AF878566

75
Tech/Elektro
'20

# Next-Generation ADCs, High-Performance Power Management, and Technology Considerations for Advanced Integrated Circuits

Andrea Baschirotto • Pieter Harpe
Kofi A. A. Makinwa
Editors

# Next-Generation ADCs, High-Performance Power Management, and Technology Considerations for Advanced Integrated Circuits

## Advances in Analog Circuit Design 2019

Springer

Mängelexemplar

*Editors*
Andrea Baschirotto
Department of Physics "G. Occhialini"
University of Milan-Bicocca
Milan, Italy

Pieter Harpe
Eindhoven University of Technology
Eindhoven, Noord-Brabant, The Netherlands

Kofi A. A. Makinwa
Delft University of Technology
Delft, Zuid-Holland, The Netherlands

ISBN 978-3-030-25266-3 ISBN 978-3-030-25267-0 (eBook)
https://doi.org/10.1007/978-3-030-25267-0

© Springer Nature Switzerland AG 2020
This work is subject to copyright. All rights are reserved by the Publisher, whether the whole or part of the material is concerned, specifically the rights of translation, reprinting, reuse of illustrations, recitation, broadcasting, reproduction on microfilms or in any other physical way, and transmission or information storage and retrieval, electronic adaptation, computer software, or by similar or dissimilar methodology now known or hereafter developed.
The use of general descriptive names, registered names, trademarks, service marks, etc. in this publication does not imply, even in the absence of a specific statement, that such names are exempt from the relevant protective laws and regulations and therefore free for general use.
The publisher, the authors, and the editors are safe to assume that the advice and information in this book are believed to be true and accurate at the date of publication. Neither the publisher nor the authors or the editors give a warranty, express or implied, with respect to the material contained herein or for any errors or omissions that may have been made. The publisher remains neutral with regard to jurisdictional claims in published maps and institutional affiliations.

This Springer imprint is published by the registered company Springer Nature Switzerland AG.
The registered company address is: Gewerbestrasse 11, 6330 Cham, Switzerland

# Preface

This book is part of the Analog Circuit Design series and contains contributions by all the speakers at the 28th Workshop on Advances in Analog Circuit Design (AACD), which was held in Milan, Italy, from April 1 to 3, 2019. The aim of the workshop was to bring together a group of expert designers to discuss new developments and future options.

The local organizers were Andrea Baschirotto (University of Milano-Bicocca) and Piero Malcovati (University of Pavia). Infineon and RedCat Devices were the platinum sponsors of the event that has been opened by the Rector of the University of Milano-Bicocca, Prof. Cristina Messa.

Each AACD workshop is followed by the publication of a book by Springer, which then becomes part of their successful series on Analog Circuit Design. A full list of the previous books and topics covered in this series can be found on subsequent pages. Each book can be seen as a reference work for students and designers interested in advanced analog and mixed-signal circuit design.

This book is the 28th in this series. It consists of three parts, each with six chapters, that cover the following topics that are currently considered of high importance by the analog and mixed-signal circuit design community:

- Next-Generation ADCs
- High-Performance Power Management
- Technology Considerations for Advanced IC

We are confident that this book, like its predecessors, will prove to be a valuable contribution to our analog and mixed-signal circuit design community.

Milan, Italy — Andrea Baschirotto
Eindhoven, The Netherlands — Pieter Harpe
Delft, The Netherlands — Kofi A. A. Makinwa

## Topics Previously Covered in the Springer Series on Analog Circuit Design

| | | |
|---|---|---|
| 2018 | Edinburgh (UK) | Low-Power Analog Techniques |
| | | Sensors for Mobile Devices |
| | | Energy-Efficient Amplifiers |
| 2017 | Eindhoven (The Netherlands) | Hybrid ADCs |
| | | Smart Sensors for the IoT |
| | | Sub-1V and Advanced-node Analog Circuit Design |
| 2016 | Villach (Austria) | Continuous-Time $\Sigma\Delta$ Modulators for Transceivers |
| | | Automotive Electronics |
| | | Power Management |
| 2015 | Neuchâtel (Switzerland) | Efficient Sensor Interfaces |
| | | Advanced Amplifiers |
| | | Low-Power RF Systems |
| 2014 | Lisbon (Portugal) | High-Performance AD and DA Converters |
| | | IC Design in Scaled Technologies |
| | | Time-Domain Signal Processing |
| 2013 | Grenoble (France) | Frequency References |
| | | Power Management for SoC |
| | | Smart Wireless Interfaces |
| 2012 | Valkenburg (The Netherlands) | Nyquist A/D Converters |
| | | Capacitive Sensor Interfaces |
| | | Beyond Analog Circuit Design |
| 2011 | Leuven (Belgium) | Low-Voltage Low-Power Data Converters |
| | | Short-Range Wireless Front Ends |
| | | Power Management and DC-DC |
| 2010 | Graz (Austria) | Robust Design |
| | | Sigma Delta Converters |
| | | RFID |
| 2009 | Lund (Sweden) | Smart Data Converters |
| | | Filters on Chip |
| | | Multimode Transmitters |
| 2008 | Pavia (Italy) | High-Speed Clock and Data Recovery |
| | | High-Performance Amplifiers |
| | | Power Management |
| 2007 | Oostende (Belgium) | Sensors, Actuators, and Power Drivers for the Automotive and Industrial Environment |
| | | Integrated PAs from Wireline to RF |
| | | Very High-Frequency Front Ends |
| 2006 | Maastricht (The Netherlands) | High-Speed AD Converters Automotive Electronics: EMC Issues Ultra Low Power Wireless |

(continued)

| | | |
|---|---|---|
| 2005 | Limerick (Ireland) | RF Circuits: Wide Band, Front Ends, DACs |
| | | Design Methodology and Verification of RF and Mixed-Signal Systems |
| | | Low Power and Low Voltage |
| 2004 | Montreux (Switzerland) | Sensor and Actuator Interface Electronics |
| | | Integrated High-Voltage Electronics and Power Management |
| | | Low-Power and High-Resolution ADCs |
| 2003 | Graz (Austria) | Fractional-N Synthesizers |
| | | Design for Robustness |
| | | Line and Bus Drivers |
| 2002 | Spa (Belgium) | Structured Mixed-Mode Design |
| | | Multi-bit Sigma-Delta Converters |
| | | Short-Range RF Circuits |
| 2001 | Noordwijk (The Netherlands) | Scalable Analog Circuits |
| | | High-Speed D/A Converters |
| | | RF Power Amplifiers |
| 2000 | Munich (Germany) | High-Speed A/D Converters |
| | | Mixed-Signal Design |
| | | PLLs and Synthesizers |
| 1999 | Nice (France) | XDSL and Other Communication Systems |
| | | RF-MOST Models and Behavioral Modelling |
| | | Integrated Filters and Oscillators |
| 1998 | Copenhagen (Denmark) | 1-Volt Electronics |
| | | Mixed-Mode Systems |
| | | LNAs and RF Power Amps for Telecom |
| 1997 | Como (Italy) | RF A/D Converters |
| | | Sensor and Actuator Interfaces |
| | | Low-Noise Oscillators, PLLs, and Synthesizers |
| 1996 | Lausanne (Swiss) | RF CMOS Circuit Design |
| | | Bandpass Sigma-Delta and Other Data Converters |
| | | Translinear Circuits |
| 1995 | Villach (Austria) | Low Noise/Power/Voltage |
| | | Mixed-Mode with CAD Tools |
| | | Voltage, Current, and Time References |
| 1994 | Eindhoven (The Netherlands) | Low-Power Low-Voltage |
| | | Integrated Filters |
| | | Smart Power |
| 1993 | Leuven (Belgium) | Mixed-Mode A/D Design |
| | | Sensor Interfaces |
| | | Communication Circuits |
| 1992 | Scheveningen (The Netherlands) | OpAmps |
| | | ADCs |
| | | Analog CAD |

# Contents

**Part I Next-Generation ADCs**

1 **Emerging ADCs** ........ 3
Gabriele Manganaro

2 **Noise-Shaping SAR ADCs** ........ 21
Shaolan Li, Jiaxin Liu, Wenjuan Guo, and Nan Sun

3 **Efficient High-Resolution Nyquist ADCs** ........ 41
P. Vogelmann, Y. Luo, M. A. Mokhtar, and M. Ortmanns

4 **Continuous-Time ADCs for Automotive Applications** ........ 59
Muhammed Bolatkale

5 **Continuous-Time Delta-Sigma Converters with Finite-Impulse-Response (FIR) Feedback** ........ 77
Shanthi Pavan

6 **High-Speed ADCs for Wireless Base Stations** ........ 91
Bob Verbruggen, Bruno Vaz, Christophe Erdmann, Diarmuid Collins, John Mcgrath, Ali Boumaalif, David Melinn, Edward Cullen, Darragh Walsh, Alonso Morgado, Conrado Mesadri, Brian Long, Rajitha Pathepuram, Ronnie De La Torre, Alvin Manlapat, Georgios Karyotis, Patrick Lynch, Peng Lim, Daire Breathnach, and Brendan Farley

**Part II High-Performance Power Management**

7 **Advanced Multiphasing: Pushing the Limits of Fully Integrated Switched-Capacitor Converters** ........ 109
Nicolas Butzen and Michiel Steyaert

8 **Highly Efficient Power Management in Wearables and IoT Devices** ........ 125
Juergen Wittmann, Francesco Cannillo, Dan Ciomaga, Mihail Jefremow, and Fabio Rigoni

9 **Current Sensing Techniques: Principles and Readouts** ........ 143
Mahdi Kashmiri

10 **Wide Bandgap Integrated Circuits for High Power Management in Extreme Environments** ........ 167
Carl-Mikael Zetterling, Saleh Kargarrazi, and Muhammad Shakir

11 **On the Limits of Driving Wide-Bandgap Transistors** ........ 179
Jef Thoné and Mike Wens

12 **Challenges in Driving New Generations of Power Switches for Motor Drive: A dV/dt Self-Adjusting Architecture for Superjunction Power Devices** ........ 193
Sergio Morini, Davide Respigo, and Martina Arosio

**Part III Technology Considerations for Advanced IC**

13 **Silicon Technologies for the Next Age of Wireless Connectivity** ........ 215
Baljit Chandhoke

14 **IC Technologies for MM-Wave Applications** ........ 225
M. Pagani, P. Rossi, and R. Lombardi

15 **Accuracy: The Next Opportunity for MEMS** ........ 243
Andrea Onetti

16 **Robustness, Reliability and Diagnostic Aspects in Sensors for Automotive Applications: The Magnetic Sensors Case** ........ 253
A. Laville, M. Pardoen, G. Close, M. Poezart, and D. Gerna

17 **Rad-Hard Mixed-Signal IC Design, Theory and Implementation** .... 273
C. Calligaro and U. Gatti

18 **1-GRad-TID Effects in 28-nm Device Study for Rad-Hard Analog Design** ........ 299
Marcello De Matteis, F. Resta, A. Pipino, F. Fary, S. Mattiazzo, C. Enz, and A. Baschirotto

**Index** ........ 317

# Part I
# Next-Generation ADCs

The first part of this book is dedicated to recent developments in the field of analog-to-digital converters. The first papers discuss new architectures such as time-domain converters and noise-shaping SAR ADCs, followed by chapters which discuss how the energy efficiency and bandwidth of existing ADC architectures can be significantly improved.

In Chap. 1, Gabriele Manganaro makes the point that most ADCs are designed to meet application needs, and so may not look good when assessed in terms of general figures of merit. As a result, some emerging classes of ADCs may not be getting the attention that they deserve. Examples of these are time-domain converters, which are compact and highly scalable, and ADCs based on compressive sampling techniques, which result in lower output data rates and thus to (sensor) systems with lower power dissipation.

Chapter 2 authored by Nan Sun et al. provides a comprehensive overview of the emerging class of noise-shaping SAR ADCs. These achieve high resolution and energy efficiency by combining the merits of SAR and ADCs. The resulting ADCs are highly scalable and amenable to low supply voltages, making them attractive candidates for emerging sensing and communication systems.

In Chap. 3, Maurits Ortmanns et al. discuss innovative Nyquist ADC architectures that can achieve both high resolution and energy efficiency. In contrast to oversampled ADCs, such ADCs do not filter their inputs, and thus do not have memory, allowing them to be multiplexed over multiple channels.

In Chap. 4, Muhammed Bolatkale describes the use of continuous-time ADCs in wideband software defined radios. The chapter gives an overview of the key architectures and circuit-level techniques employed in recently published high-speed continuous-time ADC designs.

In Chap. 5, Shanthi Pavan provides an overview of the use of finite-impulse-response (FIR) feedback in continuous-time ADCs. Apart from mitigating the effect of clock jitter, FIR feedback confers high linearity, and facilitates the use of chopping to reduce offset and 1/f noise.

In Chap. 6, Bob Verbruggen et al. describe the design of an ADC intended for direct RF-to-digital conversion in wireless base stations. The ADC employs an 8-channel time-interleaved architecture together with various foreground and background calibration schemes. Implemented in a 16 nm FinFET process, it achieves 60.9 dB SFDR and 55.3 dB SNDR at a sampling rate of 5.2 GS/s.

# Chapter 1
# Emerging ADCs

**Gabriele Manganaro**

## 1.1 Introduction

New converter architectures are introduced from time to time to address a variety of application requirements or technology challenges, and either mature alongside traditional ones or are eventually abandoned, almost like in a Darwinian selection process [1–6]. Greater awareness of some of these evolution and selection mechanics empowers the technical community to avoid unintended outcomes and possibly foster the development of useful new architectures. This topic and two examples of interesting classes of emerging analog-to-digital converters (ADCs) are discussed in this paper: time-domain converters and compressive sensing converters.

How can one objectively compare between many different converters and designs? One common way to discriminate is by quantifying the energy efficiency with which a given converter performs its function. The latter is generally assessed and tracked by means of a couple of popular figures of merit (FOMs) [1, 3, 4]. FOMs do, indeed, capture fundamental trade-offs between power consumption, signal bandwidth, and spectral purity. But, over time, FOMs have also been employed to highlight performance trends, to point to architectural strengths and shortcomings.

In some cases, FOMs have been nearly promoted to the rank of another design specification, the deliberate optimization of which may end up being rewarded with a scientific publication. Such unintended effects of FOMs are beginning to be humorously acknowledged by the technical community [7].

New points in a FOM scatter plot regularly emerge as the result of what designers are working on, which is influenced by application and business dynamics. So, the emergence of new points should not be confused as an indication of what

G. Manganaro (✉)
Analog Devices, Inc., Wilmington, MA, USA
e-mail: gabriele.manganaro@ieee.org

© Springer Nature Switzerland AG 2020

A. Baschirotto et al. (eds.), *Next-Generation ADCs, High-Performance Power Management, and Technology Considerations for Advanced Integrated Circuits*,
https://doi.org/10.1007/978-3-030-25267-0_1

converter technology could possibly do in absolute sense. Some level of correlation between technological potential and a FOM trend should not be hastily confused with causation.

Time-domain converters and compressive sensing converters address very important application and technology challenges and hence deserve attention from the technical community. But neither of them, at least till now, have demonstrated impressive FOM results; that was not their intent. But because of this, these ADCs risk being neglected or even driven into wrongful extinction. A discussion about FOMs is then inevitable. Perhaps, this may stimulate a broader debate on whether our maturing technical field is accidentally restraining new and uncommon innovative efforts or narrowly channeling them due to a single-metric approach.

As for these two classes of emerging ADCs, what is covered here is by no means exhaustive. On the contrary, publication references are provided to the reader as a starting point to explore many of the subjects. However, this paper attempts to draw the attention of the technical community to interesting cases of such emerging converters, while offering some original observations about them.

This paper is organized as follows: Sect. 1.2 discusses how data converter innovation happens as a symbiosis between application needs and technology progression and how the increasing popularity of power efficiency FOMs can introduce blinders. Section 1.3 talks about time-domain converters and provides conjectures for their future evolution. Section 1.4 deals with compressive sampling and offers a brief survey of the most recent architectural breakthroughs. Some conclusions are drawn in Sect. 1.5.

## 1.2 Technology Progression is Seldom a Straight Line

### *1.2.1 FOMs: Side-Effects Include . . .*

A commonly used ADC FOM is the so-called Schreier's FOM, measured in dB/J (although the unit "Joules" is usually dropped) and defined as follows [1]:

$$\mathrm{FOM}_{S,\mathrm{hf}} = \mathrm{SNDR} + 10\log\left(\frac{\mathrm{BW}}{P}\right) \tag{1.1}$$

where SNDR is the signal-to-noise-and-distortion ratio in dB and measured for high frequency inputs (hence the subscript hf in the FOM symbol), $P$ is the corresponding power consumption, in watts, and BW is the input signal bandwidth in Hz. BW is generally assumed to be equal to the sample rate $f_s$ divided by the oversampling ratio (OSR). This definition allows both Nyquist converters (for which BW $= f_s/2$) and oversampled converters to be compared [1]. A scatter plot based on ADCs published during the last 20 years at the ISSCC and VLSI conferences is depicted in Fig. 1.1 [8].

It is easy to see that most new data points (indicated by the squares and diamonds) correspond to the highest bandwidth ADCs, which are close to the diagonal dashed asymptote known as the "technology front." A similar distribution of new data points is found, year after year, with newer points moving the dashed line asymptotes to wider band and higher FOM. In fairness, not all papers accepted at these two conferences are required to establish a substantially better FOM, provided that valuable innovations are demonstrated in other important dimensions, as seen by the few new points located far away from the hustle and bustle along the dashed line.

However, this picture, while insightful from an energy efficiency perspective, should be used carefully. Because while quantitative and objective, it omits relevant architectural innovation that was either not submitted for publication or that while addressing other valuable problems does not stand out in terms of FOM.

For example, while high-speed ADCs are often published at ISSCC and the VLSI symposium, a number of companies developing, among others, innovative high-speed data converters embedded in complex integrated systems simply choose not to publish. That is true for both commercial applications in ultra-wide band optical, wired, and wireless infrastructure communication systems, as well as for defense and space applications. Non-CMOS technologies, such as heterogeneous or optical technologies, are also sometimes used for these applications in order to handle signal bandwidths that, can be an order of magnitude above the technology front of Fig. 1.1.

There are also cases where an ADC is allowed to use as much power as is needed to meet ambitious performance objectives. For these, the FOM or the physical size would not compare favorably with the designs shown in Fig. 1.1. It should be said that while these could be regarded as outliers, if added to the scatter plot, they would distort the regularity of the distribution in Fig. 1.1.

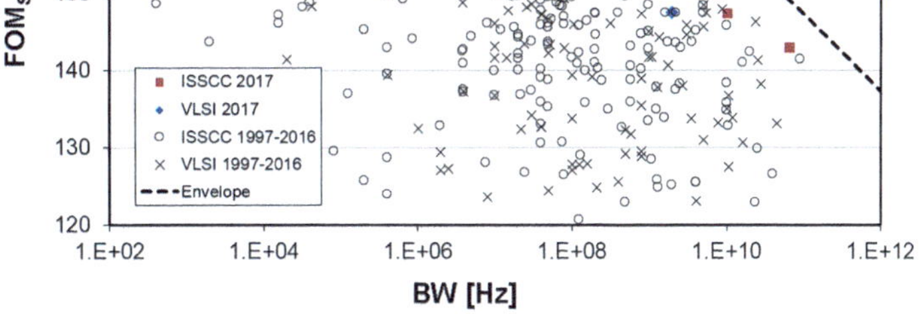


**Fig. 1.1** The Schreier's FOM versus signal bandwidth for ADCs published at the ISSCC and VLSI Symp (1997–2017)

Also, as noted earlier, the horizontal asymptote, known as the "architecture front" does not change much from year to year. This may suggest stale innovation in low bandwidth ADCs.[1] In fact, there is a lot of relevant converter innovation in narrow-band applications that doesn't necessarily aim to optimize the FOM. As a matter of fact, most commercial converters process much lower bandwidths than those close to the "technology front." Such ADCs, often called "precision converters" (low bandwidth, high dynamic range), address important application problems in innovative ways, but are intentionally seldom disclosed in publications. These converters rely on proprietary circuit and algorithmic techniques and leverage special process technology capabilities to achieve very high linearity and noise performance. All these forms of innovation are protected by trade secrets and patents and it is often deemed to be counterproductive to give them a high visibility in the open literature. Expectably, none of these specifics will intentionally be disclosed here. Although the interested reader can confirm such assertions by an in-depth study of the relevant cases, publicly available at the US and European patent office websites.

In conclusion, FOMs are very useful tools if used with care. Conversion efficiency is only one lens for looking at converters' progress. Overly emphasizing conversion efficiency at prestigious conferences will inevitably incentivize and possibly develop gregarious lines of research at the expenses of more important directions. Secondly, though not less important, FOM-based trends can miss out some important industrial innovation.

### *1.2.2 When an Application Requirement Triggers a Major Turn*

Application requirements are a main innovation driver and progression of design specifications can change quite dramatically within an application space, hence forcing technology dislocations.

For instance, high-speed converters are often required in cellular wireless infrastructure systems [9]. In fact, presently, this is probably the largest application space for this class of ADCs and hence a clear trendsetter.

About seven or so years ago, a receive signal path for a cellular base station (BTS) would have been required to process a signal such as a multi-carrier GSM channel with an RF bandwidth (BW) of 75 MHz or a CDMA channel with an RF BW of 100 MHz. The prior generation's requirement was about 40 MHz, while about 3 years later, a subsequent generation of BTSs required an RF bandwidth of BW = 200 MHz. Today, consensus is that BTS intended for so-called fifth generation (5G) mm wave systems [10] should be capable of processing RF

[1] A theoretical upper bound of 192 dB is set by the thermal noise limits. This points to the challenge of going higher [3].

bandwidths of 1–1.2 GHz. The need to be compatible with legacy systems means that the dynamic requirements of the prior generation must also be maintained in the wider bandwidths of the new systems.

So, if an ADC is used to digitize baseband in a homodyne receiver, its sample rate would need to roughly double when going from the 40 MHz generation to the 100 MHz generation and then double again to enable the 200 MHz generation. But the following ADC generation would require a sample rate that is five to six times higher than its predecessor to process a 1–1.2 GHz band. So, while in the previous cases an appropriate process technology transition of nearly the same ADC architecture could meet the requirement, in the last case a substantial architectural change is indispensable.

A closer look within the same application space shows that the evolution of ADC requirements is in fact even more non-linear. For instance, if the popular heterodyne receiver is considered, an ADC can be used to digitize a desired communication channel with band BW but centered at an intermediate frequency $f_{IF}$, rather than in baseband/zero IF. In the 100 MHz BW systems generation, this IF frequency was commonly chosen between 150 MHz and 350 MHz. In the 200 MHz systems generation, some BTS designs have moved their $f_{IF}$ to slightly higher frequencies. So, again, a sample rate doubling is very challenging but not necessarily requiring a completely new ADC architecture.

However, in some more recent cases, the requirement on the input signal for the ADC has moved to much higher frequencies. Namely, the RF to IF frequency down conversion is moved from the analog domain, in front of the ADC, to the digital domain, right after digitization. In other words, the 200 MHz wideband signal that the ADC needs to sample is not centered at a few hundred megahertz; it is now located at a few GHz. And while undersampling is a possible avenue, the demand is to use the first Nyquist band for acquisition. As for 5G cellular communication, designers distinguish between sub-6 GHz systems, where the RF channel is placed below 6GHz, and millimeter wave systems, where the channel is located between 29 GHz and 32 GHz or so [10]. So, for example, if a 10–12 GSPS ADC could be used as an RF digitizer [11] in the receive path of a sub-6 GHz system, doubling $f_s$ to 20–24 GSPS could only provide some incremental advantage in processing gain and in terms of analog filtering requirements. So a completely different approach to the millimeter wave systems is needed.

Additionally, one of the other technologies required by 5G mm wave communication systems is beamforming. The ability to establish a spatially directed receive/transmit communication link between certain mobile devices and the BTS is obtained via phased arrays of antennas, each one of which may have its own RF/mixed-signal chain. While, certainly, energy efficiency is very important (FOM), the size and weight of the electronics introduce very restrictive conditions to the system design that trickle down also to the data converters. Converter architectures that can be very compact in area, that scale well with nanometer process technologies and can then be integrated in large channel count are receiving substantial attention. That includes classic SAR ADC architectures. But it also includes emerging classes of converters such as the time-to-digital and digital-to-time converters discussed in the following sections.

### 1.2.3 When a Converter Breakthrough is an Enabler

The innovation cycle doesn't simply work in the direction of an application challenge driving an engineering solution. It works also in the opposite direction, when a technology breakthrough enables an application that was not practical or conceivable before.

For instance, while trimming has been fairly common practice in precision analog circuits for many decades, despite much research, self-calibration has only become mainstream in industrial data converter design in the last 15 years or so. Self-calibration techniques substantially loosen analog design trade-offs between matching, area, noise and linearity, power consumption or speed [1, 12]. Because of that, in the mid-2000s, there has been a rapid expansion in terms of converter architecture innovation significantly pushing the performance fronts in multiple directions, particularly in CMOS processes [1]. First, 8–10b ADCs went from sample rates of a few hundred MSPS to well into the GSPS range, thanks to a combination of substantial circuit size reduction (calibration correcting for matching limitation, hence allowing size reduction and hence speed acceleration) and simple two-way ("ping-pong") interleaving. Then further improvements in core self-calibration, plus higher-order time-interleaving (8 sub-ADCs or higher) assisted by channel mismatch calibration, allowed Nyquist rate 12–14b ADCs to also break the GSPS speed barrier [1, 2, 11]. Different self-calibration techniques were employed in continuous-time $\Delta\Sigma$ ADCs to control parametric spread in the loop filters, in the feedback delays, and to linearize the feedback DACs. This allowed such architectures to digitize hundreds of MHz of signal band at frequency centered all the way up to the low GHz range [9].

As a result, considering again the examples in the previous section, cellular wireless communication systems have been positively impacted by the ability to employ RF digitization and synthesis. That has made it possible to move a lot of the modulation/demodulation functionality from the analog/RF domain into the digital domain, with substantial benefit to integration, flexibility/programmability, development time, etc.

Similarly, the substantial reduction in size and power enabled by new self-calibration techniques has also enabled considerable miniaturization/integration in medical instrumentation systems where data converters also constituted one of the bottlenecks, hence enabling the creation of affordable portable health monitoring systems such as ultra-sound systems, etc. with an appreciable benefit for our well-being.

Finally, while the philosophy of development of analog systems has traditionally been to design for best performance, leaving to trimming and calibration the role of making up for manufacturing imperfections, recent advances in self-calibration are rapidly changing this strategy. Looking ahead, deeper analog–digital co-design is anticipated. For example, to further overcome power/speed limitations, the data converter architectural preferences may go to those that, while characterized by high but predictable and correctable nonlinearity, can enable substantial higher speed or lower power or smaller area, leaving to self-calibration and software algorithms the task of linearization [1–3, 13].

## 1.3 Time-to-Digital (TDC) and Digital-to-Time (DTC) Converters

### *1.3.1 Why Time-Domain Data Converters?*

MOS device scaling is accompanied by voltage supply scaling. Difficult trade-offs between signal headroom, noise, linearity, bandwidth, power consumption, and device matching introduce limitations to the performance of voltage-domain analog circuits; including data converters [12].

In the early 1990s, in response to the shrinking voltage headroom issue, researchers explored current-mode circuits [14]. But while a hard ceiling in the current domain isn't always immediately explicit, currents and voltages are tied to one another by finite node impedances and homologous challenges in current mode systems surfaced. Moreover, many of the signal sources, sensors, and actuators are voltage-mode devices, hence making the voltage-to-current and current-to-voltage transducers the inevitable new bottlenecks.

While the pace of reduction in supply voltages has since slowed down, the voltage headroom problem has not gone away. Analog designers have begun looking at another analog variable that could be used to represent and process information: time intervals.[2] Time-domain circuits such as phase-locked loops (PLLs) or delay-locked loops (DLLs) are very mature architectures and seminal work in time-domain data converters can be traced back to many decades ago. Time-to-digital (TDCs) and digital-to-time (DTCs) converters have been important functional blocks for digital and semi-digital timing/clock systems [1].

### *1.3.2 TDCs/DTCs Building Blocks*

Two of the most important analog circuit primitives for processing time are the CMOS inverter and the D-type edge-triggered flip-flop (DFF) [1, 15]. The voltage/current-domain signals processed by TDCs/DTCs have generally an approximately rectangular or, at high frequency, a distorted-sine shape. What really

[2]While we will refer to "time" for brevity, let it be understood that we mean "time intervals."

matters is not their shape, but the moment when such signals cross a pre-established set of thresholds, hence determining the instant of transition from 0 to 1 or from 1 to 0. That is liberally referred to as the "zero crossing" time.

In TDCs/DTCs, the CMOS inverter is often current-starved to be able to adjust its gate delay by means of a control current $I_c$ or a control voltage $V_c$ and it is employed to realize a voltage-controlled delay unit (VCDU) as in the example depicted in Fig. 1.2 [15]. The input is represented by a signal $\phi_{\text{in}}$, while the output is a signal $\phi_{\text{out}}$. The control variable in this example is $V_c$, and it can vary the net gate delay $\Delta T$. The small signal gain $G\phi$ at the VCDU's quiescent point of the voltage-to-time characteristic determines the ability of this primitive to process time [15].

VCDUs like the one in Fig. 1.2 or alternate ones, especially those implemented in differential form, are building blocks for ring-oscillator VCOs and voltage-controlled delay lines which are then used for continuous processing of time signals.

The other time-domain primitive is the D-type edge-triggered flip-flop (DFF), as the one shown in Fig. 1.3. The DFF can be used as an analog primitive to realize a comparator function since, given two pulses, say $\phi_{\text{in}}$ and $\phi_{\text{ref}}$, fed to its D input and clock input, respectively, as shown in Fig. 1.3, will return a logic 1 at its Q output when $\phi_{\text{in}}$ leads $\phi_{\text{ref}}$ ($\phi_{\text{in}} < \phi_{\text{ref}}$) and 0 otherwise ($\phi_{\text{in}} \geq \phi_{\text{ref}}$).

The VCDU and the DFF can also be used to build a variety of TDCs and DTCs using the architectures described in excellent tutorials such as [15, 16].

A rather simple one is the time-domain flash ADC depicted in Fig. 1.4. Here the VCDU's gate delay $\Delta T$ is used to set the time-domain comparator thresholds

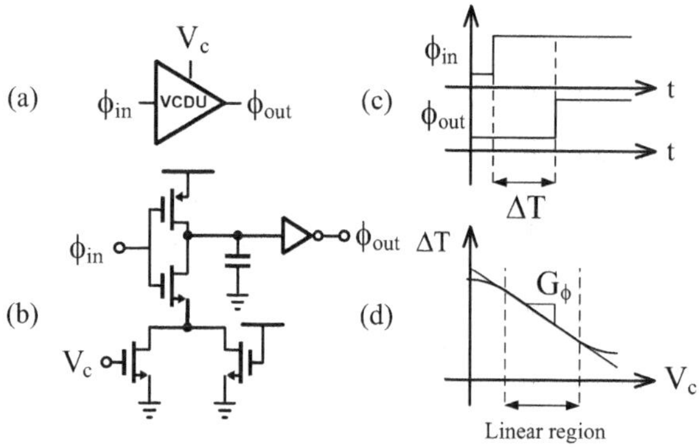


**Fig. 1.2** (**a**) A VCDU; (**b**) a circuit implementation using a current-starved inverter; (**c**) time diagram of the input $\phi_{\text{in}}$ and output $\phi_{\text{out}}$; (**d**) an example of the voltage-to-phase characteristic of a VCDU where a central linear region can be identified with a corresponding linearized voltage-to-phase gain $G\phi$

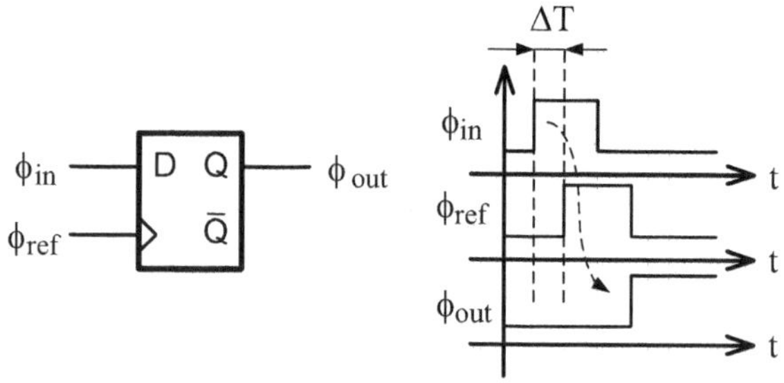


**Fig. 1.3** D-type pos. edge-trig. Flip-flop (DFF) as a time-mode comparator

Fig. 1.4 A flash TDC

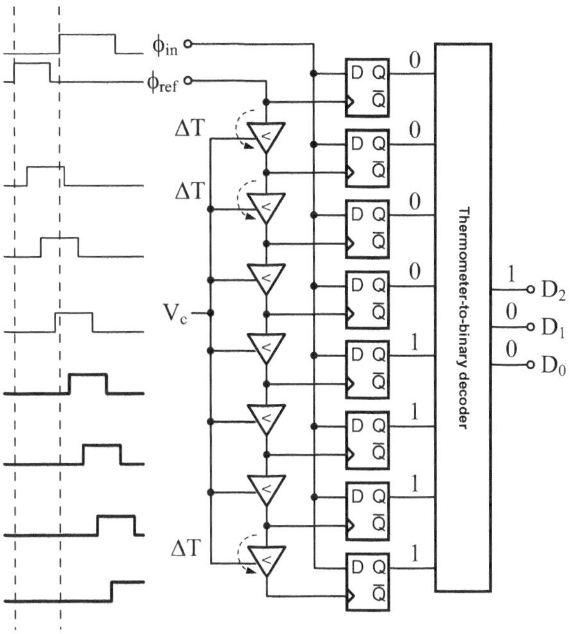


and hence the quantization step and the nominal resolution of the converter. Finer steps can be obtained by phase interpolation between two delay elements' outputs or by using a sliding "time Vernier" obtained by introducing a second time-shifted servoed voltage-controlled delay line [1, 15]. These techniques, however, introduce additional complexity, area, power consumption, noise, and linearity issues that need to be carefully managed.

Another way to realize a TDC consists of building a ring-oscillator VCO from VCDUs and then using the analog voltage input signal to control the VCDUs so that the frequency of oscillation of the VCO depends on the input to be digitized. Finally, a counter, or an array of DFFs properly connected to the output phases of the ring oscillator, is used to map the oscillator frequency to a digital representation of the analog input [1, 15, 16]. More advanced "VCO ADC" architectures use such VCOs as quantizers embedded in time-domain or classic voltage-domain delta-sigma modulators [16].

### 1.3.3 So, Do TDCs Work Then?

How well do TDCs perform compared with traditional ADCs? The aperture plot and the energy plot for all ADCs, traditional and time-domain, are shown in Fig. 1.5 for all the papers published at ISSCC and VLSI symposium during the last 20 years [8]. In these two plots, the TDCs are highlighted by using black squares. These

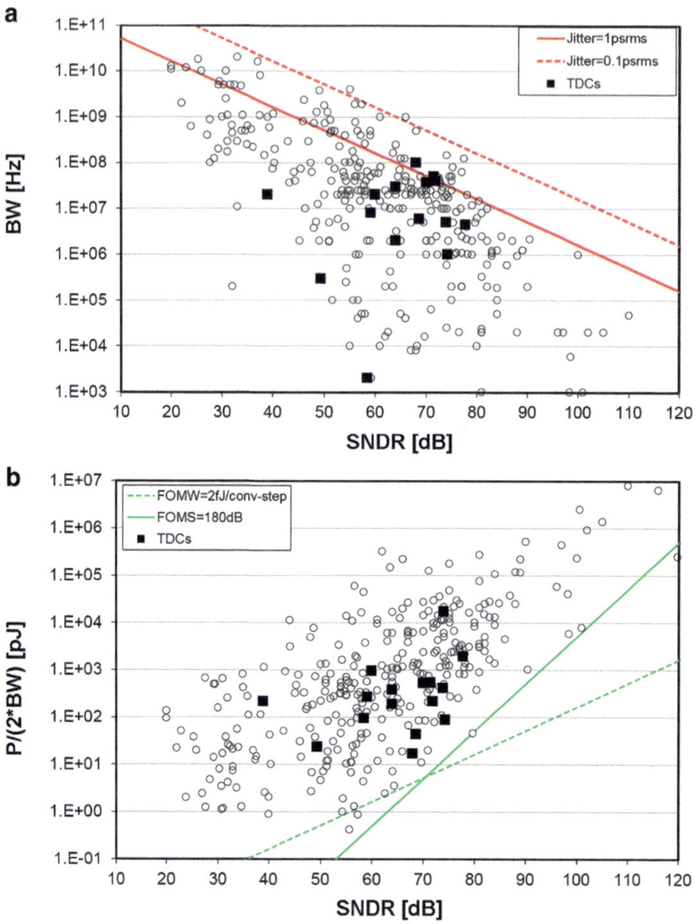


**Fig. 1.5** (**a**) Aperture plot for signal bandwidth (BW) versus SNDR at high input frequency for publications at the ISSCC and VLSI conferences between 1997 and 2017. TDCs, hybrids included, are marked with black squares. (**b**) Same data set but for the energy plot graphing conversion energy P/(2BW) versus SNDR at high input frequency

also include a few hybrid TDCs mixing traditional voltage-mode circuits with time-domain sub-blocks. The most recent data points tend to be those closer to the overall state-of-the art lines (low jitter contours, high in the aperture plot, and best figures of merit contours, low in the energy plot).

Overall, the performance of the published TDCs so far spans the medium SNDR, medium BW range. The energy efficiency isn't the most competitive, though recent data points show appreciable improvement in efficiency. Indeed, as stated in Sect. 1.2.1, one should be careful in quickly drawing conclusions only based on inspection of these plots, particularly with respect to the question about what type of performance it may be possible to attain with TDCs.

More in-depth examination of the papers behind these data points suggests that, especially in the case of industrial publications, these tend to target signal bandwidths in the tens of MHz with SNDR of around 75 dB, presently finding application as embedded ADCs in mobile handset's SoCs.

Yet a distinctive feature of TDCs is their substantial area compactness, making them very competitive with comparably performing but physically larger pipelined and SAR ADCs.

Another application space where TDCs are also finding increasing use is as part of digital temperature sensors [1, 17–19] and other low frequency/low power sensing and digitization systems, including those for Internet of Things (IoTs). That is due to the combination of very high compactness, low power, and low cost.

### *1.3.4 How Well Do TDCs/DTCs Scale?*

As stated above, one of the motivations to develop TDCs and DTCs as alternate data converter architecture relates to their scalability with CMOS process technology. Considering the primitives of Figs. 1.2 and 1.3, some observations follow.

First, their area scales approximately with Moore's law, which is expected to continue to hold up to 7 nm and likely beyond that. That is an advantage over traditional ADCs and DACs since amplifiers, for example, don't shrink that well.

The minimum gate delay $\Delta T_{\min}$ of a VCDU is process technology dependent. Based on actual data reported in [20] and accounting for the transition from planar MOS to FinFET occurring around 22 nm it is possible to estimate that $\Delta T_{\min}$ shortens from one CMOS node to the next one with an approximate geometric progression of factor 1.15–1.2. But, since a reduction in $\Delta T_{\min}$ sets the TDC's quantization capability, this is a relatively modest improvement.

The gate switching energy, though, enjoys a better scaling profile. Based on the trends shown in [20], we can estimate a relative energy reduction of about 1.52–1.55 times from a CMOS generation to the next one. This significantly impacts conversion efficiency and tends to be higher than what most traditional ADC architectures experience for the same node transition.

But while reducing $\Delta T_{\min}$ improves quantization in TDCs/DTCs, the phase noise on the zero crossing would still limit the realizable dynamic range. A concern valid until recent years was that while MOS's transconductance $g_m$ improved at a faster pace than the supply drop, reducing thermal noise, on the other hand, the flicker noise $1/f$ corner substantially increased in frequency. Beyond 90 nm, the latter could be the dominant contributor to phase noise. This required, for example, various forms of mitigations in different architectures employing CMOS delay lines and oscillators, depending on the resulting noise modulation mechanisms contributing to phase noise/jitter [21, 22].

But with the introduction of FinFETs, both the flicker and the thermal noise performance of FETs have been substantially improved over that of planar high-K gate MOSFETs (e.g., about 3 dB better in 16 nm FinFET than in 28 nm planar MOS [23, 24]). This is very encouraging news. While, to the best of the author's knowledge, a quantitative assessment of the impact on TDCs has not yet been published, it should be expected that all TDC architectures will see a larger net jitter improvement compared to the previously cited decrease in $\Delta T$. If that is indeed the case, then this points to renewed potential for developing higher dynamic range TDCs.

## 1.4 Compressive Sampling ADCs

### *1.4.1 Why Compressive Sampling ADCs?*

While many applications, such as those in communication systems or in high performance instrumentation, deal with very active signals, sensing applications in health/vital monitoring, seismic/environmental monitoring, and some industrial process control applications, among others, deal with signals that experience very little change for extended lengths of time, followed by short bursts of activity [19, 25]. There are also classes of signals (e.g., audio) that can be represented by either few significant components in the frequency domain or by limited events of activity in the time-domain. Because of that such signals are said to be "sparse": sparse in frequency domain or sparse in time-domain, respectively. Classic time-uniform Shannon sampling theory is not very efficient for sparse signals as it results in very long sample series requiring too many samples/data to deliver the desired information content. A mathematically accurate description of signal sparsity can be found in [26–28].

This issue of signal sparsity and associated processing, while well-known for several decades in many engineering disciplines (e.g., compression algorithms are ubiquitous in software design and data storage; also, wavelet theory is well established in signal processing) has recently found renewed attention in the circuit design community due to the rapid growth of the Internet of Things (IoTs). This is particularly true in the case of wireless sensor networks (WSNs), where a network of sensor nodes (SNs) senses, pre-processes, and wirelessly delivers information to a central hub/base station. Each SN comprises the sensor(s), the conditioning and data acquisition circuitry, a local DSP, and the wireless transceiver (TRX) (plus a power management unit) as shown in Fig. 1.6.

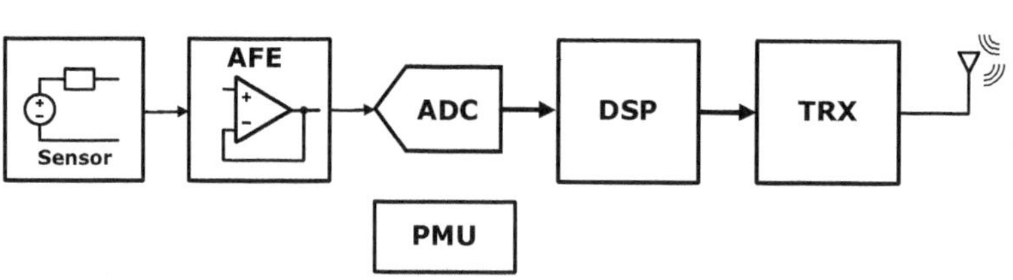


**Fig. 1.6** High level block diagram for a sensor node (SN) in a WSN

The requirements on size, weight, and power (SWAP) associated with SNs are extremely demanding. But while each circuit block in an SN must meet challenging specifications, in several cases the real bottlenecks and the most power-hungry functions are either in the data transmission from the SN to the hub (TRX) or the SN's digital signal processing (DSP) required to extract the relevant information from the data to be sent to the hub. The ADC only consumes a very modest fraction of the overall power budget (e.g., ~5% of the total SN's power consumption) [29]. But if the ADC is a classic time-uniform sample rate (Shannon) converter, it will produces a large amount of data that then causes the DSP and/or the TRX to use even more power.

On the other hand, a compressive sampling (CS) architecture processing the same information with a lower average data throughput to perform the analog-to-digital conversion can result to an overall lower power consumption budget for the SN as a whole. The compressed information is then transmitted to the hub, where, with a substantially larger computational capability and power budget, the reconstruction of the received compressed signal into the original sensed signal can be performed. This is a compelling example of a class of ADCs that despite having a mediocre FOM can deliver a far better engineering outcome than the alternative.

### 1.4.2 A Short Tour in the Land of Compressive Sampling

In Shannon's uniform sampling theory, a time-domain sampled signal can be thought of as a modulation/convolution between the original continuous input signal and a Dirac pulse train. While at a very high level, in compressive sampling the pulse train is replaced by pulse amplitude modulated signals whose amplitude is set by an independent, identically distributed noise (ideally Gaussian) vectors (usually a pseudo-random binary sequence, or PRBS) constituting an alternate representation basis. If the input signal is sparse, then after convolution with the PRBS signals (the operation of compression), the resulting signal has far fewer samples [28]. To reconstruct the original signal, the operation needs to be reversible with tolerable/controllable losses/degradation. In general, the compression operation is a matrix product (inner product) between the vector of the samples of the original input signal with an encoding matrix made of appropriate PRBS vectors.

Compression can occur at different stages in the signal chain of Fig. 1.6. It can be done in the analog continuous-time domain before the ADC, so the ADC's uniform sample rate can be reduced (to a sub-Nyquist rate), while the burden of the encoding lies with the front-end analog convolution circuit. It could also be done digitally with a DSP, after the ADC. In this case the ADC is a traditional uniform sampling converter and the burden of encoding is on the DSP. Or, it can be performed in the analog domain and combined with the ADC function (running at sub-Nyquist rate), leading to compressive sampling ADC architectures.

In [30, 31] the compression is performed in the analog domain before the ADC. The encoder uses a so-called random-modulation pre-integrator (RMPI) architecture consisting of an array of parallel signal paths each one mixed with a different random basis function, followed by a low-pass/integrator stage and a reduced sample rate (usually a SAR) ADC. While the mixers with $\pm 1$ random components are efficient analog circuits, the filtering/integration requires power/area hungry operational transconductance amplifiers (OTAs). While the ADCs have low sample rate and don't require a lot of power and area, the rest of the analog encoder can require substantial power and area. Moreover, the parallel paths require proper time-alignment, which is non-trivial.

In [29], the CS encoder is implemented digitally. The integrators are replaced by energy efficient digital accumulators, though the ADC, while using a very power efficient implementation, runs at Nyquist rate.

In [32] the mixing of the input signal with the PBRS functions and subsequent integration are replaced by a much simpler architecture where the sampler in front of the ADC is controlled by the PBRS. So, instead of sampling $N$ consecutive samples at rate $f_s$, this CS sampler picks only $M$ of them at random (with $M < N$) from each successive length-$N$ window of the input. The resulting non-uniform time sequence (NUS) corresponds to a lower sample rate of average $(M/N)f_s$ and is digitized by an ADC that is structurally a conventional asynchronous SAR ADC, but where each conversion cycle is edge-triggered by the sampler's PBRS clock.

In [33] a rather simple hardware NUS implementation of the SN shifts the burden of the decompression to the hub/base station. However, they also have a more limited performance in terms of the sparsity of signal they can process, compared to alternatives [34].

Lastly, a very effective approach is introduced in [34]. Here a SAR ADC is augmented with an extended CS front-end encoding in discrete time. The mixing with the PBRS sequences is done similarly to the RMPI implementations using four-switch passive mixers, though the discrete-time implementation has advantages over the continuous-time circuits used in the RMPI architectures. The subsequent integration operation is performed in the charge domain using a reconfigurable extension of the SAR's capacitive DAC array, avoiding the power/area hungry OTAs used in the previous RMPI architectures and only using passive switch capacitor charge-domain circuitry.

Renewed attention to level-crossing ADCs [19, 25] has led to several interesting NUS architectures which, in addition to providing some of the system-level benefits highlighted above, have a few additional important advantages. A remarkable recent example is the architecture described in [35, 36]. With this architecture the analog anti-aliasing filter required in front of a conventional uniform sampling system is substantially simplified and essentially replaced by a digital anti-aliasing filter.

In this architecture, the sampling time occurs when the analog input crosses the quantizer threshold and this time instant gets simultaneously quantized and stored. A benefit to dynamic range (DR) is so realized by the level-crossing [25, 35] as it can be also guessed as there is virtually no level-quantization error once the captured analog input level corresponds to the quantum level at the time of its acquisition. The sampled series in the digital domain is then made of data pairs of measured voltages and their corresponding quantized time instants (time-stamps). But while time-quantization of the time-stamp introduces an approximation error,[3] current nanometer process technology supports what can be a time-resolution equivalent to using a traditional uniform sample rate well into the THz range.[4] That is substantially higher than the state-of-the-art $f_s$ of traditional ADC alternatives. This higher equivalent $f_s$ results in a substantial relaxation of the anti-aliasing filter in front of the ADC.

But, to be further processed by a traditional uniform sampling DSP, the NUS sampled series needs to be resampled at a (much lower) uniform clock rate. This would cause aliasing and hence DR degradation, nulling what has been so far obtained, if nothing else is done. But, before the uniform digital re-sampling, the NUS digital series is digitally filtered using a clever adaptive FIR-like filter architecture that cuts the out of band noise, hence preserving most of the dynamic range previously gained. Another attractive aspect of this seemingly involved approach is that much of this converter system is fully digital. So, it directly benefits from CMOS scaling.

Overall, the examples reported here show very encouraging progress occurred over a limited span of time. Once again, the emphasis on ADC energy efficiency is not that important and can distract attention from greater advantages. A good reason to develop a CS ADC is for its impact to the signal chain and to the overall power of the SN. Let's not think of the ADC in isolation.

## 1.5 Conclusions

Recent developments in the innovative field of data converters have been discussed. The symbiosis between applications and technology leads to technology dislocations and trend changes. This contrasts with an otherwise regular, Moore's-like, orderly progression seen when trying to correlate too many heterogeneous data points. Special attention has here been given to the promising technologies of TDCs/DTCs [37, 38] and compressive sampling converters [39, 40]. Neither of these converter classes seem attractive given the current emphasis on converter energy efficiency. But both are demonstrating visible progress in addressing valuable engineering problems. While recognizing the value of FOMs, it is incumbent upon the technical community to look at architectural innovation in the widest possible perspective.

[3] A time-domain quantization error.

[4] And we have already emphasized how time-domain resolution promises to benefit from CMOS scaling when we talked about TDCs in the previous section.

## References


1. G. Manganaro, *Advanced Data Converters* (Cambridge University Press, Cambridge, 2011)
2. D. Robertson, A. Buchwald, M. Flynn, H.-S. Lee, U.-K. Moon, B. Murmann, Data converter reflections: 19 papers from the last ten years that deserve a second look, in *2016 European Solid-State Circuits Conference (ESSCIRC)* (2016), pp. 161–164
3. B. Murmann, The race for the extra decibel. IEEE Solid-State Circuits Mag. **Summer**, 58–66 (2015)
4. B.E. Jonsson, A survey of A/D-converter performance evolution, in *IEEE Int. Conf. Electronics, Circuits and Systems* (2010), pp. 766–769
5. R.H. Walden, Analog-to-digital converter survey and analysis. IEEE J. Sel. Areas Commun. **17**(4), 539–550 (1999)
6. G. Manganaro, Emerging data converter architectures and techniques, in *2018 IEEE Custom Integrated Circuits Conference (CICC), San Diego (CA), USA* (2018)
7. G. Manganaro, F. Tavernier, B. Dobkin, A. de Graauw, J. Stauth, L. Loh, P. Andreani, Figures-of-merit on trial, in *2018 IEEE Int. Solid-State Circuits Conference*, Evening Session (2018)
8. B. Murmann, ADC performance survey 1997–2017 [Online]. Available http://web.stanford.edu/~murmann/adcsurvey.html
9. G. Manganaro, D. Leenaerts, *Advances in Analog and RF IC Design for Wireless Communication Systems* (Academic Press, Orlando, 2013)
10. J. Thompson, Z. Ge, Z.-C. Wu, R. Irmer, H. Jiang, G. Fettweis, S. Alamouti (Eds.), Special issue on "5G wireless communication systems: prospects and challenges, in *IEEE Communications Magazine* (2014)
11. S. Devarajan, L. Singer, D. Kelly, T. Pan, J. Silva, J. Brunsilius, D. Rey-Losada, F. Murden, C. Speir, J. Bray, E. Otte, N. Rakuljic, P. Brown, T. Weidgandt, Q. Yu, D. Paterson, C. Petersen, J. Gealow, G. Manganaro, A 12b 10GS/s interleaved pipeline ADC in 28nm CMOS technology, in *IEEE J. Solid-State Circuits* (2017)
12. K. Bult, in The effect of technology scaling on power dissipation in analog circuits, ed. By M. Steyaert, A.H. van Roermund, J.H. Huijsing. *Analog Circuits Design* (Springer, Berlin, 2006), pp. 251–290
13. N. Rakuljic, C. Speir, E. Otte, J. Bray, C. Petersen, G. Manganaro, In-situ nonlinear calibration of a RF signal chain, in *2018 IEEE Int. Symp. on Circuits and Systems (ISCAS), Florence, Italy* (2018)
14. C. Toumazou, F.J. Lidgey, D.G. Haigh (Eds.), Analog IC Design: the current-mode approach, in *IEE Circuits, Devices and Systems Series* (1990)
15. G.W. Roberts, M. Ali-Bakhshian, A brief introduction to time-to-digital and digital-to-time converters. IEEE Trans. Circuits Syst. **57**(3), 153–157 (2010)
16. P.K. Hanumolu, Time-based $\Delta\Sigma$ ADCs, in *CICC* (2017)
17. X. Tang, W.T. Ng, K.-P. Pun, A resistor-based sub-1-V CMOS smart temperature sensor for VLSI thermal management. IEEE Trans. VLSI **23**(9), 1850091 (2015)
18. S. Jeong et al., A fully-integrated 71 nW CMOS temperature sensor for low power wireless sensor nodes. IEEE J. Solid State Circuits **49**(8), 1682–1693 (2014)
19. N. Sayiner, H. Sorensen, T. Viswanathan, A level-crossing sampling scheme for A/D conversion. IEEE Trans. Circuits Syst. **43**(4), 335–339 (1996)
20. W. Holt, Moore's law: a path going forward, in *ISSCC* (2016), pp. 8–13
21. A. Hajimiri, S. Limotyrakis, T.H. Lee, Jitter and phase noise in ring oscillators. IEEE J. Solid State Circuits **34**(6), 790–804 (1999)
22. D. Ponton et al., Assessment of the impact of technology scaling on the performance of LC-VCOs. in *ESSCIRC* (2009), pp. 364–367
23. S.-Y. Wu et al., A 16nm FinFET CMOS technology for mobile SoC and computing applications, in *IEDM* (2013)

24. S.-Y. Wu et al., An enhanced 16nm CMOS technology featuring 2nd generation FinFET transistors and advanced Cu/low-k interconnect for low power and high performance applications, in *IEDM* (2014)
25. Y. Tsividis, Event-driven data acquisition and digital signal processing - a tutorial. IEEE Trans. Circuits Syst. **57**(8), 577–581 (2010)
26. E.J. Candès, J. Romberg, T. Tao, Robust uncertainty principles: exact signal reconstruction from highly incomplete frequency information. IEEE Trans. Inf. Theory **52**(2), 489–509 (2006)
27. D.L. Donoho, Compressed sensing. IEEE Trans. Inf. Theory **52**(4), 1289–1306 (2006)
28. M. Mangia, R. Rovatti, G. Setti, Rakeness in the design of analog-to-information conversion of sparse and localized signals. IEEE Trans. Circuits Syst. **59**(5), 1001–1014 (2012)
29. F. Chen, A.P. Chandrakasan, V.M. Stojanovic, Design and analysis of a hardware-efficient compressed sensing architecture for data compression in wireless sensors. IEEE J. Solid State Circuits **47**(3), 744–756 (2012)
30. X. Chen, Z. Yu, S. Hoyos, B.M. Sadler, J. Silva-Martinez, A sub-Nyquist rate sampling receiver exploiting compressive sensing. IEEE Trans. Circuits Syst. **58**(3), 507–520 (2011)
31. D. Gangopadhyay, E.G. Allstot, A.M.R. Dixon, K. Natarajan, S. Gupta, D.J. Allstot, Compressed sensing analog front-end for bio-sensor applications. IEEE J. Solid State Circuits **49**(2), 426–438 (2014)
32. M. Trakimas, R. D'Angelo, S. Aeron, T. Hancock, S. Sonkusale, A compressed sensing analog-to-information converter with edge-triggered SAR ADC Core. IEEE Trans. Circuits Syst. **60**(5), 1135–1148 (2013)
33. V.R. Pamula et al., A 172 $\mu$W compressively sampled photoplethysmographic (PPG) readout ASIC with heart rate estimation directly from compressively sampled data. IEEE Trans. Biomed. Circuits Syst. **11**(3), 487–496 (2017)
34. W. Guo, Y. Kim, A.H. Tewfik, N. Sun, A fully passive compressive sensing SAR ADC for low-power wireless sensors. IEEE J. Solid State Circuits **52**(8), 2154–2167 (2017)
35. T.-F. Wu, S. Dey, M.S.-W. Chen, A nonuniform sampling ADC architecture with reconfigurable digital anti-aliasing filter. IEEE Trans. Circuits Syst. **63**(10), 1639–1651 (2016)
36. T.-F. Wu, C.-R. Ho, M.S.-W. Chen, A flash-based non-uniform sampling ADC with hybrid quantization enabling digital anti-aliasing filter. IEEE J. Solid State Circuits **52**(9), 2335–2349 (2017)
37. S.-J. Huang, et al., A 125MHz-BW 71.9dB-SNDR VCO-Based CT $\Delta\Sigma$ ADC with segmented phase-domain ELD compensation in 16nm CMOS, in *ISSCC* (2017), pp. 470–471
38. C.-H. Weng, C.-K. Wu, T.-H. Lin, A CMOS thermistor-embedded continuous-time delta-sigma temperature sensor with a resolution FoM of 0.65pJoC2. IEEE J. Solid State Circuits **50**(11), 2491–2500 (2015)
39. P.K. Yenduri et al., A low-power compressive sampling time-based analog-to-digital converter. IEEE J. Emerg. Top. Circuits Syst. **2**, 502–515 (2012)
40. H. Mamaghanian, N. Khaled, D. Atienza, P. Vandergheynst, Design and exploration of low-power analog to information conversion based on compressed sensing. IEEE J. Emerg. Top. Circuits Syst. **2**, 493–501 (2012)

# Chapter 2
# Noise-Shaping SAR ADCs

**Shaolan Li, Jiaxin Liu, Wenjuan Guo, and Nan Sun**

## 2.1 Introduction

The rise of 5G communication and internet of things (IoTs) is bringing an unprecedented scale of connectivity and sensing needs to emerging electronic systems. Meanwhile, the continuing miniaturization of computing devices (Bell's Law [1]) necessitates higher level of integration requirements and energy constraints. These trends drive analog-to-digital converters (ADCs) towards higher energy efficiency and lower cost yet without losing performance. Among various existing ADC solutions, the successive-approximation-register (SAR) architectures are deemed promising candidates for their simple, mostly digital, implementation and the resulting superior power efficiency. As depicted in Fig. 2.1, a typical SAR ADC consists of an array of switches and capacitors, which serves as a reference level generation digital-to-analog converter (DAC), a fully digital SAR logic and a comparator. It obtains the conversion result by converging the DAC voltage to the input in a binary search fashion. Owing to the absence of analog-intensive building blocks and a fully dynamic nature, the SAR ADC is highly compatible with duty-cycled applications and, importantly, can achieve better performance naturally with technology scaling.

However, on top of the stringent power efficiency requirements, many emerging systems also have an increasing demand for higher ADC resolution. A high-resolution/dynamic-range (DR) ADC provides tolerance to strong interference, such as blocker signals in receivers and motion artifacts in sensors. Thus, it relaxes the analog front-end (FE) requirements and enables higher flexibility with more processing shifted to the digital side. In the context of SAR ADC, however,

S. Li · J. Liu · W. Guo · N. Sun (✉)
The University of Texas at Austin, Austin, TX, USA
e-mail: nansun@mail.utexas.edu

© Springer Nature Switzerland AG 2020

A. Baschirotto et al. (eds.), *Next-Generation ADCs, High-Performance Power Management, and Technology Considerations for Advanced Integrated Circuits*,
https://doi.org/10.1007/978-3-030-25267-0_2

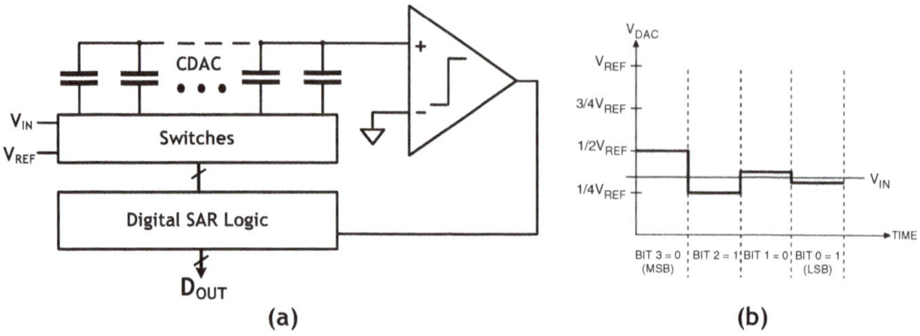


**Fig. 2.1** (**a**) Generic block diagram of a SAR ADC. (**b**) Example 4b conversion

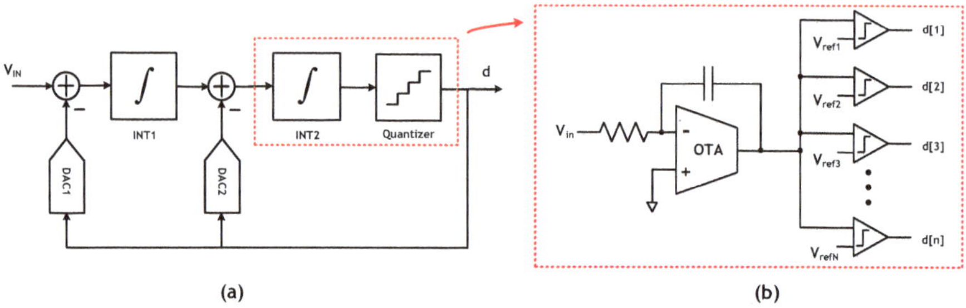


**Fig. 2.2** (**a**) Example block diagram of a second-order ΔΣ ADC. (**b**) Classic integrator and quantizer implementation

it becomes quite challenging to design for high resolution. While the hardware complexity of the SAR ADC remains relatively constant in the mid-resolution range (e.g., 6–10 bit), it grows drastically as resolution increases. In high-resolution SAR ADCs, performance is limited by thermal noise (including comparator noise and sampling $kT/C$ noise) and capacitor mismatch. Suppressing these requires increased comparator loading as well as DAC size, leading to speed degradation. Thus, stronger gate driving is needed. The aforementioned advantage in power efficiency therefore rapidly degrade in the high-resolution regime. To address this challenge, several techniques have been proposed. Harpe et al. [2] and Chen et al. [3] demonstrated majority voting and statistical estimation, respectively, to suppress comparator noise using the hidden information in the residue voltage. Nevertheless, Harpe et al. [2] requires a carefully tuned meta-stability detector, and Chen et al. [3] suffers from a severe speed penalty as many least significant bit (LSB) comparisons need to be repeated. Pipelining is another solution to relax the noise constraint by leveraging inter-stage gain [4–8]. However, it is sensitive to inter-stage gain inaccuracy, which causes quantization error leakage from the most significant bit (MSB) conversions, leading to non-linearity and limited resolution improvement.

Traditionally, high-resolution ADCs are implemented using the ΔΣ architecture, whose conceptual diagram is shown in Fig. 2.2. The ΔΣ ADC leverages oversampling to relax the sampling capacitance by attenuating the $kT/C$ noise by the oversampling ratio (OSR), while at the same time shaping the quantizer noise (including comparator noise and quantization error), thus allowing high resolution to be realized even with low quantizer accuracy. Nevertheless, the classic implementation of the $\Delta\Sigma$ ADC involves operational transconductance amplifiers (OTAs) and precision comparators. These building blocks are power hungry and incompatible with device and voltage scaling. From the power efficiency perspective, the traditional design of $\Delta\Sigma$ ADCs is hence suboptimal for many energy-stringent emerging applications.

While the $\Delta\Sigma$ ADC lacks the desired power efficiency, its underlying principle however suggests a strong complement to the issues in high-resolution SAR ADC design. In light of this, recent research has proposed the concept of the noise-shaping SAR (NS-SAR) ADC. This emerging architecture seeks to hybridize the energy-efficient SAR operation with the $\Delta\Sigma$ mechanism. By doing so, the issues of excessive comparator noise and DAC area can be overcome by oversampling and noise shaping, while the $\Delta\Sigma$ modulation can be realized in a much more efficient manner, with great potential for the simultaneous-realization of high resolution and low power. With such merit, the NS-SAR ADC has attracted much attention in recent years. Figure 2.3 compares state-of-the-art NS-SAR ADCs with standard SAR and $\Delta\Sigma$ ADCs reported in major conferences and journals using the energy-resolution plot [9]. It can be seen that NS-SAR ADCs are at the efficiency frontier.

The generic implementation of the NS-SAR ADC can be abstracted as Fig. 2.4. It extends the operation of a standard SAR ADC with a few extra steps that filter and feed back previous conversion residues to enable the noise-shaping effect. These

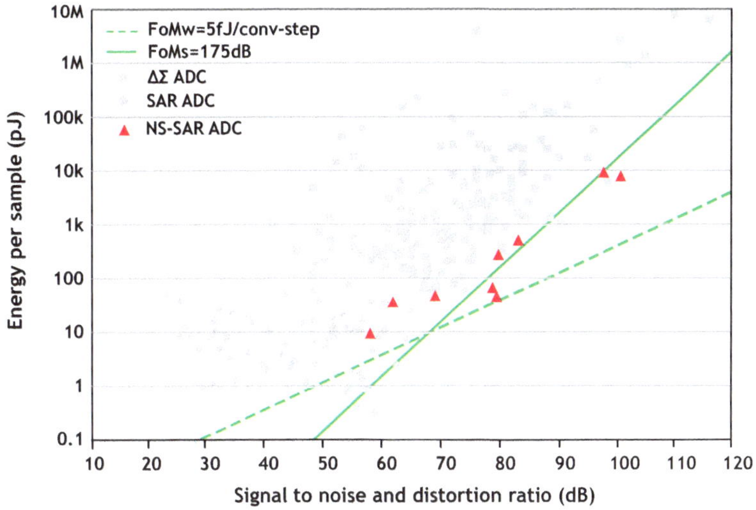


**Fig. 2.3** ADC energy vs SNDR survey for SAR, $\Delta\Sigma$, and NS-SAR ADCs. Credit: Boris Murmann ADC Survey

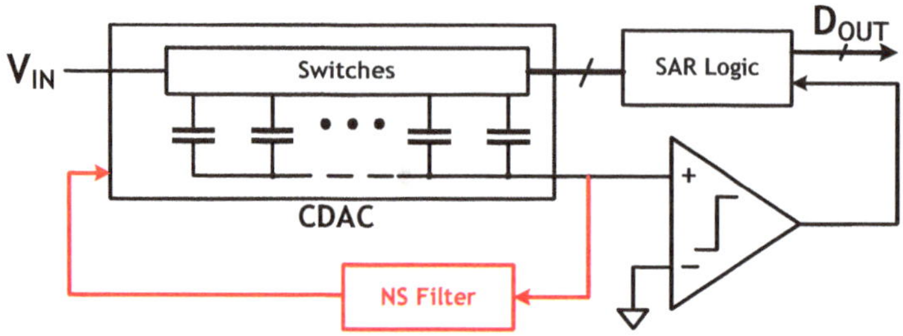


**Fig. 2.4** Generic block diagram of an NS-SAR ADC

steps, in analogy to the loop filters in $\Delta\Sigma$ ADCs, are the major factors that govern the overall performance of the NS-SAR ADC and determines if it can fully extract its aforementioned potential. This has thus fostered vigorous research efforts with diverse interesting techniques being proposed. In this chapter, we aim to provide a comprehensive survey of state-of-the-art NS-SAR ADC research. Specifically, this survey will first cover two major NS-SAR ADC implementation methodologies: the cascaded-integrator-feed-forward (CIFF) structure and the error-feedback (EF) structure, with insightful discussion on the contributions and drawbacks of existing designs. We will then review how the NS-SAR ADC itself can be further hybridized with other ADC architectures to achieve new technical advantages. Lastly, the survey will touch upon the effect of capacitor mismatch in the NS-SAR ADC and the methods to mitigate it.

## 2.2 NS-SAR ADCs Using the CIFF Structure

The CIFF structure is a widely adopted noise-shaping method in classic $\Delta\Sigma$ ADCs. As illustrated in Fig. 2.5b, this structure consists of a loop filter built with analog integrators and feed-forward paths. It processes the residue voltage obtained by subtracting the input with the converter output. The CIFF structure can be easily mapped to implement an NS-SAR ADC as indicated in Fig. 2.5a. The SAR operation assumes the role of the input feed-forward path, quantizer, and the feedback subtraction. The residue voltage can be extracted directly from the capacitor DAC (CDAC) at the end of each conversion, then processed through an integrator-based loop filter. The feed-forward path summing block can be realized using a multi-input comparator.

An important characteristic of the CIFF structure is that the effectiveness of the noise shaping highly depends on the ideality of the integrator in the loop filter. A $z$-domain model of a discrete-time (DT) analog integrator can be expressed as $\frac{\kappa_i}{1-\kappa_i z^{-1}}$. The ideality of the integrator is captured by the quality factor $\kappa_i$ ($0 < \kappa_i < 1$). Physically, the more ideal an integrator is, the more it holds the accumulation result (less leakage), which leads to a higher $\kappa_i$. Without loss of generality, assuming a

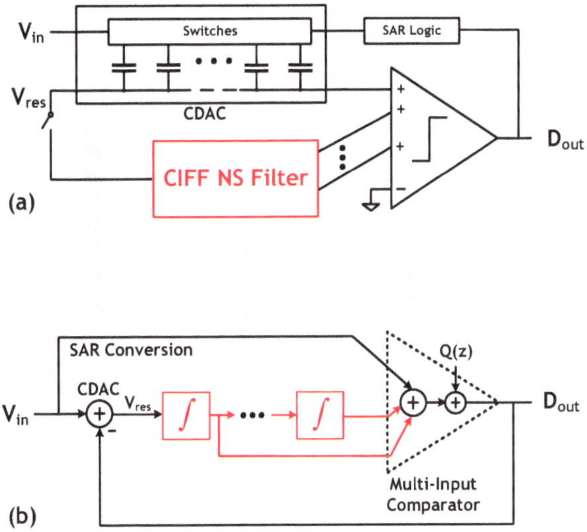


**Fig. 2.5** (**a**) General structure of a CIFF NS-SAR and (**b**) its signal flow

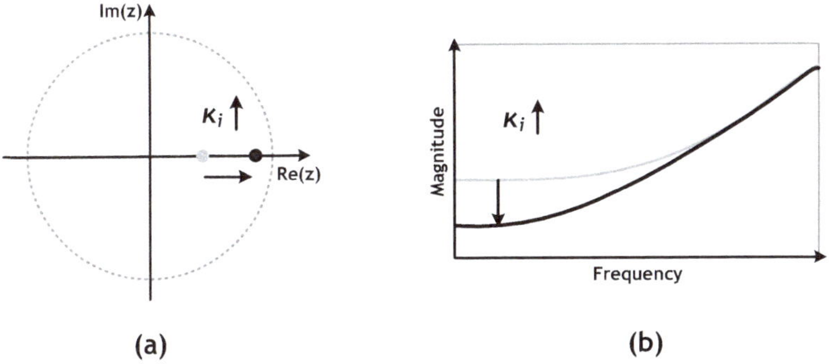


**Fig. 2.6** NTF zero (**a**) and magnitude (**b**) as a function of quality factor $\kappa_i$

first-order noise-shaping implementation, the noise transfer function (NTF) of the system is given by:

$$\mathrm{NTF}(z) = \frac{1}{1 + \mathrm{INT}(z)} = 1 - \kappa_i z^{-1} \tag{2.1}$$

With higher $\kappa_i$, the zero of the NTF will be placed closer to the unit circuit, thus enabling a sharper noise-shaping effect, as illustrated in Fig. 2.6. Achieving a higher $\kappa_i$, requires accurate charge transfer in the integrator, which typically relies on the use of high-gain, high-bandwidth OTAs. For a NS-SAR ADC, this is undesirable as the high power and hardware complexity needed for an accurate integrator will strongly conflict with the efficiency intent of this hybrid architecture. For this reason, a majority of the research effort in CIFF-based NS-SAR ADC design has been focused on optimizing the trade-off between noise-shaping effect and power efficiency.

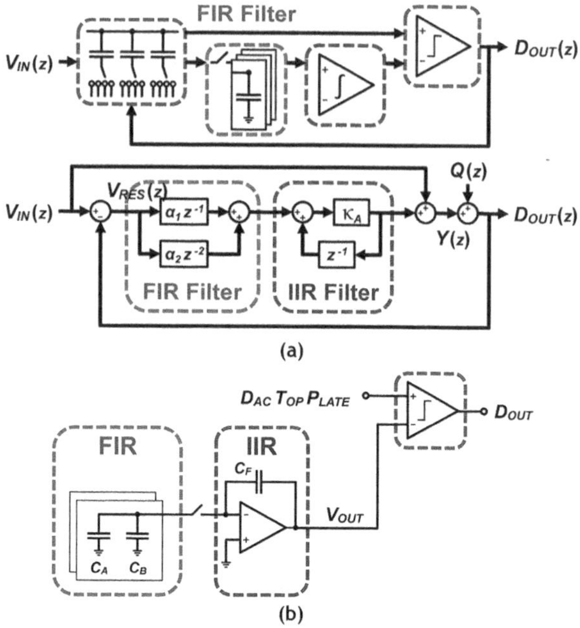


**Fig. 2.7** (**a**) Signal flow diagram of the FIR-IIR NS-SAR in [10]. (**b**) Circuit implementation

The very first NS-SAR ADC design, proposed by Fredenburg and Flynn [10], adopted an active-switched-capacitor (SC) implementation for its integrator. Figure 2.7 depicts its system and circuit diagram. To alleviate the power and complexity of the integrator, this work intentionally implemented a low $\kappa_i$ of 0.64, allowing the use of a low-gain amplifier and improving its robustness to parasitic charge sharing. To reconcile the noise-shaping degradation due to low $\kappa_i$, this work proposed using an additional finite-impulse-response (FIR) filter in conjunction with the integrator. By doing so, the loop filter can use a higher loop gain to compensate for the NTF flattening, while keeping the loop stable. The resulting NTF is expressed as follows:

$$\mathrm{NTF}(z) = \frac{1 - 0.64z^{-1}}{1 - 1.28z^{-1} + 0.64z^{-2}} \tag{2.2}$$

Figure 2.8 compares the NTF of this work with that of a standard first-order noise shaping. Noteworthily, for low OSR applications, the NTF of this work is in fact more advantageous owing to the rapid suppression. Due to the strong loop gain, a sharp out-of-band peaking occurs in the NTF of this FIR-IIR implementation. While such peaking is deemed hazardous in classic $\Delta\Sigma$ ADC, it is of less concern in a NS-SAR as long as the SAR quantizer provides sufficient resolution, as it highly reduces the risk of quantizer overload.

As shown in Fig. 2.7, this work implemented the FIR filter in a fully passive switched-capacitor manner. The residue voltage is transferred from the CDAC

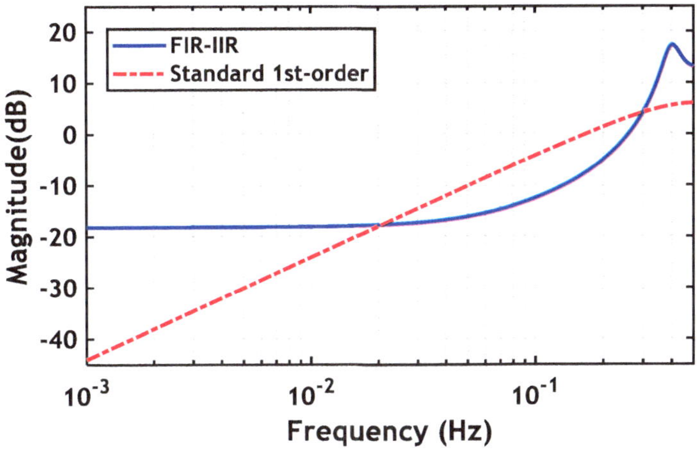


**Fig. 2.8** NTF comparison between the FIR-IIR design [10] and a standard first-order shaping

to the FIR simply through passive charge sharing. The attenuation effect can be compensated by the gain of the active IIR stage. This arrangement minimizes the active components required and thus is beneficial from a low complexity perspective. On the other hand, it exhibits a drawback in terms of thermal noise. During the residue transfer and FIR operation, extra $kT/C$ noise will be created. This noise will see an amplification when input referred due to the absence of preceding gain. As a result, larger capacitance is required to suppress the additive noise, leading to an area and power penalty.

To address this issue, in another NS-SAR design adopting a similar FIR-IIR loop filter, Shu et al. [11] placed a buffer between the CDAC and the FIR to provide active residue copying with a gain of 2. Its implementation is illustrated in Fig. 2.9. Leveraging the fact that the residue voltage is very small (in the mV range), this buffer is implemented as an open-loop amplifier operating in incomplete settling mode to minimize the power consumption. Incomplete settling brings two benefits. It relaxes the bias current requirement, and it also reduces the noise bandwidth of the amplifier, limiting its noise contribution. The small input also allows the buffer's input pair to be biased in sub-threshold region to achieve high efficiency. Yet the use of a buffer increases the static analog components in the circuit, making the design less amenable to process and voltage scaling.

Apart from the single-amplifier FIR-IIR approach discussed above, another notable CIFF NS-SAR work demonstrating a full active-SC implementation is reported by Obata et al. [12]. This work implemented a third-order system using a classic three-integrator network, as depicted in Fig. 2.10. The highlight of this work is that the amplifiers used in the integrator adopt a simple two-stage resistive-load design. These require no common-mode feedback and only need a minimum number of bias points, therefore they can be turned off most of the time and quickly powered up during the short integration phase. In other words, the power of the

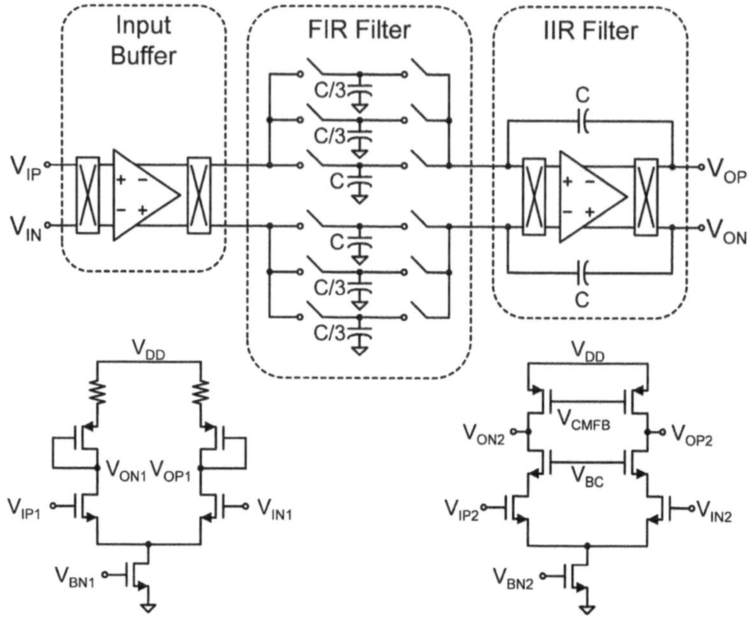


Fig. 2.9 FIR-IIR filter with input buffer proposed in [11]

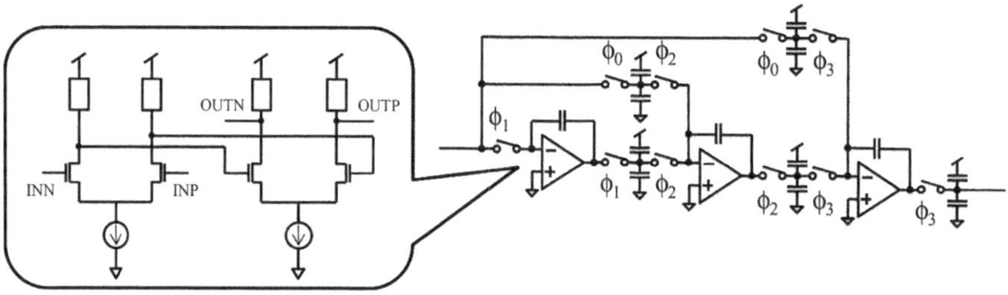


**Fig. 2.10** Full active-SC NS-SAR filter proposed in [12]

amplifiers is saved through duty cycling. However, this technique is less suitable for high speed applications, as the benefit of duty cycling diminishes when the start-up time of the amplifier becomes significant.

While the NS-SAR ADCs we have discussed so far share a common theme on how to minimize the power overhead due to the use of active-SC integrators, there is another category of NS-SAR design that, on the other hand, completely obviates the need for active amplifiers in the loop filter and adopts a fully passive implementation. Guo and Zhuang et al. demonstrated a first-order and second-order NS-SAR ADC using fully switched-capacitor techniques in [13] and [14, 15], respectively. The circuit and signal flow diagram of the second-order design is presented in Fig. 2.11. In Guo's work, the residue voltage is first sampled to an

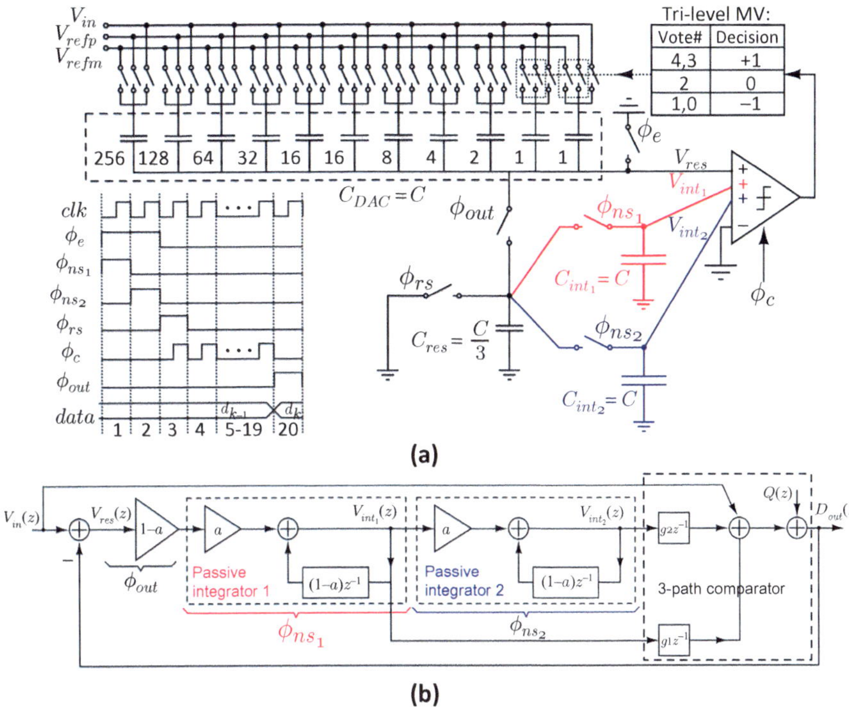


**Fig. 2.11** (**a**) Circuit and (**b**) signal flow of the second-order fully passive CIFF NS-SAR proposed in [14, 15]

intermediate capacitor $C_{res}$. $C_{res}$ will then subsequently share charge with two non-reset capacitors $C_{int1}$ and $C_{int2}$. This process realizes two leaky integration operations, as suggested in the signal flow diagram. Note that passive charge sharing will result in an integration gain lower than unity. To realize the loop gain, this work leverages the relative strength of the input pairs in the multi-input comparator, where the inputs connected to the CDAC, first and second integrator will be assigned with increasing weights, for example, 1:4:16. An input with stronger weight contributes more to the comparison, and is thus equivalent to a gain block. This work implements an NTF of $(1 - 0.75z^{-1})^2$, providing 18 dB of noise attenuation at low frequency. Compared to the active-SC employing NS-SAR ADCs, the fully passive NS-SAR ADCs offers several unique advantages. First, their fully dynamic operation allows them to be easily duty cycled to save power. Second, they can naturally take advantage of device scaling for better switches and denser capacitors, and can work well with lower supply voltages. Last but not least, all coefficients are defined by ratios. Specifically, the integrator coefficients are related to capacitor ratios, while the gain is defined by the size ratios of the comparator's input pairs. Hence, it is robust against process-voltage-temperature (PVT) variations.

Although using comparator input ratio to realize gain is hardware efficient, this method also brings drawbacks to the fully passive NS-SAR ADC. The large-ratio input pair will inject a significant amount of noise, making this gain-embedded multi-input comparator much noisier than that of a standard SAR ADC. The comparator noise in this case can even be the limiting factor on the SAR quantizer's achievable resolution. To address this limitation, a tri-level voting technique is developed in [14, 15], where the last two LSB conversions will repeatedly fire the comparator for 4 times for each bit, then determine the result according to the table shown in Fig. 2.11. By doing so, the comparator noise is suppressed through an averaging effect. A similar implementation is reported by a higher-speed passive NS-SAR design in [16]. To realize higher speed, this work uses a split comparator method.

While the comparator noise issue can be alleviated by the voting technique, the passive NS-SAR approach still suffers from unattenuated $kT/C$ noise injection as mentioned above in the FIR-IIR case. In Guo's work [14], the total thermal noise contribution of the passive loop filter in terms of in-band power spectral density (PSD) is $18kT/C_{\mathrm{DAC}}$, where $C_{\mathrm{DAC}}$ is the lumped capacitance of the SAR ADC's capacitor array. A key reason for this large noise overhead is the use of a small $C_{res}$. Liu et al. [17, 18] improved on this by getting rid of $C_{res}$. By doing so, the noise contribution reduces to $3.6kT/C_{\mathrm{DAC}}$. This also relaxes the total area and comparator noise contribution (due to lower ratio). But the integration operation cannot be combined in the sampling phase.

Seeking to achieve a better balance between design complexity, power consumption, and noise, a third category of NS-SAR ADCs exploits a combination of dynamic amplifier and passive switched-capacitor techniques, as reported by Liu et al. [19] and Miyahara et al. [20]. Unlike classic amplifiers that require static current biasing during operation, dynamic amplifiers work like a digital gate charging (or discharging) a capacitor, thus providing superior efficiency in processing discrete-time voltage quantities. The work reported by Liu et al. [19] adopts an FIR-IIR loop filter implementation. In this work, both the FIR and IIR filters are realized by fully switched-capacitor methods, as illustrated in Fig. 2.12. A dynamic amplifier is used at the front-end of the loop filter, amplifying the residue voltage before the filtering operation and supporting the loop gain. This

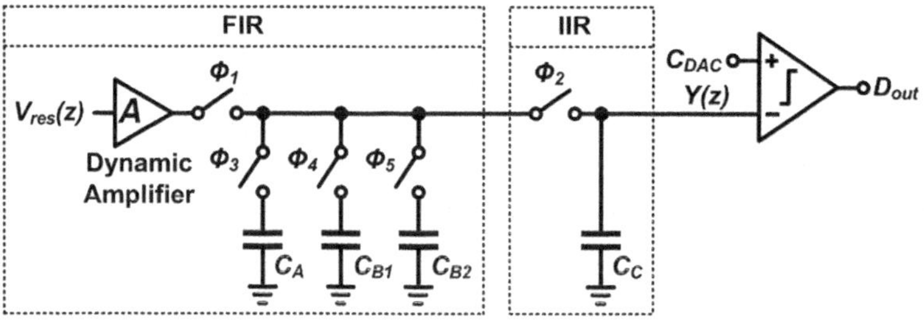


**Fig. 2.12** Dynamic amplifier and passive-SC implementation of FIR-IIR filter in [19]

effectively suppresses the noise contribution from the FIR and IIR stages, allowing small capacitors to be used for better area and speed. Miyahara's work [20] demonstrates a third-order full-integrator (IIR) based design. But instead of using the conventional OTA-based active-SC integrator, the authors proposed a design that uses two dynamic amplifiers and three capacitors to realize the ideal integrator response. The downside of using dynamic amplifiers, on the other hand, is that their gain is sensitive to the timing waveform. This renders the system less robust to PVT variations compared to OTA-based or fully passive implementations, but it requires extra complexity in their timing generation.

## 2.3 NS-SAR ADCs Using the EF Structure

While the maturity of CIFF-based $\Delta\Sigma$ realization provides a good foundation for NS-SAR ADCs, from the discussion above it can be seen that the CIFF NS-SAR implementations have several limitations in common. Firstly, due to the direct link between the NTF zero location and integrator quality as shown in the previous section, there exists a steep trade-off between noise-shaping effect and power efficiency. Although the FIR-IIR approach can alleviate this trade-off using loop gain to compensate for a less ideal integrator design, this method becomes less effective for high-resolution high-OSR design as the NTF flattens out at low frequencies. Owing to the same trade-off, NTF optimization[1] is considered costly for CIFF NS-SAR ADCs, because it demands high quality integrators to ensure close proximity of NTF zeros to the unit circle. Existing NS-SAR ADCs in the CIFF structure have yet to report NTF optimization. Secondly, although using multi-input comparator obviates the need of an explicit path summing block, the extra input pairs are extra sources for noise and mismatch, which can lead to potential performance loss.

Researchers have thus investigated alternative methods to implement NS-SAR ADCs that can circumvent the drawbacks of the CIFF structure. Recent research has demonstrated that the error-feedback (EF) structure can be a promising solution. Figure 2.13 depicts a generic block diagram of the EF structure. In contrast to the CIFF methods that process the subtraction result of the input and converter output, the EF path extracts and filters the unshaped quantization error $Q(z)$. The filtered prior errors are then combined with the input signal during subsequent quantization. This structure produces a set of neat signal and noise transfer functions (STF and NTF) as shown in Fig. 2.13. The NTF expression indicates an attractive feature, that using an FIR loop filter, rather than low-loss integrators, will suffice for creating

[1]NTF optimization is a common practice in classic $\Delta\Sigma$ ADC design. It splits the NTF zeros into complex conjugate pairs other than placing them all on the real axis, such that they will create "stop-band" notches in the NTF. This will further minimize the energy of the shaped quantization noise.

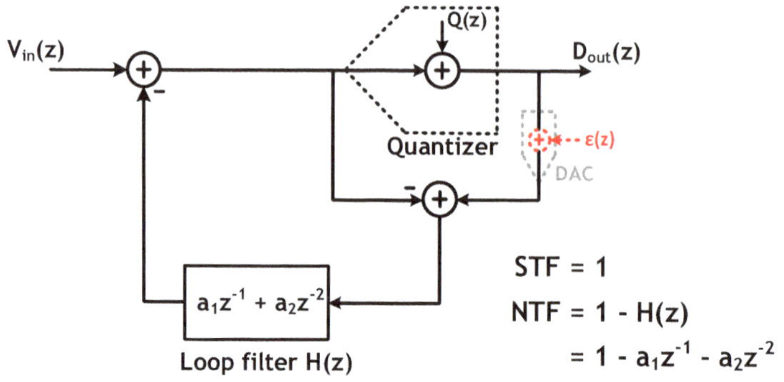


**Fig. 2.13** Example block diagram of a second-order EF structure

highly effective optimization notches. To elucidate with a 2nd-order example, a two-tap FIR filter $(a_1z^{-1} + a_2z^{-2})$ results in an NTF of $1 - a_1z^{-1} - a_2z^{-2}$, whose zeros are at $z = \frac{a_1 \pm \sqrt{a_1^2 + 4a_2}}{2}$. Provided that $a_2 = -1$ and $a_1 \leq 2$, the NTF zeros will be placed on the unit circle, thus realizing a strong notch. FIR filters are simply combinations of delay elements. They are relatively easy to design, thus allowing the EF structure to achieve aggressive noise shaping more efficiently.

Thanks to its structural simplicity, the EF structure has been a dominant method in digital $\Delta\Sigma$ modulators. In classic $\Delta\Sigma$ ADCs, however, the EF structure is traditionally deemed less attractive. The major reason is that the EF structure is vulnerable to inaccurate residue extraction. Classic $\Delta\Sigma$ ADCs often employ a flash ADC as the quantizer. To extract the quantization error, a separate extraction DAC is needed. In practice, it is difficult to guarantee that the quantizer and the extraction DAC will have the same reference levels and full scale. In the presence of discrepancy, an error will be added to the extraction result as indicated by $\varepsilon(z)$ in Fig. 2.13. This error will not be shaped and thus will corrupt the ADCs performance. Owing to this drawback, the EF structure has typically been used as an order booster at the back-end of a high-order integrator-based loop where its non-idealities causes little impact [21, 22], thereby restricting its potential. In the context of NS-SAR, on the other hand, this limitation is naturally resolved. The SAR operation concurrently quantizes and extracts using the same DAC, thus inherently providing accurate residues at the end of each conversion, allowing the EF structure to be adopted straightforwardly and reliably. This key observation led to the revitalization of the EF structure in NS-SAR ADCs.

The first attempt to use the EF structure for NS-SAR was reported by Chen et al. [23]. In this work, the authors proposed a modified EF scheme, in lieu of a direct implementation of Fig. 2.13, to suit a fully passive implementation. The circuit and signal flow diagram is presented in Fig. 2.14. Nevertheless, this fully passive modification renders a relative shallow NTF design of $\frac{1-0.5z^{-1}}{1+0.5z^{-1}}$. Due to the first-order nature and far zero placement, their scheme only provides a maximum of

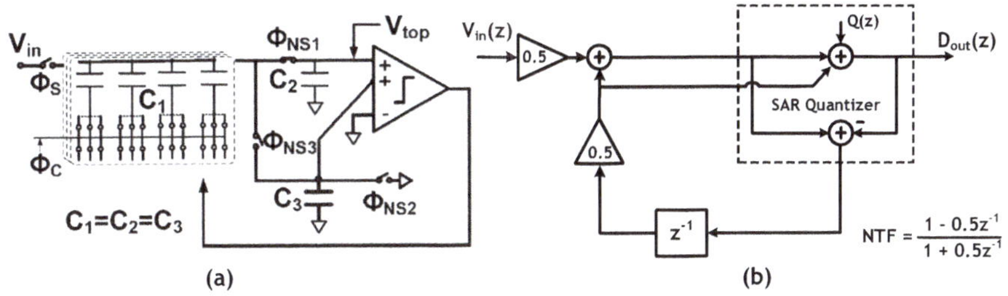


**Fig. 2.14** (**a**) Circuit and (**b**) signal flow of the fully passive EF NS-SAR proposed in [23]

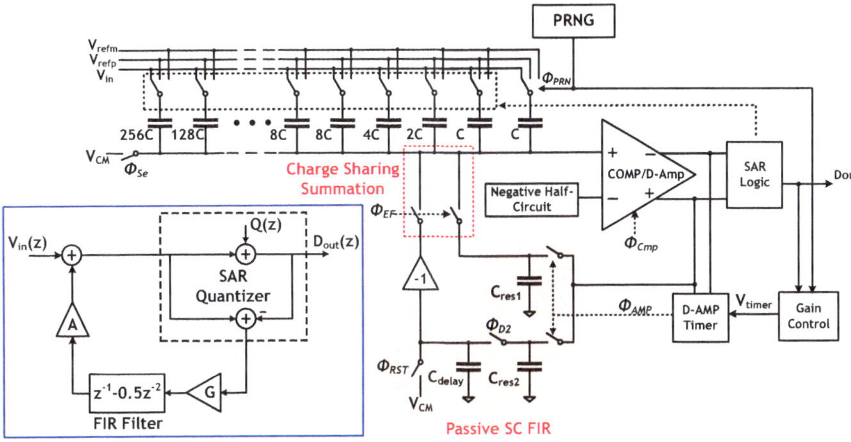


**Fig. 2.15** An EF NS-SAR featuring a comparator reused dynamic amplifier, passive FIR, and charge sharing feedback, as proposed in [25]

9.5 dB in-band attenuation. It also suffers from significant signal attenuation due to internal charge sharing, which results in limited DR performance.

Seeking to further extract the potential of the EF structure, Li et al. [24, 25] proposed a NS-SAR ADC featuring a more direct EF implementation, whose signal flow and circuit diagram are shown in Fig. 2.15. This work presents a reduced-variable EF scheme, where the EF path is implemented by a tunable gain followed by a two-tap fixed-coefficient FIR. This scheme is able to generate optimized NTF with high flexibility by simply using the gain as the single tuning knob. To visualize this effect, Fig. 2.16 plots the magnitude of the NTF as a function of the EF path gain $GA$. By effectively optimizing the NTF, this work demonstrates a 7 dB noise-shaping improvement compared to the state-of-the-art performance in [14] and [19]. The design realizes the reduced-variable EF scheme in a way that only requires minimum modifications to a standard SAR ADC are required. As shown in Fig 2.15, it implements the tunable gain by reusing the comparator as a dynamic amplifier, and the FIR and feedback summation through fully passive switched-capacitor network and charge sharing. Similar to [19], the use of residue pre-amplification highly

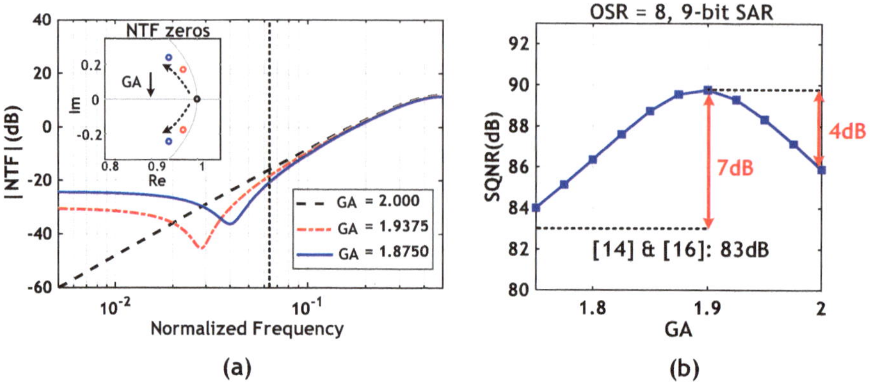


**Fig. 2.16** (**a**) NTF as a function of EF path gain. (**b**) SQNR as a function of EF path gain

relaxes the noise from switched-capacitor FIR, thus reducing its area overhead. Through the charge sharing summation, it obviates the need for a multi-input comparator, which brings an additional advantage.

As mentioned in the previous chapter, using a dynamic amplifier brings the drawback of PVT sensitivity. This work addresses this issue through a dither-based least-mean-square background calibration, which continuously monitors the gain of the amplifier and regulates the timing generation. Noteworthily, background calibration is traditionally deemed non-trivial for $\Delta\Sigma$ ADCs, as it relies on accurately replicating the analog transfer function in the digital domain. This work takes advantage of the simplicity of the EF structure, that its analog transfer function is well-controlled and easy to be digitally represented, thus enabling precise gain tracking using low-power digital techniques. In comparison, background calibration can be more challenging for CIFF NS-SAR ADCs for the same reason as classic $\Delta\Sigma$ ADCs. On the other hand, it should be noted that the accuracy requirement for the EF path gain becomes highly stringent for ultra-high-resolution high-OSR applications, as the NTF notch becomes more sensitive to the gain. This results in a trade-off with the convergence speed of the background calibration, which can be disadvantageous for applications that require fast start-up.

## 2.4 ADCs Hybridized with an NS-SAR

In recent years, there has been a popular trend in using the SAR ADC to replace the flash quantizer in classic architectures such as the pipelined ADC and $\Delta\Sigma$ ADC. Compared to flash ADC, which generally limited to 4-bit resolution (due to exponentially increased hardware complexity), the SAR ADC can provide higher resolution while consuming less power/area and exhibiting better tolerance to low voltage. However, the successive conversion introduces a speed penalty to the design, making the advantages of the SAR quantizer diminish in high-bandwidth applications. The development of NS-SAR ADCs brings a solution to this drawback. From our previous analysis, we can see that the STF of an NS-SAR ADC remains unity like a standard SAR. Thus, the NS-SAR can readily assume the role of a sub-quantizer. With noise shaping, it equivalently requires less raw SAR resolution, thus compensating for the speed penalty. Especially for ΔΣ ADCs, the extra orders of noise shaping provided by the NS-SAR quantizer can reduce the ΔΣ loop filter complexity, hence improving power and stability. The following are several notable existing ADC designs that embed the NS-SAR as part of the system.

Recently, Chen et al. [26] and Liu et al. [17, 18] demonstrated the use of a NS-SAR ADC as the quantizer of a discrete-time (DT) and continuous-time (CT) ΔΣ ADC, respectively. The architecture diagrams of these two works are presented in Figs. 2.17 and 2.18. The NS-SAR quantizers in these two works both adopt a fully passive design. Noteworthily, Liu's work [17, 18] embeds a second-order NS-SAR ADC in a first-order CT loop filter, it achieves a third-order noise shaping in total but requiring only a single OTA. This ΔΣ ADC can also be viewed as a CT-DT hybrid with the one-order noise shaping realized in CT domain and two orders realized in DT domain. It combines the merits of CT ΔΣ ADCs, which provides intrinsic anti-alias filtering, with the merit of DT ΔΣ ADCs, which is PVT robustness. The overall stability of the third-order CT ΔΣ ADC is similar to that of a first-order CT ΔΣ ADC when considering tolerance to RC variation.

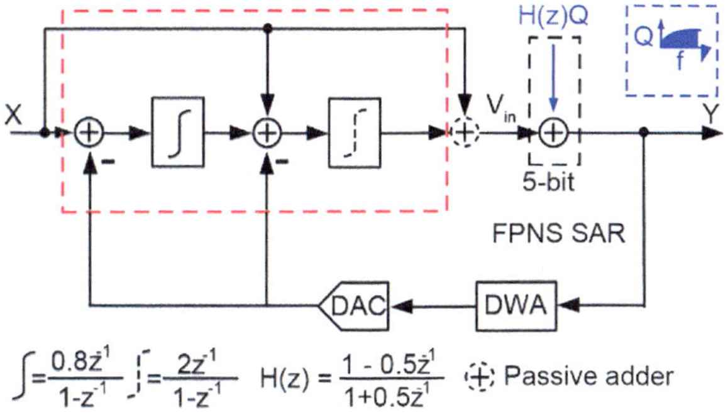


**Fig. 2.17** A single-OTA DT ΔΣ ADC using a first-order NS-SAR quantizer, as proposed in [26]

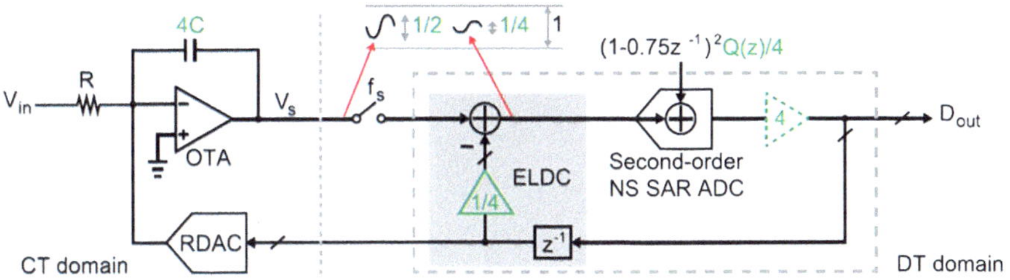


**Fig. 2.18** A single-OTA, CT-DT hybrid ΔΣ ADC using a second-order NS-SAR quantizer, as proposed in [18]

In another design, Song et al. [27] present a two-stage pipeline-SAR ADC featuring a NS-SAR ADC as the second stage. The NS-SAR adopts a first-order EF implementation. This work simultaneously leverages the inter-stage amplifier to extract the previous conversion NS-SAR residue and combines it with the first-stage residue. By doing so, the EF operation can be guaranteed lossless while saving the need for an extra residue amplifier.

## 2.5 Capacitor Mismatch in NS-SAR

While the quantization error and comparator noise are suppressed in NS-SAR ADCs, in practice the resolution is determined by another factor that we have not touched upon so far: the mismatch effect or non-linearity in the CDAC. To get a more intuitive understanding of this mismatch effect, we can represent the behavior of a SAR conversion and output reconstruction using Fig. 2.19. The SAR's behavior can be regarded as subtracting the input with a sequence of analog DAC weights. Due to mismatches, the analog DAC weights will drift away from their intended values. This error can be captured by an error term $E_{\mathrm{DAC}}$. Figure 2.19 suggests that the reconstructed output of the SAR, $D_{OUT} = D_{\mathrm{MSB}} + D_{\mathrm{LSBs}}$, is related to the input, quantization error $Q$ and DAC mismatch error $E_{\mathrm{DAC}}$ in the relationship of $D_{OUT}(n) = V_{IN}(n) + Q(n) - E_{\mathrm{DAC}}(n)$. This behavior can be captured in the signal flow of the NS-SAR ADC shown in Fig. 2.20. Note that during extraction, the DAC error will be intrinsically applied to the analog representation of $D_{OUT}$. For this reason, $E_{\mathrm{DAC}}$ will appear directly at the output without experiencing any transfer function. This suggests that the DAC mismatch error in the NS-SAR ADC can be mitigated in the same way as a standard SAR ADC. Existing solutions in foreground or background calibration can be adopted directly.

Apart from using calibration, the oversampling nature of NS-SAR ADCs in fact allows them to leverage mismatch error shaping techniques similar to those used in a $\Delta\Sigma$ ADCs. The classic mismatch error shaping (MES) technique in $\Delta\Sigma$ ADCs is dynamic element matching (DEM) [28, 29], which requires the DAC to

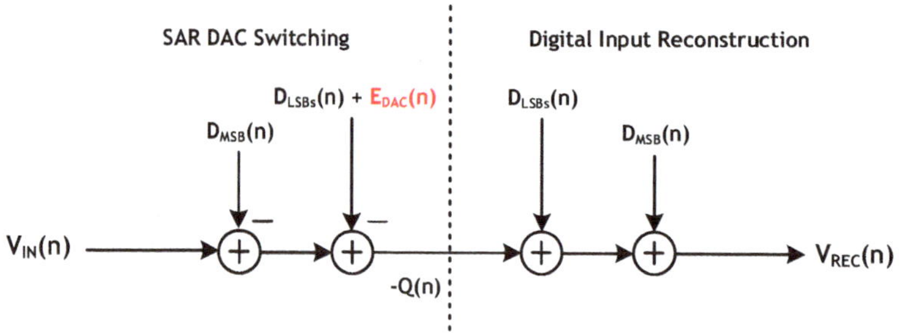


**Fig. 2.19** Behavior of the SAR operation and the effect of DAC mismatch

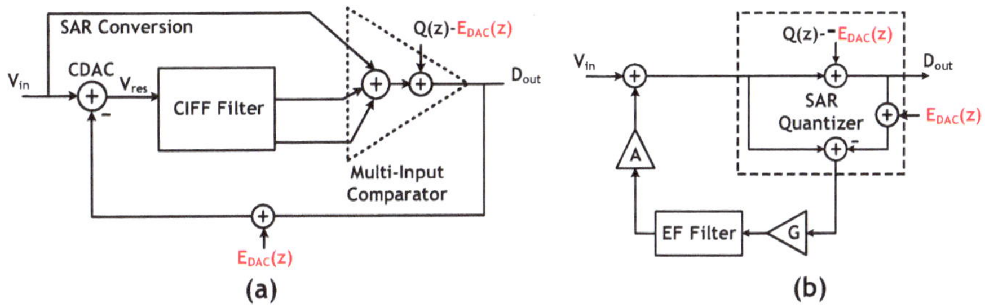


**Fig. 2.20** Mismatch effect modeling in the (**a**) CIFF and (**b**) EF NS-SAR ADCs

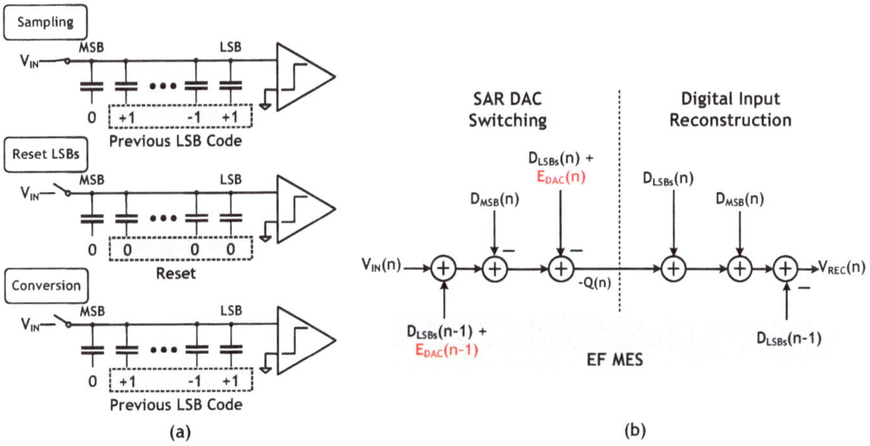


**Fig. 2.21** Error-feedback MES technique proposed in [11]: (**a**) Operation breakdown; (**b**) Error model

be decomposed into equally weighted unit elements. However, a critical limitation for DEM is that the number of independently addressable unity-weighted DAC elements grows exponentially with DAC resolution, making the hardware cost too high when it is directly applied in NS-SAR ADCs, where the number of bit of DAC is usually much larger than that of $\Delta\Sigma$ ADCs.

In [11], Shu et al. developed a new MES technique that well suits the nature of NS-SAR ADCs, as is shown in Fig. 2.21. The idea is to reapply the conversion result of the previous cycle during sampling. By doing so, it explicitly feeds back the mismatch error from the previous conversion cycle, thus creating an EF operation on the mismatch error, making it first-order shaped. In implementation, it is realized by simply delaying the reset of the LSB capacitors. The digital output will be subtracted by the last result to recover the conversion result. This MES technique can directly operate on a power-of-two-weighted DAC, therefore its hardware cost grows only linearly with the number of bits and is substantially lower than DEM. However, the MES of [11] comes with the limitations of limited shaping capability and reduced dynamic range. Recently, Liu et al. systematically generalized this MES technique to second-order [30] and other more advanced forms [31], enhancing its shaping capability and enabling it to be applied to various types of low-pass, band-pass, and high-pass converters. Moreover, the work of [31] addresses the critical dynamic range loss problem with digital prediction and names this new class of MES technique as EF MES based upon its operation principle.

## 2.6 Conclusions

In this chapter, we have presented a comprehensive review of NS-SAR ADCs. Our discussion begins from the motivation of the architecture, followed by a comprehensive survey covering existing architectures and design techniques for existing NS-SAR ADCs. We have reviewed standalone NS-SAR ADC designs in both CIFF and EF structures, the use of embedded NS-SAR ADCs as a subquantizer in other architectures, as well as techniques for addressing the CDAC mismatch for NS-SAR ADCs. NS-SAR ADCs are versatile, simple to design, and scaling-friendly. Their performance is expected to improve as CMOS technology scaling continues. This is still very much an active area of research with many research groups, including the authors, contributing to this area. It is also expected that there will be more innovative architectures that will further extract the merit of this framework and push the envelope of bandwidth and power to unprecedented levels.

## References


1. C.G. Bell, R. Chen, S. Rege, Effect of technology on near term computer structures. Computer **5**(2), 29–38 (1972)
2. P. Harpe, E. Cantatore, A. van Roermund, An oversampled 12/14b SAR ADC with noise reduction and linearity enhancements achieving up to 79.1dB SNDR, in *2014 IEEE International Solid-State Circuits Conference Digest of Technical Papers (ISSCC)* (IEEE, Piscataway, 2014), pp. 194–195
3. L. Chen, X. Tang, A. Sanyal, Y. Yoon, J. Cong, N. Sun, A 0.7-V 0.6- $\mu$W 100-kS/s low-power SAR ADC with statistical estimation-based noise reduction. IEEE J. Solid-State Circuits **52**(5), 1388–1398 (2017)
4. C.C. Lee, M.P. Flynn, A SAR-assisted two-stage pipeline ADC. IEEE J. Solid-State Circuits **46**(4), 859–869 (2011)
5. H.Y. Lee, B. Lee, U.K. Moon, A 31.3fJ/conversion-step 70.4dB SNDR 30MS/s 1.2V two-step pipelined ADC in 0.13$\mu$m CMOS, in *2012 IEEE International Solid-State Circuits Conference* (IEEE, Piscataway, 2012), pp. 474–476
6. B. Verbruggen, M. Iriguchi, J. Craninckx, A 1.7 mW 11b 250 MS/s 2-times interleaved fully dynamic pipelined SAR ADC in 40 nm digital CMOS. IEEE J. Solid-State Circuits **47**(12), 2880–2887 (2012)
7. F. van der Goes et al., A 1.5 mW 68 dB SNDR 80 Ms/s 2$\times$ interleaved pipelined SAR ADC in 28 nm CMOS. IEEE J. Solid-State Circuits **49**(12), 2835–2845 (2014)
8. Y. Lim, M.P. Flynn, A 1 mW 71.5 dB SNDR 50 MS/s 13 bit fully differential ring amplifier based SAR-assisted pipeline ADC. IEEE J. Solid-State Circuits **50**(12), 2901–2911 (2015)

9. B. Murmann, ADC performance survey 1997–2018, http://web.stanford.edu/murmann/adcsurvey.html
10. J.A. Fredenburg, M.P. Flynn, A 90-MS/s 11-MHz-bandwidth 62-dB SNDR noise-shaping SAR ADC. IEEE J. Solid-State Circuits **47**(12), 2898–2904 (2012)
11. Y.S. Shu, L.T. Kuo, T.Y. Lo, An oversampling SAR ADC with DAC mismatch error shaping achieving 105 dB SFDR and 101 dB SNDR over 1 kHz BW in 55 nm CMOS. IEEE J. Solid-State Circuits **51**(12), 2928–2940 (2016)
12. K. Obata, K. Matsukawa, T. Miki, Y. Tsukamoto, K. Sushihara, A 97.99 dB SNDR, 2 kHz BW, 37.1 $\mu$W noise-shaping SAR ADC with dynamic element matching and modulation dither effect, in *2016 IEEE Symposium on VLSI Circuits (VLSI-Circuits)* (IEEE, Piscataway, 2016), pp. 1–2
13. W. Guo, N. Sun, A 12b-ENOB 61$\mu$W noise-shaping SAR ADC with a passive integrator, in *Proceedings of 42nd European Solid-State Circuits Conference (ESSCIRC)* (IEEE, Piscataway, 2016), pp. 405–408
14. W. Guo, H. Zhuang, N. Sun, A 13b-ENOB 173dB-FoM 2nd-order NS SAR ADC with passive integrators, in *Proceedings of IEEE Symposium VLSI Circuits* (IEEE, Piscataway, 2017), pp. C236–C237
15. H. Zhuang, W. Guo, J. Liu, H. Tang, Z. Zhu, L. Chen, N. Sun, A second-order noise-shaping SAR ADC with passive integrator and tri-level voting. IEEE J. Solid-State Circuits **54**(6), 1636–1647 (2019)
16. Y.Z. Lin, C.H. Tsai, S.C. Tsou, R.X. Chu, C.H. Lu, A 2.4-mW 25-MHz BW 300-MS/s passive noise shaping SAR ADC with noise quantizer technique in 14-nm CMOS, in *Proceedings of IEEE Symposium on VLSI Circuits* (IEEE, Piscataway, 2017), pp. C234–C235
17. J. Liu, S. Li, W. Guo, G. Wen, N. Sun, A 0.029MM2 17-FJ/Conv.-Step CT $\Delta\Sigma$ ADC with 2nd-order noise-shaping SAR quantizer, in *IEEE Symposium on VLSI* (IEEE, Piscataway, 2018), pp. 201–202
18. J. Liu, S. Li, W. Guo, G. Wen, N. Sun, A 0.029-mm2 17-fJ/conversion-step third-order CT $\Delta\Sigma$ ADC with a single OTA and second-order noise-shaping SAR quantizer. IEEE J. Solid-State Circuits **54**(2), 428–440 (2019)
19. C.C. Liu, M.C. Huang, 28.1 A 0.46mW 5MHz-BW 79.7dB-SNDR noise-shaping SAR ADC with dynamic-amplifier-based FIR-IIR filter, in *IEEE International Solid-State Circuits Conference (ISSCC) Digest of Technical Papers* (IEEE, Piscataway, 2017), pp. 466–467
20. M. Miyahara, A. Matsuzawa, An 84 dB dynamic range 62.5–625 kHz bandwidth clock-scalable noise-shaping SAR ADC with open-loop integrator using dynamic amplifier, in *Proceedings of IEEE Custom Integrated Circuits Conference (CICC)* (IEEE, Piscataway, 2017), pp. 1–4
21. S. Pavan, R. Schreier, G.C. Temes, *Understanding Delta-Sigma Data Converters*, 2nd edn. (IEEE Press, Piscataway, 2017)
22. K. Lee, M.R. Miller, G.C. Temes, An 8.1 mW, 82 dB delta-sigma ADC with 1.9 MHz BW and −98 dB THD. IEEE J. Solid-State Circuits **44**(8), 2202–2211 (2009)
23. Z. Chen, M. Miyahara, A. Matsuzawa, A 9.35-ENOB, 14.8 fJ/conv.-step fully-passive noise-shaping SAR ADC, in *Proceedings of IEEE Symposium on VLSI Circuits* (2015), pp. C64–C65
24. S. Li, B. Qiao, M. Gandara, N. Sun, A 13-ENOB 2nd-order noise-shaping SAR ADC realizing optimized NTF zeros using an error-feedback structure, in *IEEE International Solid-State Circuits Conference (ISSCC) Digest of Technical Papers* (IEEE, Piscataway, 2018), pp. 234–236
25. S. Li, B. Qiao, M. Gandara, D.Z. Pan, N. Sun, A 13-ENOB second-order noise-shaping SAR ADC realizing optimized NTF zeros using the error-feedback structure. IEEE J. Solid-State Circuits **53**(12), 3484–3496 (2018)
26. Z. Chen, M. Miyahara, A. Matsuzawa, A stability-improved single-opamp third-order $\Delta\Sigma$ modulator by using a fully-passive noise-shaping SAR ADC and passive adder, in *European Solid-State Circuits Conference* (IEEE, Piscataway, 2016), pp. 249–252

27. Y. Song, Y. Zhu, C. Chan, L. Geng, R.P. Martins, A 77dB SNDR 12.5MHz bandwidth 0–1 MASH $\Sigma\Delta$ ADC based on the pipelined-SAR structure, in *IEEE Symposium on VLSI* (IEEE, Piscataway, 2018), pp. 201–202
28. R.J.V.D. Plassche, Dynamic element matching for high-accuracy monolithic D/A converters. IEEE J. Solid-State Circuits **11**(6), 795–800 (1976)
29. B.H. Leung, S. Sutarja, Multibit $\Sigma - \Delta$ A/D converter incorporating a novel class of dynamic element matching techniques. IEEE Trans. Circuits Syst. II Analog Digit. Signal Process. **39**(1), 35–51 (1992)
30. J. Liu, G. Wen, N. Sun, Second-order DAC MES for SAR ADCs. Electron. Lett. **53**(24), 1570–1572 (2017)
31. J. Liu, C. Hsu, X. Tang, S. Li, G. Wen, N. Sun, Error-feedback mismatch error shaping for high-resolution data converters, in *IEEE Transactions on Circuits and Systems I: Regular Papers* (IEEE, Piscataway, 2018)

# Chapter 3
# Efficient High-Resolution Nyquist ADCs

**P. Vogelmann, Y. Luo, M. A. Mokhtar, and M. Ortmanns**

## 3.1 Introduction

In many sensor applications, such as in high-channel-count biomedical applications, analog-to-digital converters (ADCs) with rather high resolution, are needed. Simultaneously such ADCs must be capable of being multiplexed between individual channels without memory and thus without inter-sample interference. Therefore, they need to provide true sample-to-sample conversion at Nyquist rate. Successive-approximation register (SAR) ADCs offer the best efficiency for a huge range of sampling frequencies and resolutions [1]. However, to achieve high resolution (e.g., $\geq$14 bits) and high linearity (e.g., $\geq$90 dB), noise shaping and mismatch error shaping have been used almost exclusively [2, 3]. These techniques introduce memory to the system and thus such ADCs cannot be multiplexed. Similarly, freely running delta-sigma modulators (DSMs) dominate the state of the art in efficiency for high-resolution designs. Likewise, they use noise shaping and have memory within their loop and decimation filters, and thus cannot be multiplexed nor do they offer true Nyquist-rate sample-to-sample conversion.

In Fig. 3.1a, an overview of recently published ADCs is presented based on [1]. Firstly, we see that the terms high efficiency or high resolution are obviously relative. Depending on conversion speed, "High" resolution corresponds to a completely different number of bits. The intention of this chapter is to focus on high resolution in absolute terms, meaning more than, e.g., 14 bits of resolution and linearity. Obviously, this is an area where over the last decades very few designs have been published: there exists an almost Nyquist-ADC free gap, marked in red in Fig. 3.1a. Similarly, Fig. 3.1b shows an overview of published ADCs, but displaying

P. Vogelmann · Y. Luo · M. A. Mokhtar · M. Ortmanns (✉)
Institute of Microelectronics, University of Ulm, Ulm, Germany
e-mail: maurits.ortmanns@uni-ulm.de

© Springer Nature Switzerland AG 2020

A. Baschirotto et al. (eds.), *Next-Generation ADCs, High-Performance Power Management, and Technology Considerations for Advanced Integrated Circuits*,
https://doi.org/10.1007/978-3-030-25267-0_3

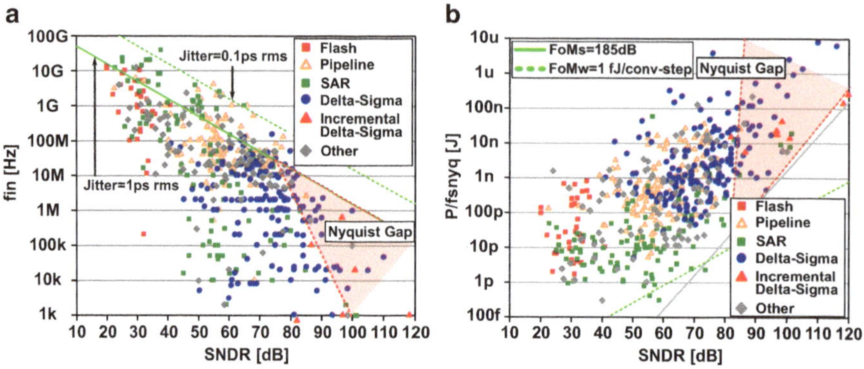


**Fig. 3.1** State-of-the-art comparison for ADCs. (**a**) State-of-the-art ADCs: Speed vs. SNDR. (**b**) State-of-the-art: Energy vs. SNDR

energy over SNDR. Again, it is clear that high efficiency is very rarely achieved without the use of oversampling and noise shaping, and true Nyquist-rate converters are again almost exclusively found outside the "Nyquist-gap." In the following sections, we will therefore review some of the advances in the field of efficient high-resolution Nyquist-rate converters, thereby focusing on SAR and incremental DSM that are capable of true Nyquist-rate conversion.

## 3.2 SAR ADCs

The main building blocks of a SAR based ADC are shown in Fig. 3.2, which are the comparator, the DAC, and the SAR logic. The major error sources for limited resolution are $kT/C$ noise of the sampling and switched-capacitor DAC (CDAC), the comparator noise, and the DAC mismatch. The major sources of power dissipation and thus of limited efficiency are the comparator, the switched CDAC, and—very minorly in advanced technologies—the digital control logic circuitry. The power consumption of the comparator is mainly determined by noise; the CDAC size and its power consumption are determined by technology, matching, and noise specification, and advanced switching schemes have helped to significantly suppress the CDAC power consumption [4–6].

Notably, in most published work, the power consumption of the necessary preceding anti-aliasing filter, the driver, and the reference-voltage buffer is not included in power consumption or non-ideality considerations, even though these can easily consume considerably more power than the ADC itself—especially for high resolution.

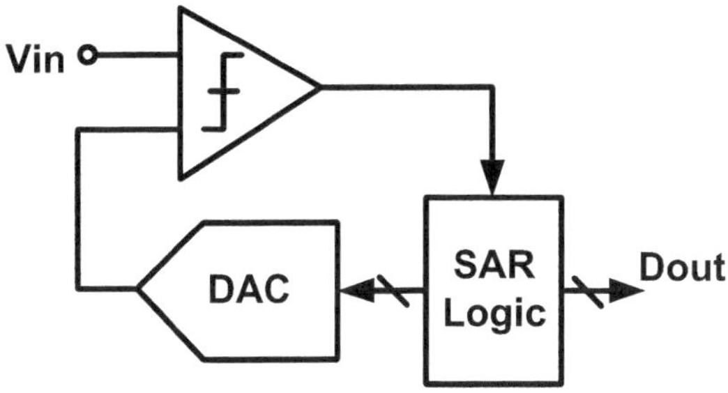


**Fig. 3.2** Block diagram of a SAR ADC

### *3.2.1 Oversampling, Noise Shaping, and Nyquist-Rate SAR ADCs*

SAR ADCs have gained high popularity due to their superior power efficiency for moderate resolution and moderate bandwidth A/D conversion [4, 6], especially in advanced CMOS technologies. For high resolution, however, the stringent accuracy requirement imposed on the circuit blocks of a SAR ADC degrades its efficiency. To achieve high linearity in the CDAC, a straightforward way is to increase the element size of the DAC, leading to an increased power consumption. Since at high resolution the required size of the DAC elements to meet noise specifications is typically smaller than that to meet the matching specifications [7], the energy-efficient solution to the DAC sizing is based on their noise contribution; thereafter, various calibration schemes are employed to ensure the adequate linearity of the DAC.

The comparator is usually the noise bottleneck of high-resolution SAR ADCs, since it needs to feature a resolution as small as the LSB of the ADC. A 6 dB noise reduction increases the power consumption at least by a factor of 4. In order to cope with comparator noise efficiently, oversampling and noise shaping are often used for SAR ADCs.

For example, the noise-shaping SAR ADC presented in [8] utilizes a cascaded FIR-IIR filter as a loop filter to perform noise shaping for the DAC residue voltage generated at the end of the SAR conversion. By noise shaping, the comparator noise and quantization noise in the signal band get suppressed, obviating the need for a power-hungry precision comparator. By oversampling, also the error due to $kT/C$ noise and DAC element mismatch can be suppressed as well. Such an ADC can be regarded as a hybrid between a SAR and a $\Delta\Sigma$ ADC. By doing so, the work in [9] achieved a Schreier FoM of 180.1 dB in 2017.

Together with the comparator and quantization noise, the DAC mismatch error also can be shaped [3]. By postponing the reset of the LSB CDAC array until the sampling switches open, the relative capacitor mismatch error, with respect to the MSB cell, gets first-order high-pass filtered, thus an SFDR = 105 dBFS, SNDR = 101 dB are achieved over 1 kHz bandwidth, leading to a Schreier FoM of 178.9 dB without any further calibration.

Those are perfect examples of how the state of the art has driven efficiency in high-resolution ADCs. Unfortunately this comes at the cost of losing true Nyquist-rate operation. When no oversampling or noise or mismatch error shaping can be employed to the whole SAR ADC, DAC mismatch can still be digitally calibrated; this leaves comparison noise as the key obstacle to implementing an efficient high-resolution SAR ADC. Given that only very few comparisons in one conversion cycle of a SAR ADC are noise critical, a data-driven noise-reduction method was proposed in [10]. There, based on the comparator decision time, noise-critical decisions are detected, for which the comparator decision is repeated several times and the final result is determined by voting. This technique, which basically employs comparator oversampling when it is needed, ensures that the extra power in the comparator is spent on the exact bit trials where it is needed.

To achieve a fully adaptive noise reduction based on the comparator input, an eye-opening VCO-based comparator is proposed in [11]. As shown in Fig. 3.3, the comparator consists of two identical VCO rings controlled by the two input signals and a phase detector. The input voltage difference is converted to a difference in oscillating frequency, which integrates into phase difference. The integration continues until the phase difference exceeds a pre-set threshold in the phase detector. Based on the lag/lead phase relation, the comparison result can be determined. Assuming that $N$ cycles of oscillation are performed before the decision is made, the comparator's input-referred noise is improved by $\sqrt{N}$. For a certain threshold, different comparator inputs lead to different $N$. Thus, the comparator input-referred noise and thus the power consumption are scaled with respect to the actual input voltage difference, resulting in a comparator with "power on demand." Although the VCO-based comparator is used in a 13-b SAR ADC in [11], it can be used for higher resolution. Furthermore, since this VCO-based comparator consists of inverters and some logic circuit, it is a good candidate for advanced technology.

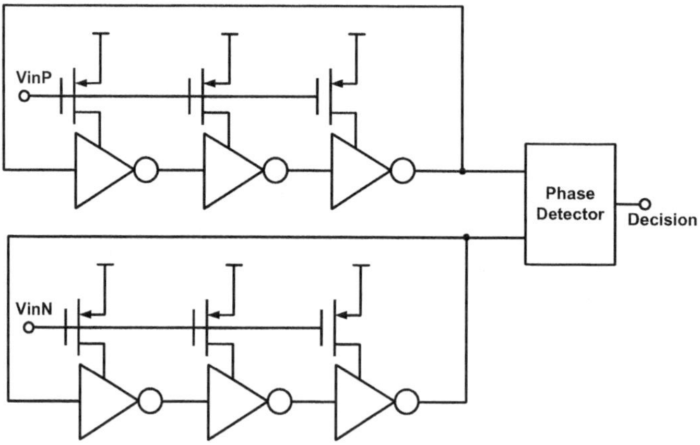


**Fig. 3.3** Eye-opening VCO-based comparator [11]

### 3.2.2 *Design Examples for High-Resolution Nyquist-Rate SAR*

To implement efficient high-resolution SAR ADCs at Nyquist rate, an efficiency enhanced comparator is essential, but still architectural innovation is needed for further error attenuation. In this section, three state-of-the-art designs are presented as examples of architectural innovations and solutions.

#### SAR ADCs with Coarse and Fine Phases

In a SAR ADC, the noise-critical decisions tend to occur towards the LSB decisions. Thus, the A/D conversion can be divided into two phases: coarse phase and fine phase. With redundancy between these two phases, the accuracy requirement for the coarse phase is relaxed [12]. Hence, a low noise comparator is only required for the fine phase, while a low power comparator can be adopted in the coarse phase. The coarse-fine architecture based SAR ADC has proven to be an efficient solution for resolution improvement [12, 13]. The design in [13] is elaborated here.

The block diagram of the 16-b SAR ADC from [13] is shown in Fig. 3.4. The first 7 MSBs are resolved by a coarse-phase SAR ADC consisting of a 7-b CDAC and a low-power dynamic comparator; then, the remaining 9 bits are resolved by a fine-phase SAR ADC consisting of an incremental $\Delta\Sigma$ DAC (I$-\Delta\Sigma$DAC) and a VCO-based comparator [11]. The I$-\Delta\Sigma$ DAC consists of an intrinsically linear 1.5-b, third-order digital $\Delta\Sigma$ modulator followed by a semi-digital FIR D/A interface and a third-order reconstruction filter. Non-binary radix is used in the fine phase, which helps to suppress settling errors. To address the offset between the coarse and fine phase comparators, one bit redundancy is introduced at the beginning of the fine phase. The input signal operates from the I/O supply (2.5 V), while the fine phase and the digital logic work at 1.1 V supply, which enhances efficiency.

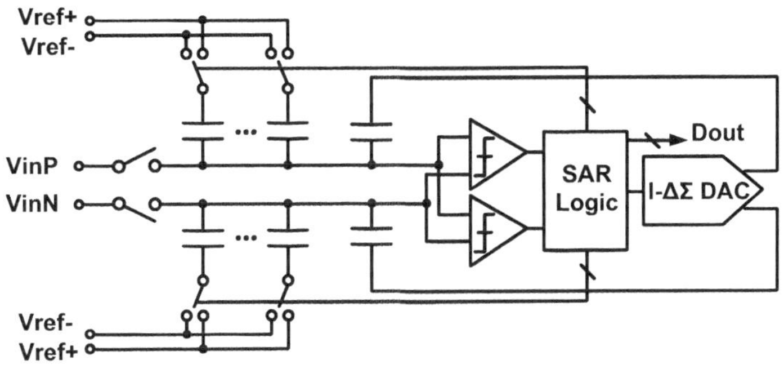


**Fig. 3.4** Block diagram of a SAR ADC with coarse and fine phases [7]

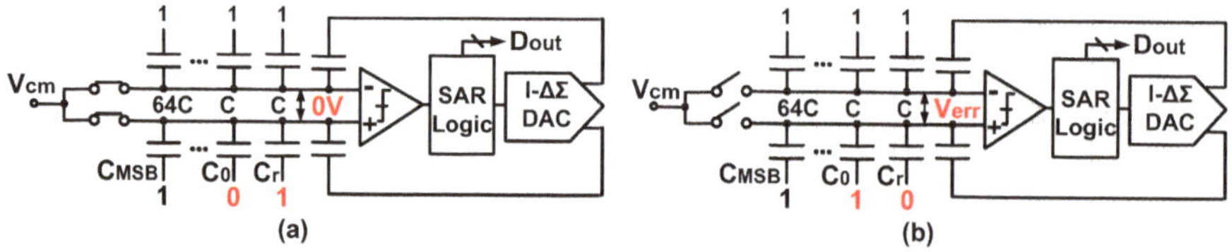


**Fig. 3.5** Example of detecting capacitor error [7]

The fine phase of the SAR ADC is also used to calibrate the 7-b coarse-phase CDAC during start-up. In the calibration scheme, the relative capacitance error of each binary capacitor in the coarse CDAC is estimated by the fine-phase SAR ADC. For example, to find out the relative error of $C_0$ with respect to $C_r$ in the positive capacitive array, as shown in Fig. 3.5a, first, the top plates of both the positive and negative capacitive arrays are set to $V_{\mathrm{cm}}$, and the bottom plates of all the capacitors are set to "1" except that of $C_0$, which is set to "0." Subsequently, the top plates of the two capacitive arrays disconnect from $V_{\mathrm{cm}}$; next, the bottom plate of $C_0$ is set to "1," and that of $C_r$ is set to "0," as seen in Fig. 3.5b. Ideally, the differential voltage of the top plates should be zero. Due to the mismatch between $C_0$ and $C_r$, a non-zero differential voltage appears at the top plates, which is a representation of the relative capacitance error of $C_0$ with respect to $C_r$. This voltage is digitized by the fine-phase SAR ADC. In a similar way, the relative capacitance error of all the capacitors in the CDAC arrays can be found. This information is used to linearize the coarse CDAC in the digital domain.

This design achieved an SNDR above 84.8 dB and an SFDR of 104 dBFS for an input frequency of up to 20 kHz at a Nyquist rate of 80 kS/s [13] and was limited by the unexpectedly high 1/f-noise contribution of the VCO-based comparator.

## A Pipelined SAR ADC with Digital Calibration

In [14], an 18-b pipelined SAR ADC with 98.6 dB SNDR and 100.2 dB dynamic range at 5 MS/s is presented, whose combination of resolution and speed sets the current state of the art in Fig. 3.1a. Instead of working on improving the comparator noise performance, this design inserts a residue amplifier (RA) during the SAR bit trials to decouple the ADC performance from the comparator noise.

The block diagram of the ADC is shown in Fig. 3.6 [14]. It consists of three sub-SAR ADCs with their individual CDACs. The MSB ADC resolves the first 4 MSBs. After the first 4 bits are determined, ADC1 resolves another 9 bits. At the end of the conversion of ADC1, the RA amplifies the residual quantization error of ADC1, which is eventually converted by ADC2.

Employing the low-noise, high-bandwidth RA with a gain of 64 is the highlight of this design. The RA is in auto-zero mode during ADC1 operation; when the conversion of ADC1 is completed, it amplifies the residue. Once ADC2 samples

Fig. 3.6 Block diagram of a pipelined SAR ADC [14]

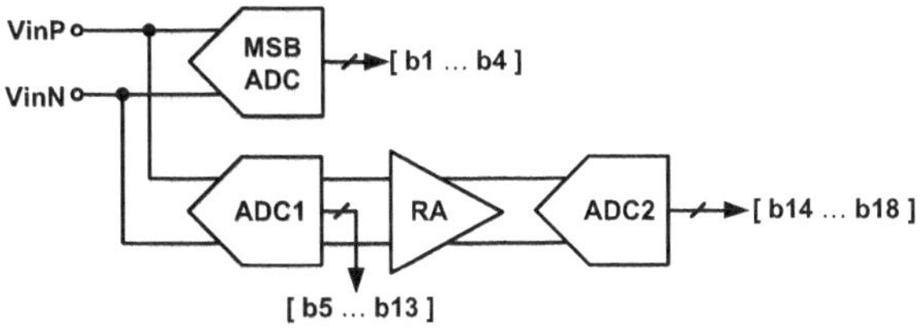


its output, the RA is powered down. The RA allows the accuracy of the individual bit trial among all the 18 bits to be no better than approximately 10 bits. Thus, the comparator noise requirement and the CDAC settling accuracy are significantly relaxed, which is an important reason why the ADC achieves above 100 dB SNDR at 5 MS/s.

The differential input signal swing is as large as 10 Vpp, thanks to the used technology node, while the active circuitry operates from 1.8 V core supply. The dual supply scheme improves the power efficiency as before in [13]. Due to the redundancy in ADC1, only a 4-bit accuracy is required in the MSB ADC, which is small, fast, and low power.

Since the CDAC element mismatch in the MSB ADC and in ADC2 is suppressed by the redundancy and the RA, respectively, the CDAC mismatch in ADC1 dominates the overall static non-linearity. One-time calibration of the CDAC in ADC1 is used to determine the actual bit weight, which is used to digitally correct the output code. Besides, the already known voltage dependency of the capacitor in the CDAC is employed to further linearize the conversion results.

It is worth noting that this design is implemented in a 0.18 μm technology, which is beneficial in terms of the RA implementation; however, as a digital intensive data converter, high-efficiency SAR ADCs tend to be implemented in advanced technology nodes; in this design though, as the RA is the key to the achieved high SNDR, the technology is of great advantage and the same architecture should become challenging in advanced technology nodes. Moreover, the design features a large digital correction circuitry, which makes the overall core area as large as 4 $mm^2$.

For the same architecture, instead of one-time calibration of the CDAC in ADC1, a signal-independent background calibration scheme is proposed in [15]. It determines the element mismatch error of the CDAC in ADC1 by a heuristic process with certain constraints and convergence conditions. The background calibration is all digital and implemented in an FPGA and Matlab. In addition to this background calibration, polynomial correction is applied to calibrate the static low-order error, which is not caused by CDAC1 mismatch, such as capacitor voltage dependency and signal-dependent sampling errors. The coefficients of the polynomial correction are obtained by measuring the harmonics in the frequency domain. With the proposed calibration techniques, the SAR ADC is able to achieve an SNDR above 100 dB and an SFDR above 110 dBc for an input frequency up to 100 kHz at a Nyquist rate of 1 MS/s, leading to a Schreier FoM of 175.5 dB.

## 3.3 Incremental ΔΣ ADCs

One of the most prominent architectures for high-resolution ADCs is the ΔΣ modulator, which makes use of oversampling and noise shaping to mitigate noise and matching requirements to large extent. However, these ADCs have memory due to the structure of their loop and decimation filters, which hinders them from operating as true Nyquist-rate converters. To combine the idea of ΔΣ modulation with the Nyquist-rate operation, incremental ΔΣ (I−ΔΣ) converters have been introduced, which periodically reset the entire memory within the loop filter and the digital decimation filter. While offering Nyquist-rate conversion capabilities, I−ΔΣs are commonly found less power efficient than freely running ΔΣ modulators; e.g., I−ΔΣ ADCs pay a noise penalty and usually work with higher oversampling ratios (OSR) than their freely running counterparts. Therefore, there is a continuous interest in all system-level and circuit-level techniques that help to close the efficiency gap between freely running and incremental ΔΣ ADCs.

### 3.3.1 *Operation Principle of Incremental ΔΣ ADCs*

Conventional ΔΣ ADCs are widely known for their ability to permit high resolution at low- to medium-speed ranges. Circuit non-idealities can be traded in for speed. This is due to the two fundamental principles: oversampling and noise shaping. These shall shortly be explained in the following before the incremental operation is introduced. The discrete-time, first-order modulator depicted in Fig. 3.7 shall serve as an example. For analysis, the non-linear quantizer is replaced by the white quantization noise model having an effective gain $k_{eff}$ and a quantization error $e_q[k]$. To simplify the following calculations, the quantizer gain is further simplified to be $k_{eff} = 1$. Under these assumptions, the Z-domain signal and noise transfer functions (STF and NTF) become

$$\mathrm{STF}(z) = \frac{D(z)}{U(z)} = z^{-1} \tag{3.1}$$

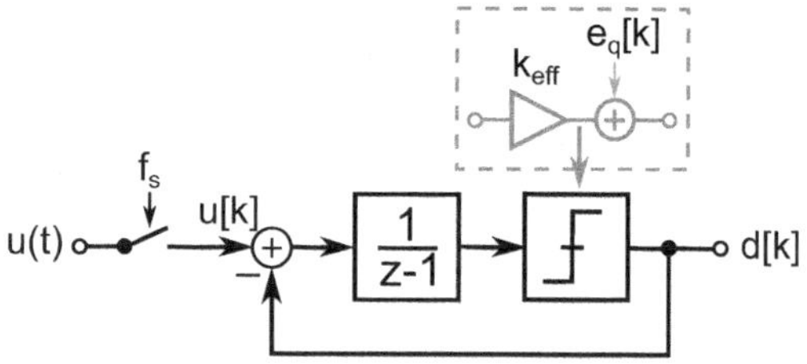


**Fig. 3.7** A freely running, discrete-time, first-order ΔΣ modulator

$$\mathrm{NTF}(z) = \frac{D(z)}{E_q(z)} = 1 - z^{-1}. \tag{3.2}$$

The first principle of $\Delta\Sigma$ modulators is oversampling: by overfulfilling the Nyquist criterion ($f_S \gg 2f_B$), sampling and quantization noise will be spread to a larger frequency band and can be filtered before they alias into the baseband by decimation.

The second principle is noise shaping. Looking at Eq. (3.2), the NTF possesses a high-pass characteristic. As the oversampling makes it possible to have a larger frequency range that can be divided in in-band and out-of-band, the NTF can be used to high-pass filter spectrally spread quantization noise (QN) and shape it into higher frequency bands. Such out-of-band QN can easily be filtered in the digital domain, which heavily increases the resolution. This noise-shaping is depicted in Fig. 3.8.

Combining both principles offers the possibility to use a low-resolution quantizer while still achieving high resolutions with superb efficiency, which is proven, e.g., by Fig. 3.1a, b and the Nyquist gaps discussed in Sect. 3.1.

In order to provide true sample-to-sample correspondence at Nyquist rate, any memory from one sample to another has to be erased within the system. This can be achieved by introducing a periodic reset after the conversion of every sample as shown in Fig. 3.9.

For such an incremental modulator it is not possible anymore to use a conventional low-pass filter for reconstruction as they come again with memory causing inter-symbol interference. When operated in an I$-\Delta\Sigma$ modulator, the filter itself needs to be reset together with the analog integrators in the modulator loop.

Unfortunately, the STF and NTF—calculated for the freely running modulator in Eqs. (3.1) and (3.2)—are not valid anymore as the reset signal transforms the linear

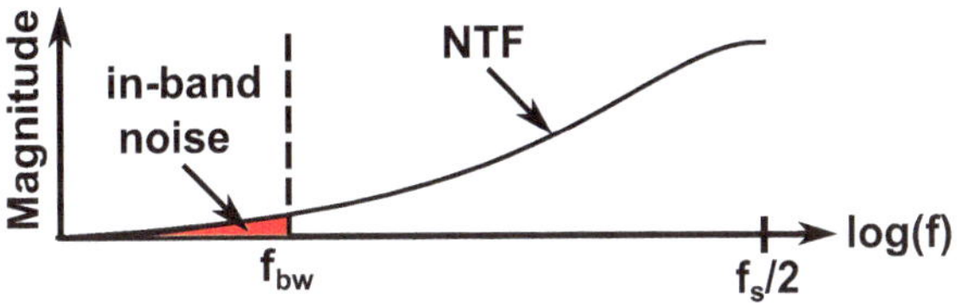


**Fig. 3.8** The principle of noise shaping

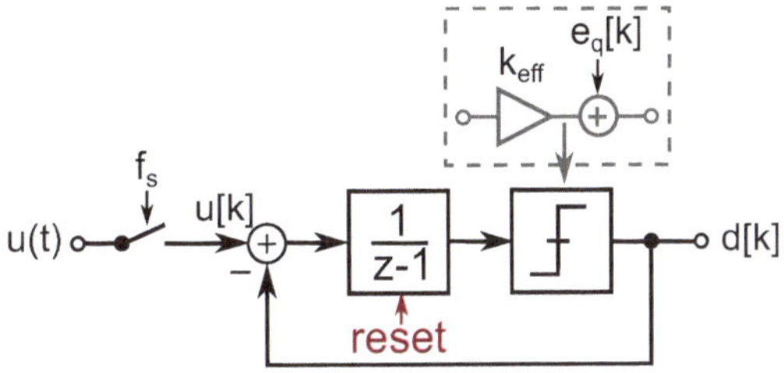


**Fig. 3.9** Resetting a freely running $\Delta\Sigma$ modulator to obtain an incremental modulator

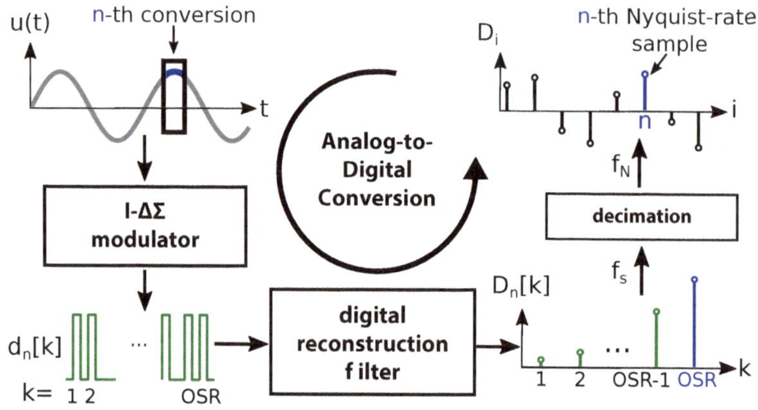


**Fig. 3.10** Working principle of an I−ΔΣ ADC

time-invariant system of a conventional ΔΣ modulator into a linear time-varying system; this does not allow for simple Fourier transformation anymore. Therefore, it is convenient to describe the system in the time domain by using the impulse responses of the individual components.

The working principle of an I−ΔΣ ADC (modulator+filter) is illustrated in Fig. 3.10. The continuously varying input signal u(t) is observed for a certain time frame as indicated by the black box in the top left. For the given example it is assumed that the current conversion is the n-th conversion, where n can be an arbitrary integer number. During this time, the input signal may slightly vary. A dedicated S&H amplifier can be used to keep the signal constant, but this leads to a significant increase in power consumption. It was, e.g., shown in [16] that the S&H upfront can be omitted. In consequence, higher frequency signal components (towards the band edge and higher) are slightly attenuated; this can usually be tolerated.

This part of the input signal is now fed to the I−ΔΣ modulator. During the whole conversion period, the modulator produces $OSR$ quantizer decisions. Also in I−ΔΣ modulators, $OSR$ is usually referred to as the oversampling ratio. The output sequence of the I−ΔΣ modulator is then fed to the digital reconstruction filter with an impulse response $h[k]$. Usually, for the sake of simplicity, a chain-of-integrators (CoI) filter is chosen as it yields a very good reconstruction quality for a comparably low complexity. Nevertheless, there are implementations making use of sinc filters, which provide line noise suppression at the cost of a worsened reconstruction quality that requires an increased OSR to restore the targeted SQNR as in [17]. The output of the CoI filter still runs at the full sampling rate $f_S$ of the I−ΔΣ. Therefore, only the last sample of every Nyquist-rate conversion is taken and the rest is discarded. As it can be seen in Fig. 3.10, the decimated samples $D_i$ are the digital representation of the analog input signal running at Nyquist rate $f_N$. The DFT of this signal can be calculated to evaluate the performance of the I−ΔΣ converter.

I$-\Delta\Sigma$ modulators can be realized in DT and CT architectures, offering similar advantages as their freely running counterparts. The calculation of the resulting STF and also a definition of an NTF for an I$-\Delta\Sigma$ and the respective prediction of their QN-performance were recently shown in [18].

## 3.3.2 *Design Examples for High-Resolution* I$-\Delta\Sigma$ *ADCs*

In the following, state-of-the-art designs are presented that make use of incremental $\Delta\Sigma$ modulators to achieve high-resolution and high-efficiency Nyquist-rate performance.

### The "Zoom ADC"

The architecture presented in [17] is widely known as the "zoom ADC." It is designed in a 160 nm technology node and operates from a 1.8 V supply. This ADC is able to achieve an outstanding peak SNR of 119.8 dB at a power consumption of 6.3 μW, thereby achieving an effective bandwidth of 12.5 Hz. This results in the best Schreier FoM of 182.7 dB for a Nyquist-rate converter at the time of its publication. A system overview of this ADC is depicted in Fig. 3.11. Even though this converter looks like a two-step ADC and it even operates in two operation cycles, it introduces a new operation principle. First, a SAR ADC converts the input signal X. This SAR ADC offers an effective resolution of 6 bits. After this coarse conversion, the approximate range is known where the input signal should be found. This information is used to adjust the feedback references of an I$-\Delta\Sigma$ modulator, which performs the fine conversion. A simplified circuit implementation of the I$-\Delta\Sigma$ is shown in Fig. 3.12. Thereby, the SAR ADC helps the I$-\Delta\Sigma$ modulator to zoom into the signal range, where it will perform the fine conversion. After filtering the output of the I$-\Delta\Sigma$ by means of a reconstruction filter—implemented as a resettable sinc-filter—the results of the coarse and the fine conversion are added, respectively.

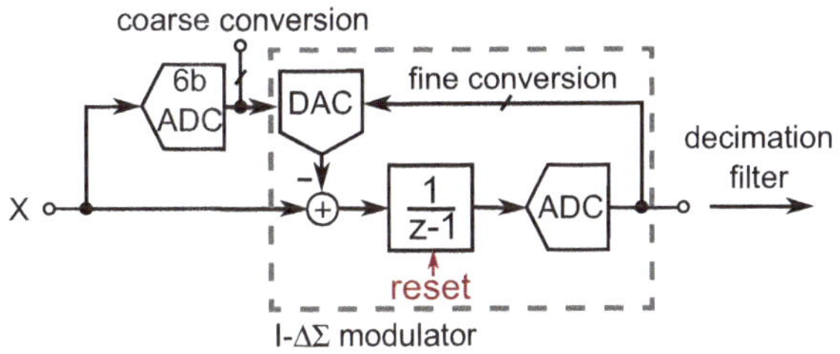


**Fig. 3.11** System-level overview of the "zoom ADC"

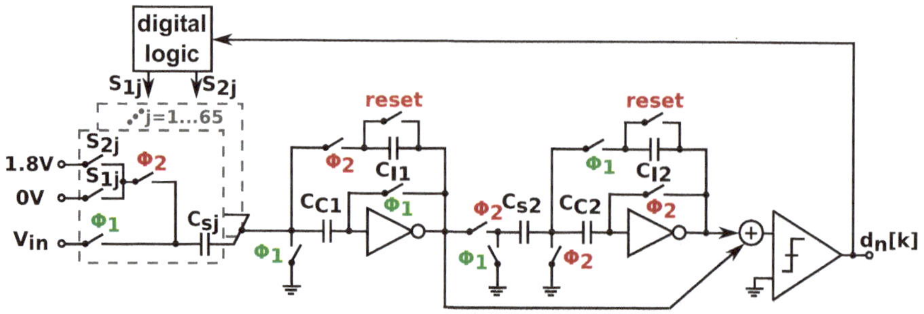


**Fig. 3.12** Circuit implementation of the I−ΔΣ used in the fine-conversion phase

As the effective input signal to the I−ΔΣ modulator is very small, the requirements on the input stage are greatly relaxed. Furthermore, as the coarse and the fine ADC share a big part of the same hardware, this can be reused for the I−ΔΣ resulting in an increased area and power efficiency, respectively. However, the selection of the feedback references in the I−ΔΣ modulator comes similar to using a multibit converter, such that mismatch in those feedback levels would cause major non-linearity. Thus, data-weighted averaging (DWA) is employed. As the OSR of this architecture is as large as 2000, DWA can be very effectively used. Furthermore, correlated double sampling (CDS) and system-level chopping are used to reduce 1/f-noise, as it is crucial for the performance of this low bandwidth ADC.

### The "Two-Step" ADC

Another highly efficient architecture using an I−ΔΣ modulator was presented in [19]. It is implemented in a 65 nm technology node and the circuit is operated from a 1.2 V supply voltage. The reported SNDR/DR are 90.8 dB and 99.8 dB, respectively. Given a power consumption of 10.7 μW and an effective bandwidth of 250 Hz, this results in a Schreier FoM of 173.5 dB. This is achieved using a two-step conversion.

The I−ΔΣ ADC, as shown in Fig. 3.13, operates on the input signal for 128 cycles. After this phase, the first integrator is reset by "reset1" and the modulator is reconfigured to perform the second conversion step. As has often been shown in literature, the residual quantization error is available at the output of the last integrator. By performing another A/D conversion of this error and by adding it—correctly scaled—to the output, the effective resolution can be increased; in literature, this is known as extended counting in I−ΔΣ ADCs. To avoid the necessity for another ADC, Chen et al. [19] reuses the hardware by simply reconfiguring the I−ΔΣ modulator as shown in Fig. 3.14. The last integrator is now used as a hold amplifier storing the residual error—that is left after the first conversion step—and feeding it to the first integrator in the loop. Thereby, the order of the I−ΔΣ is reduced from two to one, and the modulator runs for another 64

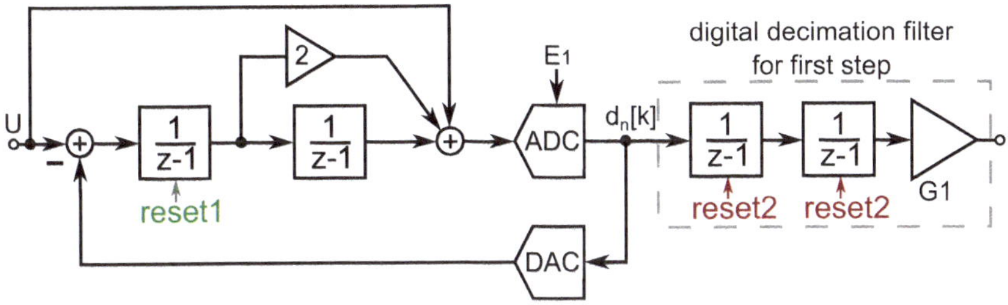


**Fig. 3.13** A second-order I$-\Delta\Sigma$ ADC performs the first conversion step

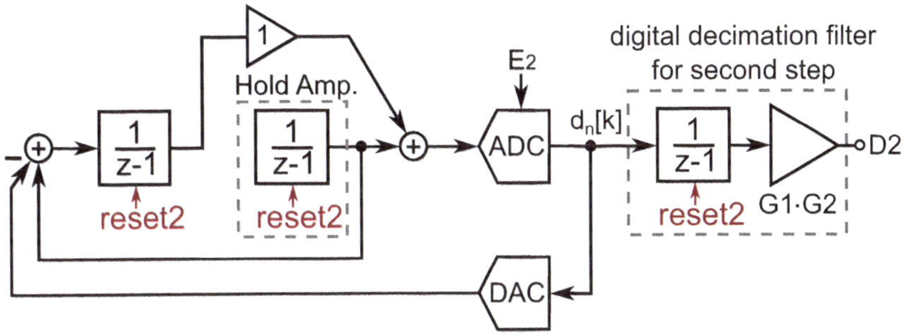


**Fig. 3.14** A first-order I$-\Delta\Sigma$ ADC performs the second conversion step

cycles on the residual error. After this conversion step, the digital outputs D1 and D2 are added and all elements are reset by the signal "reset2."

By making use of this implementation of extended counting, the reduction of the quantization noise is greatly improved. To achieve an SNDR/DR of 90.8 dB and 99.8 dB in a 250 Hz bandwidth, respectively, chopping techniques to remove the 1/f-noise contributions are again used. The OTA in the first integrator is chopped at a speed of $fs/2$, thereby greatly reducing the 1/f-noise contribution. In order to relax the requirements on the first integrator and to increase the maximum stable amplitude (MSA) as much as possible, a five-level quantizer is used in the presented work. This comes with the drawback of increased harmonic distortion, which is reduced by using DWA.

### The "Sliced" ADC

The architecture presented in [20] makes use of the dynamic increase of non-idealities during the operation of the I$-\Delta\Sigma$ modulator to increase the power efficiency. The modulator is implemented in a 180 nm technology node and the circuit is operated from a 3 V supply, thereby approximately consuming 1.1 mW. The reported SNDR/DR are 86.6 dB and 91.5 dB, respectively. Achieving an effective bandwidth of 100 kHz results in a Schreier FoM of 171.1 dB.

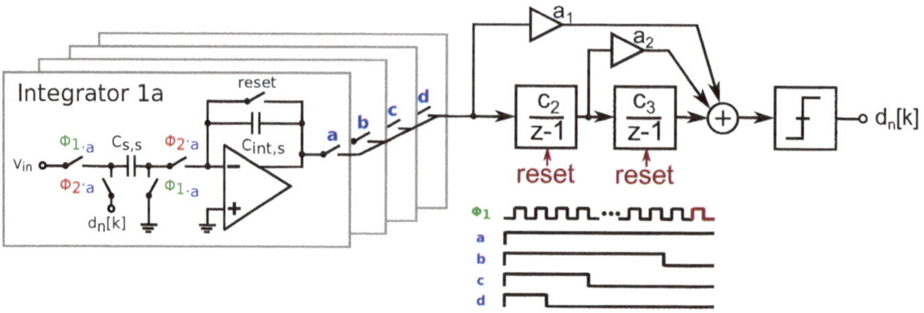


**Fig. 3.15** Implementation of the I$-\Sigma\Delta$ modulator including the "slicing technique"

This work makes use of the unequal weighting of the individual samples of the I$-\Delta\Sigma$ in the reconstruction filter. It can be shown that samples towards the end of a conversion are much less weighted by the digital reconstruction filter than samples at the beginning of a conversion [20]. This strongly depends on the order of the filter as well as on the OSR. This theoretically allows to increase circuit non-idealities towards the end of every Nyquist-conversion cycle, since there non-idealities are of less importance.

To achieve this, the work introduces the "slicing technique," in which the entire first integrator consisting of sampling network and OTA is split into four equal parts (or slices) as shown in Fig. 3.15. Every single slice can be deactivated during the runtime to save power. As the power savings linearly depend on the number of cycles that a slice is activated, but the contribution of the noise decreases superlinearly over the conversion cycles, power savings are possible by reducing the number of used slices during the runtime of each Nyquist-conversion cycle as indicated by the clocking scheme and the control signals a–d. By making use of this technique, the power consumption of this modulator can be reduced by 30%, while the SNR is barely affected.

### The "Linear-Exponential" Incremental ADC

Another architecture was recently presented in [21]; it makes use of a single-loop two-phase exponential incremental ADC. The presented circuit is implemented in a 65 nm CMOS technology with a 1.2 V supply. The ADC is able to achieve an SNDR of 100.8 dB as well as a dynamic range of 101.8 dB in a bandwidth of 20 kHz. The ADC consumes 550 $\mu$W and achieves a Schreier FoM of 176.4 dB. The consumed area is only 0.134 mm$^2$. The basic principle of this ADC is depicted in Fig. 3.16. In the first phase, the part indicated in red is inactive. The modulator works as a first-order I$-\Delta\Sigma$ modulator with a first-order CoI as reconstruction filter for 246 cycles. Instead of a single-bit quantizer, a multibit quantizer is used to increase the achievable resolution.

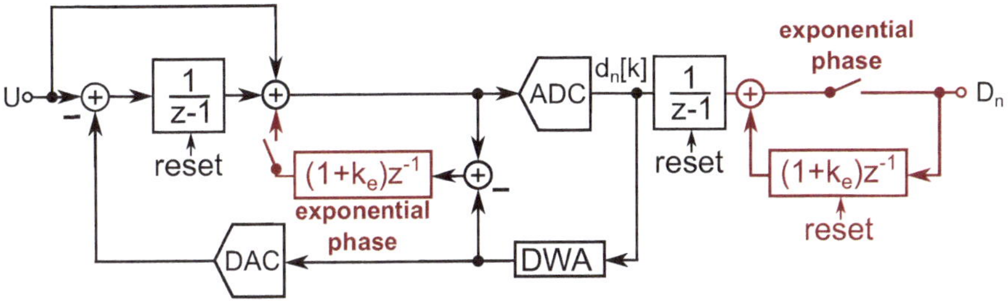


**Fig. 3.16** System overview of the linear-exponential incremental ADC

During the second phase, the path as shown in red is switched in for the last 10 cycles; this noise-coupling loop is fed back with a gain factor of $1 + k_e$, where $k_e$ is a gain factor. This feedback exponentially decreases the quantization error during the last cycles of the conversion. But in order to precisely reconstruct the signal, any change within the modulator has to be imitated by the reconstruction filter. Therefore, an accumulator block is switched in. This block modifies the transfer function of the digital reconstruction filter such that the change in the modulator is imitated in the digital domain. During this phase, the SQNR is increased by 66 dB.

As a multibit quantizer is used to boost the SQNR during the whole conversion phase, the non-linearity from the feedback DAC has to be attenuated. This is achieved by means of DWA. As a first-order modulator with large OSR is used, a simple DWA can be used in contrast to more sophisticated DEM for higher-order I$-\Delta\Sigma$ modulators [22]. To reduce the 1/f-noise contribution, chopping is again used in the first integrator. However, the noise penalty that is usually paid by using higher-order modulators (and reconstruction filters) is avoided in this design as a first-order modulator does not suffer from this noise penalty as all samples are weighted equally in the reconstruction filter.

## 3.4 Conclusions

High-efficiency data converters are of continuous research interest. When high efficiency is combined with high resolution, this is mostly achieved by oversampling and noise-shaping techniques, which makes such converters dominate the state of the art for such applications. If at the same time, true Nyquist-rate conversion is required, e.g., for multiplexing or true sample-to-sample conversion, those architectures fail to achieve this. Thus, an increasing research interest is again dedicated to efficient, high-resolution Nyquist-rate ADCs. This chapter has reviewed some of the prominent examples in the state of the art, mostly looking into SAR ADCs and incremental $\Delta\Sigma$ converters, both of which already today promise to fulfill those requirements and any further architectural or circuit level innovation of those ADCs is highly needed to further close the efficiency gap to the oversampling converters.

## References


1. B. Murmann, ADC performance survey 1997–2018, http://web.stanford.edu/~murmann/adcsurvey.html
2. K. Obata, K. Matsukawa, T. Miki, Y. Tsukamoto, K. Sushihara, A 97.99 dB SNDR, 2 kHz BW, 37.1 $\mu$W noise-shaping SAR ADC with dynamic element matching and modulation dither effect, in *2016 IEEE Symposium on VLSI Circuits (VLSI-Circuits)* (IEEE, Piscataway, 2016), pp. 1–2
3. Y.-S. Shu, L.-T. Kuo, T.-Y. Lo, An oversampling SAR ADC with DAC mismatch error shaping achieving 105 dB SFDR and 101 dB SNDR over 1 kHz BW in 55 nm CMOS. IEEE J. Solid-State Circuits **51**(12), 2928–2940 (2016)
4. S. Hsieh, C. Hsieh, A 0.44-fJ/conversion-step 11-bit 600-kS/s SAR ADC with semi-resting DAC. IEEE J. Solid-State Circuits **53**(9), 2595–2603 (2018)
5. C.-C. Liu, S.-J. Chang, G.-Y. Huang, Y.-Z. Lin, A 10-bit 50-MS/s SAR ADC with a monotonic capacitor switching procedure. IEEE J. Solid-State Circuits **45**(4), 731–740 (2010)
6. S. Hsieh, C. Hsieh, A 0.4V 13b 270kS/S SAR-ISDM ADC with an opamp-less time-domain integrator, in *2018 IEEE International Solid - State Circuits Conference - (ISSCC)* (2018), pp. 240–242
7. A. AlMarashli, J. Anders, J. Becker, M. Ortmanns, A Nyquist rate SAR ADC employing incremental sigma delta DAC achieving peak SFDR= 107 dB at 80 kS/s. IEEE J. Solid-State Circuits **53**(5), 1493–1507 (2018)
8. J.A. Fredenburg, M.P. Flynn, A 90-MS/s 11-MHz-bandwidth 62-dB SNDR noise-shaping SAR ADC. IEEE J. Solid-State Circuits **47**(12), 2898–2904 (2012)
9. C.-C. Liu, M.-C. Huang, A 0.46 mW 5MHz-BW 79.7 dB-SNDR noise-shaping SAR ADC with dynamic-amplifier-based FIR-IIR filter, in *2017 IEEE International Solid-State Circuits Conference (ISSCC)* (IEEE, Piscataway, 2017), pp. 466–467
10. P. Harpe, E. Cantatore, A. van Roermund, A 2.2/2.7 fJ/conversion-step 10/12b 40kS/s SAR ADC with data-driven noise reduction, in *2013 IEEE International Solid-State Circuits Conference Digest of Technical Papers (ISSCC)* (IEEE, Piscataway, 2013), pp. 270–271
11. K. Yoshioka, H. Ishikuro, A 13b SAR ADC with eye-opening VCO based comparator, in *European Solid State Circuits Conference (ESSCIRC), ESSCIRC 2014-40th* (IEEE, Piscataway, 2014), pp. 411–414
12. H. Tai, Y. Hu, H. Chen, H. Chen, A 0.85fJ/conversion-step 10b 200kS/s subranging SAR ADC in 40nm CMOS, in *2014 IEEE International Solid-State Circuits Conference Digest of Technical Papers (ISSCC)* (IEEE, Piscataway, 2014), pp. 196–197
13. A. AlMarashli, J. Anders, J. Becker, M. Ortmanns, A 107 dB SFDR, 80 kS/s Nyquist-rate SAR ADC using a hybrid capacitive and incremental $\Sigma\Delta$ DAC, in *2017 Symposium on VLSI Circuits* (IEEE, Piscataway, 2017), pp. C240–C241
14. A. Bannon, C.P. Hurrell, D. Hummerston, C. Lyden, An 18 b 5 MS/s SAR ADC with 100.2 dB dynamic range, in *2014 Symposium on VLSI Circuits Digest of Technical Papers* (IEEE, Piscataway, 2014), pp. 1–2
15. H. Li, M. Maddox, M.C. Coin, W. Buckley, D. Hummerston, N. Naeem, A signal-independent background-calibrating 20b 1MS/S SAR ADC with 0.3 ppm INL, in *2018 IEEE International Solid-State Circuits Conference-(ISSCC)* (IEEE, Piscataway, 2018), pp. 242–244
16. S. Tao, A. Rusu, A power-efficient continuous-time incremental sigma-delta ADC for neural recording systems. IEEE Trans. Circuits Syst. I Reg. Papers **62**(6), 1489–1498 (2015)
17. Y. Chae, K. Souri, K.A.A. Makinwa, A 6.3 $\mu$W 20 bit incremental zoom-ADC with 6 ppm INL and 1 $\mu$V offset. IEEE J. Solid-State Circuits **48**(12), 3019–3027 (2013)
18. J. Wagner, P. Vogelmann, M. Ortmanns, On the signal filtering property of CT incremental sigma-delta ADCs, in *IEEE Transactions on Circuits and Systems II: Express Briefs* (2019), p. 1
19. C. Chen, Y. Zhang, T. He, P.Y. Chiang, G.C. Temes, A micro-power two-step incremental analog-to-digital converter. IEEE J. Solid-State Circuits **50**(8), 1796–1808 (2015)

20. P. Vogelmann, M. Haas, M. Ortmanns, A 1.1mW 200kS/s incremental $\Delta\Sigma$ ADC with a DR of 91.5dB using integrator slicing for dynamic power reduction, in *2018 IEEE International Solid - State Circuits Conference - (ISSCC)* (IEEE, Piscataway, 2018), pp. 236–238
21. B. Wang, S. Sin, U. Seng-Pan, F. Maloberti, R.P. Martins, A 550-$\mu$ W 20-kHz BW 100.8-dB SNDR linear-exponential multi-bit incremental $\Sigma\Delta$ ADC with 256 clock cycles in 65-nm CMOS, in *IEEE Journal of Solid-State Circuits* (IEEE, Piscataway, 2019), pp. 1–12
22. Y. Liu, E. Bonizzoni, F. Maloberti, High-order multi-bit incremental converter with smart-DEM algorithm, in *2013 IEEE International Symposium on Circuits and Systems (ISCAS 2013)* (IEEE, Piscataway, 2013), pp. 157–160

# Chapter 4
# Continuous-Time ADCs for Automotive Applications

**Muhammed Bolatkale**

## 4.1 Introduction

Automotive applications such as AM/FM/DAB car radios, vehicle-to-anything (V2X) transceivers, multi-standard GPS receivers, and UWB based car key transceivers require data converters with bandwidths ranging from 25 to 250 MHz and dynamic ranges between 50 and 80 dB. Such data converters enable software-defined car products that can be deployed worldwide. Although many types of data converters can digitize BW >25 MHz and achieve DR larger than 50 dB, a continuous-time delta-sigma ADC architecture with its inherent anti-alias filtering and excellent linearity, spurious-free dynamic range (SFDR), and interference immunity results in the best in class receiver architecture [1].

Figure 4.1 plots the Schreier figure of merit ($FOM_S$) vs the Nyquist bandwidth of the data converters published over the last two decades. The trend line for bandwidth limited designs (BW <10 MHz), which is also known as the architectural front, has improved approximately by 20 dB over the last decade. Architectures utilizing incremental converters, noise-shaped SARs, and continuous-time delta-sigma modulators achieve $FOM_S$ better than 180 dB [3–6]. These architectures, also known as hybrid converters, minimize thermal noise sources and circuit non-linearities which limit the maximum achievable SNDR in a given power dissipation target. On the other hand, the trend line set by high speed data converters, which is known as the technology front, has improved by 25–30x.

In [7], the authors for the first time demonstrated an implementation of a multi-GHz sampling continuous-time $\Delta\Sigma$ modulator. Figure 4.2 illustrates the maximum sampling rate and bandwidth of the oversampled data converters which were

M. Bolatkale (✉)
NXP Semiconductors, Delft, The Netherlands
e-mail: muhammed.bolatkale@nxp.com

© Springer Nature Switzerland AG 2020

A. Baschirotto et al. (eds.), *Next-Generation ADCs, High-Performance Power Management, and Technology Considerations for Advanced Integrated Circuits*,
https://doi.org/10.1007/978-3-030-25267-0_4

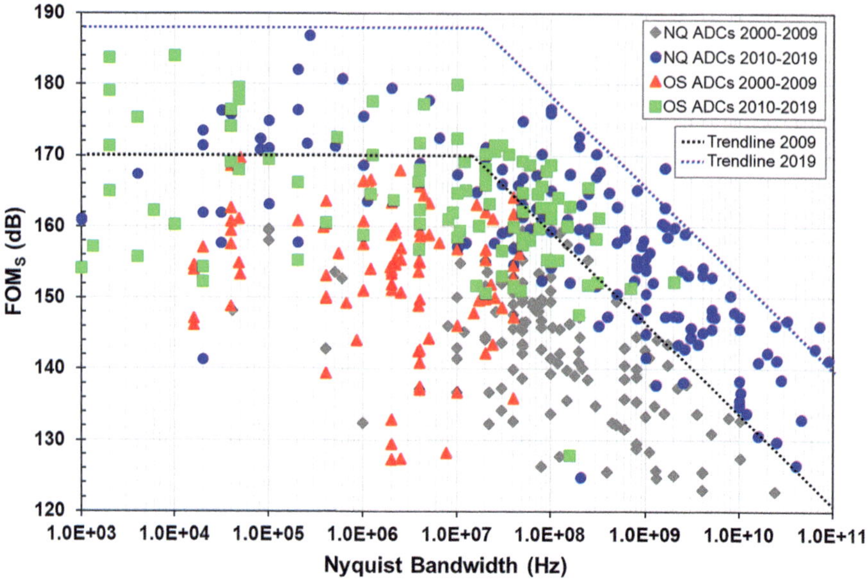


**Fig. 4.1** Schreier figure of merit, $\mathrm{FOM}_S = \mathrm{SNDR} + 10\log_{10}\left(\frac{\mathrm{BW}}{P}\right)$ [2]

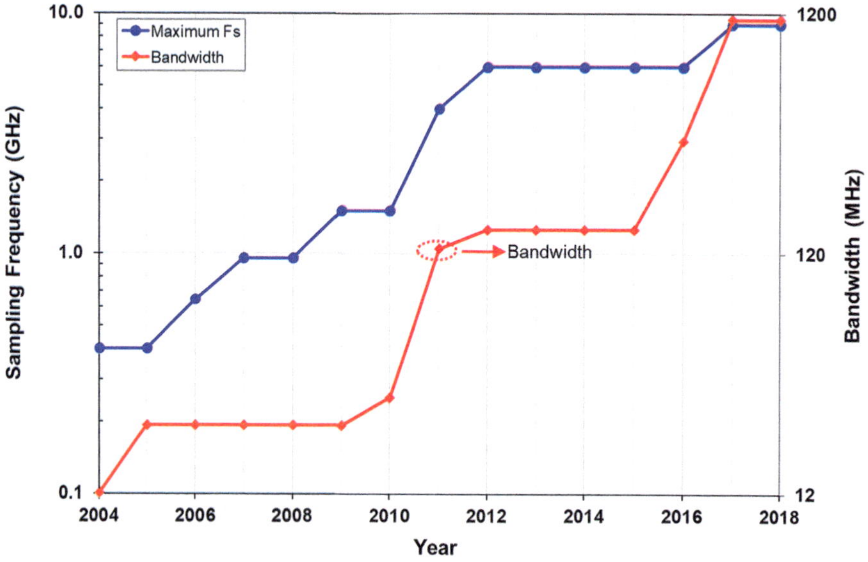


**Fig. 4.2** Maximum sampling rate and bandwidth of continuous-time ADCs

published since 2002. The highest sampling rate is nearly 10 GHz and is used in a continuous-time pipeline converter. Over a decade ago, the bandwidths of oversampled (OS) data converters were limited to 20 MHz, whereas currently a state-of-the-art design has a bandwidth of 1.125GHz [8].

In the remainder part of this chapter, various high-speed continuous-time ADC architectures are discussed. Then a brief summary of architectural constraints is introduced and the main building blocks and circuit design techniques are summarized.

## 4.2 Continuous-Time ADC Architectures

A continuous-time delta-sigma (CT$\Delta\Sigma$) modulator has an inherent anti-alias filtering, a resistive input stage, and should have a flat signal transfer function (STF) as will be discussed later on in this chapter. For automotive applications, a receiver architecture employing a CT$\Delta\Sigma$ modulator can achieve better area efficiency since the AAF filtering stages can be simplified. It is also more energy efficient to drive a resistive load rather than a switched-capacitor load.

The dynamic range (DR) of a wide bandwidth ADC is often defined by a near-band blocker that is often adjacent to the signal bandwidth. Filtering such blockers with analog filters is very costly. Therefore, Nyquist converters rely on oversampling to reduce the impact of such near-band blockers on the anti-alias filter (AAF), whereas the AAF requirement of a CT$\Delta\Sigma$ modulator is defined by blockers located around its sampling frequency. The amplitude of these blocker signals should be within the input range of the CT$\Delta\Sigma$ in order not to saturate its filter stages.

An oversampled data converter employs a decimation filter to achieve the desired resolution in the band of interest. The power dissipation of a decimation filter is proportional to its clock frequency ($F_{\text{clk}}$). The power dissipation of digital circuits and clocked circuits, such as comparators, clock buffers, will scale down for each new generation of process node. Therefore, implementing a CT$\Delta\Sigma$ modulator in advanced process nodes can reduce its power dissipation.

### *4.2.1 Design Space of Single-Loop Continuous-Time Delta-Sigma Converters*

Figure 4.3 shows a basic model of a single-loop CT$\Delta\Sigma$ modulator. It has three main building blocks: a loop filter, an ADC (also known as a quantizer), and a digital-to-analog converter (DAC). To explore the design space of a delta-sigma modulator, it is crucial to understand the fundamental working principle of each block and their trade-off between each other. The maximum clock frequency of a multi-GHz modulator is often related to the $f_T$ of the process technology. Based on the choice of

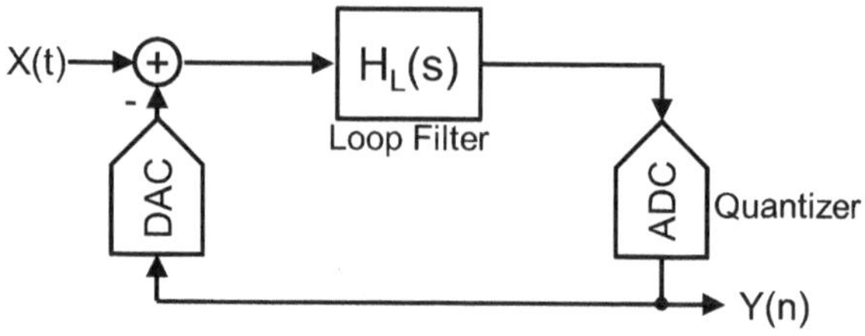


**Fig. 4.3** A basic single-loop continuous-time delta-sigma modulator

circuit architectures, each block that is clocked (i.e., comparator, DAC, etc.) requires a well-defined clock edge and amplitude. The maximum sampling rate is defined by the highest operating frequency of a CMOS clock buffer, which is proportional to the rise and fall times that can be implemented on a chip. Furthermore, the quantizer must convert its analog input signal to a full scale digital signal. Effectively, the quantizer can be modeled as a gain block which trades time for gain. For very small input signals, the quantizer will not have enough time to resolve its input to a digital output signal. Therefore, it will introduce metastability errors. Effectively, the sampling rate of a multi-bit quantizer which employs a rail-to-rail CMOS clock is approximately limited to 10 GHz. However, the maximum sampling rate could be further increased by employing low-swing clocking schemes.

Table 4.1 lists CT$\Delta\Sigma$ modulators which have bandwidths larger than 25 MHz. Table 4.1 also gives an overview of the architectural techniques employed in state-of-the-art modulator designs. Furthermore, it lists the sampling rate and the resolution of the quantizer for each design.

Increasing the resolution of the quantizer reduces the quantization error. For a given signal-to-quantization noise ratio (SQNR), the sampling rate of the modulator can be reduced, which enables the realization of many wide-bandwidth delta-sigma modulators at the expense of calibrating offset voltages of comparators as listed in Table 4.1.

Choosing a higher order loop filter reduces the oversampling ratio (OSR = Fs/2/BW), where Fs is the sampling rate and BW is the signal bandwidth. However, it is more effective at high oversampling ratios. Therefore, wideband $\Delta\Sigma$ modulators often employ high order loop filter such as between third and sixth order (Table 4.1).

### Loop Stability and Excess Loop Delay Compensation

As shown in Fig. 4.3, an ideal $\Delta\Sigma$ modulator instantaneously performs sampling, quantization, and digital-to-analog conversion. The stability theory of a $\Delta\Sigma$M has been developed from the observation that a first-order $\Delta\Sigma$M is inherently stable. To design a stable modulator, the phase margin of the loop filter must be as close as possible to 90° at $F_s/2$ [25]. However, due to the limited switching speed of the transistors in the quantizer and the DAC, there exists an excess loop delay (ELD) in the feedback of the $\Delta\Sigma$M.

**Table 4.1** Recently published wideband GHz continuous-time $\Delta\Sigma$ ADCs

| Ref. | Arch. | LF | BW (MHz) | Fs (GHz) | Int. | Amplifier | DAC | Q (#b) |
|---|---|---|---|---|---|---|---|---|
| Shibata et al. [8] | Pipe | – | 1000 | 9 | LPF | Multi-stage inv. FF comp | CS | 4 |
| Dong et al. [9] | Mash | FB | 465 | 8 | RC | Multi-stage inv. FF comp | CS | 4 |
| Wu et al. [10] | SL | FF/FB | 160 | 2.9 | RC | Single-stage | CS | 5 |
| Dayanik et al. [11] | SL | FF | 156 | 5 | RC | Multi-stageFF comp | CS | 3.7 |
| Shibata et al. [12] | SL | FB | 150 | 4 | RC | Multi-stageFF comp | CS | 4 |
| Bolatkale et al. [7] | SL | FF | 125 | 4 | RC/GmC | Multi-stageFF comp | CS | 4 |
| Huang et al. [13] | SL | FF | 125 | 2.15 | RC | – | CS | VCO |
| Ho et al. [14] | SL | FF | 80 | 2.8 | RC | Multi-stageMiller&FF | CS | 3 |
| Vidrios et al. [15] | SL | FF | 75 | 3.2 | RC, Biquad | – | CS | 4 |
| Srinivasan et al. [16] | SL | FB | 60 | 6 | Passive RC, GmC | Gm | CS | 1 |
| Jain et al. [17] | SL | FF/FB | 60 | 6 | RC, GmC | Multi-stageFF comp | FIR | 2xTI-1b |
| Yoon et al. [18] | Mash | FF | 50 | 1.8 | RC, GmC | Multi-stageMiller&FF | CS | 4&3 |
| Young et al. [19] | SL | FF/FB | 50 | 1.3 | VCO, RC | Multi-stageMiller comp | CS | 3 |
| Reddy et al. [20] | SL | FF | 50 | 1.2 | RC | Multi-stageFF comp | CS | VCO |
| He et al. [21] | SL | FF | 50 | 2 | RC | – | CS | 4 |
| Dong et al. [22] | Mash | FB | 46 | 3.2 | RC | Multi-stageFF comp | CS | 4 |
| Loeda et al. [23] | SL | FF | 40 | 2.4 | RC | – | FIR | 1 |
| Shettigar et al. [24] | SL | FF/FB | 36 | 3.6 | RC | Multi-stageFF comp | Rdac | 1 |
| Breems et al. [1] | SL | FF/FB | 25 | 2.2 | RC | Multi-stageMiller comp. | Rdac | 1 |

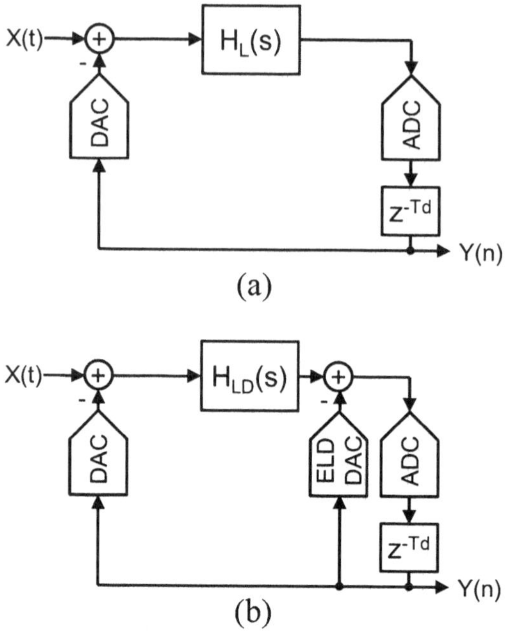


**Fig. 4.4** A basic single-loop continuous-time delta-sigma modulator with excess loop delay model (**a**) and ELD introduced around the quantizer (**b**)

Figure 4.4a illustrates a $\Delta\Sigma$M with a quantizer followed by a delay block which models the excess loop delay. Effectively, the unwanted delay in the $\Delta\Sigma$ loop increases the order of the modulator and causes instabilities. The basic goal of ELD compensation techniques is to preserve the original NTF of the modulator and thus the stability of the modulator. To compensate for ELD, a feedforward path around the quantizer must be introduced. For example, one clock cycle ELD can be compensated by introducing a direct feedforward path around the quantizer as shown in Fig. 4.4b. To match the NTF of the modulator before and after ELD compensation, the compensated loop filter $H_{\mathrm{LD}}(s)$ should have the following relation to the original loop filter $H_L(s)$ [26]:

$$H_{\mathrm{LD}}(z) = H_L(z) \cdot z^1 - c \tag{4.1}$$

where $c$ is the coefficient around the quantizer and also shown as an ELD DAC in Fig. 4.4b. By satisfying (4.1), the NTF of a modulator with one clock cycle excess delay can be matched to the original NTF.

ELD compensation schemes are employed in all the GHz sampled modulators listed in Table 4.1. In summary, ELD compensation techniques can be classified into three main categories: analog, mixed-signal, and digital ELD. Some examples from the literature are summarized as follows:

- An analog ELD technique uses a feedforward compensation path within the loop filter. The coefficient $k$ as shown in Fig. 4.5 implements an analog ELD

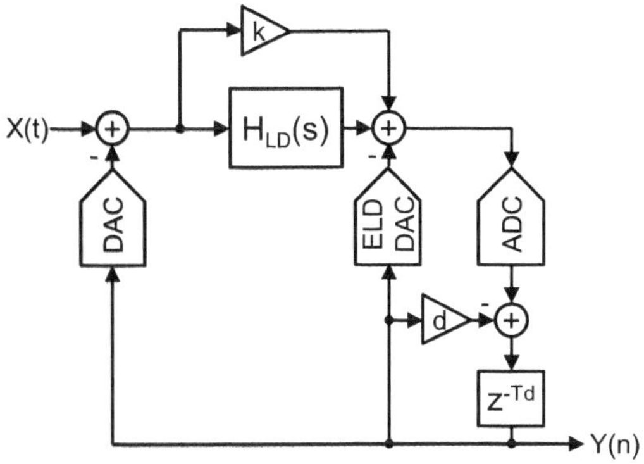


**Fig. 4.5** Overview of the ELD compensation techniques. The coefficients "*k*," "*d*," and ELD DAC represent analog, digital, and mixed-signal ELD compensation techniques, respectively

compensation path [27–30]. For example, an RC integrator can be configured to have a feedforward path by introducing a resistor in series with the integration capacitor. Since it only uses passive components, this technique is potentially low power. However, the bandwidth of the feedforward path is limited by the amplifier's bandwidth.

- Mixed-signal ELD techniques implement a feedforward path by applying the output of the quantizer to a digital-to-analog converter [1, 7, 31, 32]. The requirements of such a DAC are much more relaxed when compared to the main feedback DAC. Mixed-signal ELD is more suitable for GHz sampling $\Delta\Sigma$M since it directly accesses the output of the quantizer latch and the propagation delay of a current steering DAC is only limited to a few gates.
- Digital ELD is more suitable for low frequency applications where the delay of the digital adder does not introduce additional excess loop delay [33, 34]. In the case of a VCO based quantizer, the phase domain rotation can be utilized to implement the digital ELD domain in a hardware efficient way and also achieve high sampling rates [13].

## Signal Transfer Function and Power Efficiency

A CT$\Delta\Sigma$ modulator has an inherent signal transfer function (STF), which can be expressed as:

$$\mathrm{STF}\,(\mathrm{j}\omega) = H\,(\mathrm{j}\omega)\cdot \mathrm{NTF}\left(e^{\mathrm{jw}}\right) \tag{4.2}$$

where $H$(jw) is the continuous-time transfer function of the loop filter and NTF is the noise transfer function of the modulator. Compared to other data converter architectures which have a unity STF, a $\Delta\Sigma$ modulator can be designed to achieve a filtering STF. Figure 4.6 illustrates a $\Delta\Sigma$ modulator with a feedforward and a feedback architecture. The STF of a $\Delta\Sigma$M with feedforward architecture often exhibits out-of-band peaking because integrator outputs are added through

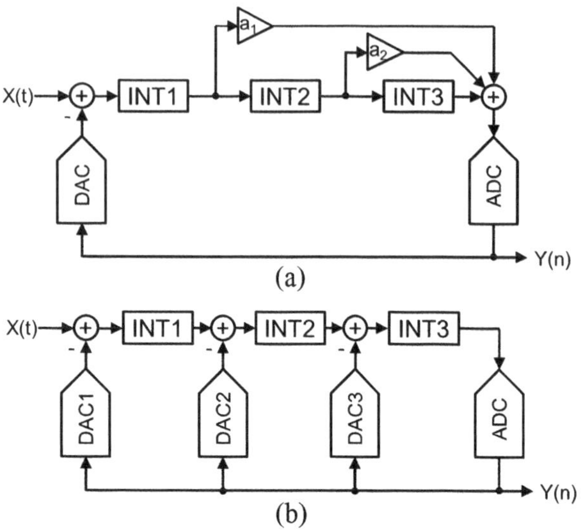


**Fig. 4.6** A $\Delta\Sigma$ modulator with feedforward (**a**), with feedback architectures (**b**)

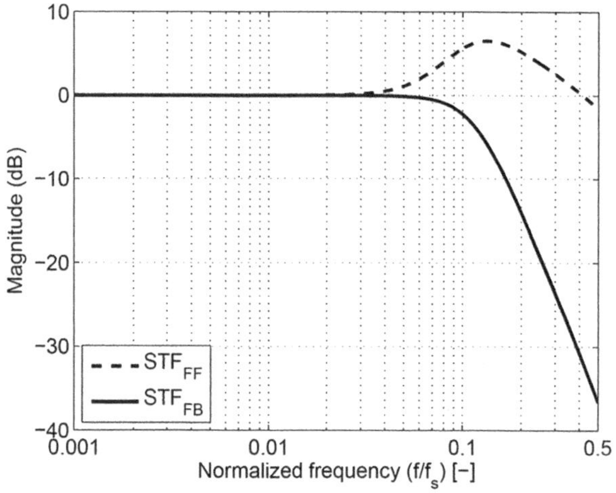


**Fig. 4.7** Signal transfer function of a CT$\Delta\Sigma$ modulator with a feedforward loop filter (dashed line) and a feedback loop filter (solid line)

feedforward coefficients ($a_1 \ldots a_n$). On the other hand, the STF of a $\Delta\Sigma$M with feedback architecture has a low-pass filtering response since the input signal goes through cascaded integrator stages. Figure 4.7 compares the STF of a CT$\Delta\Sigma$M with feedforward and feedback architectures.

The energy efficiency of a $\Delta\Sigma$M depends on its loop filter architecture. By choosing a feedforward architecture, the number of feedback DACs are reduced and the unity gain frequency of the first integrator can be set larger than signal bandwidth [10, 14, 15]. Consequently, thermal noise contribution and non-idealities of the following filter stages can be minimized. On the other hand, a feedback architecture will require a lower unity gain frequency from its first integrator [12, 16]. This general trend can be expressed as:

$$\omega_{\mathrm{int}\,1-\mathrm{FF}} > \mathrm{BW} > \omega_{\mathrm{int}\,1-\mathrm{FB}} \tag{4.3}$$

In recent years, $\Delta\Sigma$Ms with a flat STF have been published [9], where input feedforward coefficients are introduced to remove the STF peaking in CT$\Delta\Sigma$Ms with feedforward loop filter architecture [23, 34] and to remove the STF filtering in a CT$\Delta\Sigma$M with a feedback loop filter [35]. As a result, the general trend in (4.3) only holds for modulators without signal feedforward coefficients. The integrator unity gain frequencies can be independently designed for $\Delta\Sigma$ modulators with input feedforward coefficients. This technique can be applied to any $\Delta\Sigma$ modulator and is independent of the OSR.

### *4.2.2 Multi-Stage Continuous-Time Converters*

According to the published works, the maximum sampling rate of a single loop CT$\Delta\Sigma$ modulator is limited to approximately 6 GHz and its BW is less than 160 MHz. Figure 4.8 depicts the maximum SQNR for a fourth order CT$\Delta\Sigma$ modulator with 4-bit quantization. Most of the single loop designs that target DR larger than 70 dB employ an OSR larger than 8 in order to suppress the quantization error much below the thermal noise floor. In order to break the BW versus DR trade-off, multi-stage noise shaping (MASH) architecture can be adapted. In principle, this architecture enables cascading of low-order CT$\Delta\Sigma$ modulators and as a result achieves wider bandwidth for a given sampling rate [25] (Fig. 4.9).

Figure 4.10 shows a system diagram of a 1–2 MASH CT$\Delta\Sigma$ modulator [9]. The first stage is a first order CT$\Delta\Sigma$ modulator and the second stage is a second order CT$\Delta\Sigma$ modulator with a feedback loop filter architecture. The transfer function of the first and the second stages can be expressed as:

$$Y_1 = X \cdot \mathrm{STF}_1 + E_{Q1} \cdot \mathrm{NTF}_1 \tag{4.4}$$

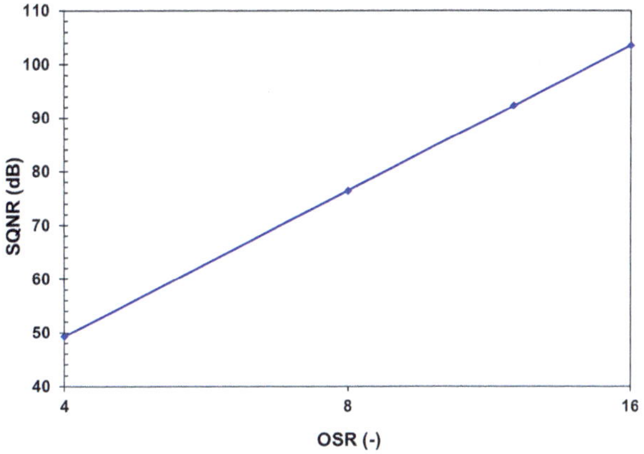


**Fig. 4.8** Maximum SQNR limit for a fourth order CT$\Delta\Sigma$ modulator with 4-bit quantization [36]

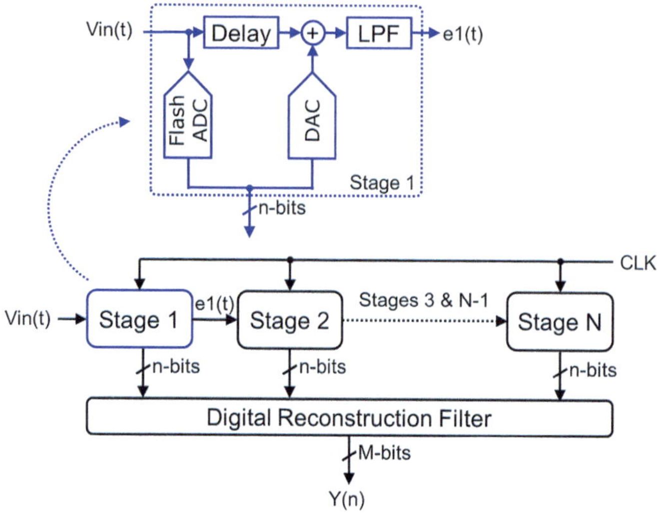


**Fig. 4.9** An example of a continuous-time pipelined ADC adapted from [8]

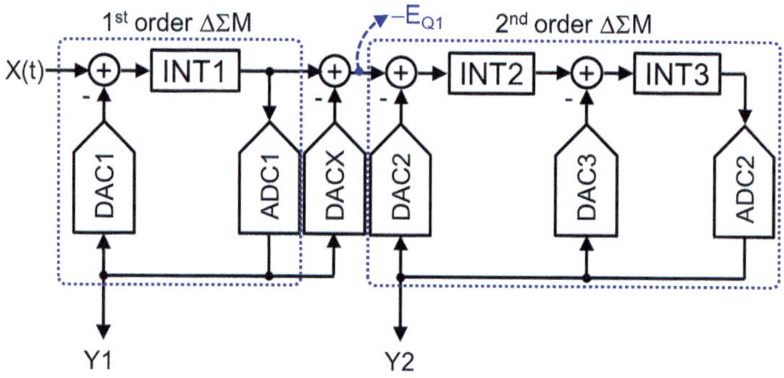


**Fig. 4.10** An example of a 1–2 MASH CT$\Delta\Sigma$ modulator [9]

$$Y_2 = -E_{Q1} \cdot G \cdot \mathrm{STF}_2 + E_{Q2} \cdot \mathrm{NTF}_2 \tag{4.5}$$

where $G$ is the inter-stage gain which is defined as $G = \mathrm{LSB}_{\mathrm{DACX}}/\mathrm{LSB}_{\mathrm{DAC2}}$. Assuming $\mathrm{STF}_1$ and $\mathrm{STF}_2$ are unity, and an accurate representation of $\mathrm{NTF}_1$ can be implemented in digital ($\mathrm{NTF}_{1d}$), the final transfer function of the MASH converter can be expressed as:

$$Y_{\mathrm{out}} = Y_1 + Y_2 \frac{\mathrm{NTF}_{1d}}{G} = X \cdot \mathrm{STF}_1 + E_{Q2} \cdot \frac{\mathrm{NTF}_{1d}}{G} \mathrm{NTF}_2 \tag{4.6}$$

The coefficients of $NTF_{1d}$ can then be calibrated over PVT to minimize the leakage of the quantization error of the first stage. The most common techniques use either the variance of the in-band noise after decimation filter or the cross-correlation technique before the decimation filter which then update the coefficients of the digital noise cancellation filter [37, 38].

As shown in Table 4.1, the minimum OSR of the single-loop and MASH $\Delta\Sigma$ modulators is limited to around 8 which is set by the 70 dB DR requirement. For applications that do not require very high DR such as an UWB car key receiver, an ultra-low OSR of 4 can be employed to further increase the BW and reduce the power of the clocked circuits [35, 39].

To further lower the OSR ratio, a CT pipelined converter can be implemented [8, 40]. Figure 4.9 shows a simplified block diagram of a pipelined continuous-time converter. It consists of N stages, and each stage converts its input signal with n-bit resolution and forwards its residue ($e(t)$) to the following stage. The outputs of the stages are then added and reconstructed in the digital domain to achieve the target resolution of M-bit.

A basic system level model of each stage is shown in Fig. 4.9. The input signal is sampled by an n-bit flash ADC, and its output is converted to analog through an n-bit DAC. The DAC output is subtracted from a delayed version of the input signal. At this point of the signal chain, the input of the low pass filter is mostly dominated by the quantization error of the flash converter and the images of the DAC. The main function of the low pass filter is to attenuate the DAC images [8]. As a result, the following stage only processes a residue signal ($e(t)$) which is a low pass filtered version of the flash quantization error.

Compared to continuous-time $\Delta\Sigma$ modulators, the CT pipeline converter has a limited out-of-band signal handling capability [8]. Due to the sampling operation of the flash converter, the out-of-band signals located between $n_* F_S \pm$ BW alias back into the signal bandwidth. However, the delayed version of the input signal does not have any aliased tones. As a result, the residue signal might saturate and clip the following stages. Therefore, an AAF is required.

## 4.3 Main Building Blocks

A CT$\Delta\Sigma$ modulator has three main building blocks: a quantizer, a DAC, and a loop filter. We will first introduce each building block and briefly describe the circuit techniques that are required to achieve wideband conversion at GHz sampling rates.

### *4.3.1 Quantizer*

The quantizer defines the maximum sampling rate of a $\Delta\Sigma$ converter. The flash ADC is the most common architecture and can achieve resolutions of up to 5-bits

[9, 10, 12, 22]. Each comparator often uses a pre-amplifier and a latch which can be implemented using either in current mode or dynamic logic schemes. For a multi-bit flash ADC, the offset voltage ($V_{\text{offset}}$) of each comparator must be calibrated. While calibrating the offset voltage of a latch, the offset trim circuits must not load the latch critical nodes [41]. The comparator slices can also be sized for matching and the offset voltage can be reduced at the cost of area and power [7].

For CTΔΣ modulators which target below GHz sampling rate, a SAR based quantizers can also be employed [42]. Since, a 1-bit per cycle SAR converter has only one comparator, the comparator offset calibration might not be required.

## Metastability

Figure 4.11a shows a basic model of a high speed quantizer. The input signal is sampled by an edge triggered comparator, then further sampled by a D-FF. The time allocated to the comparator ($T_{\text{comp}}$) is less than the clock period of the modulator ($T_{\text{CLK}}$). As the clock rate increases, the dynamic gain that a comparator can achieve degrades. The gain of a comparator can be modeled as:

$$\text{Latch Gain} = A_{\text{pre}} \cdot e^{T_{\text{comp}}/\tau} \tag{4.7}$$

where $A_{\text{pre}}$ is the gain during pre-amplification phase and $\tau$ is the time constant of the latch. To reduce the metastability of the modulator, the time allocated to the latches must be maximized. This can be achieved by using cascaded stages of latches [7], or allocating most of the clock period to the latch which has the smallest time constant [9].

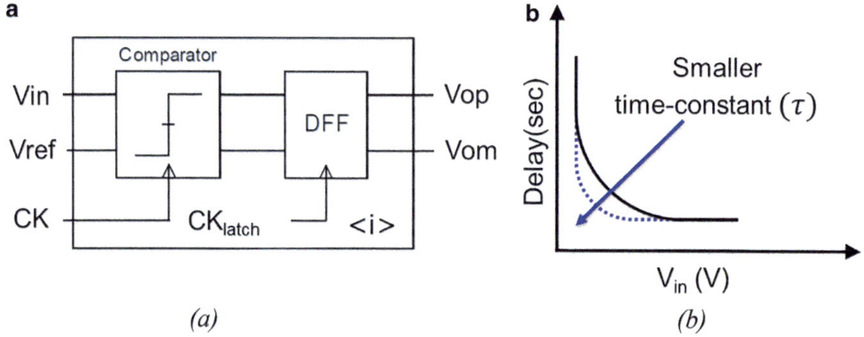


**Fig. 4.11** A basic model of a high speed comparator (**a**) and delay vs input signal transfer function of a high speed comparator (**b**)

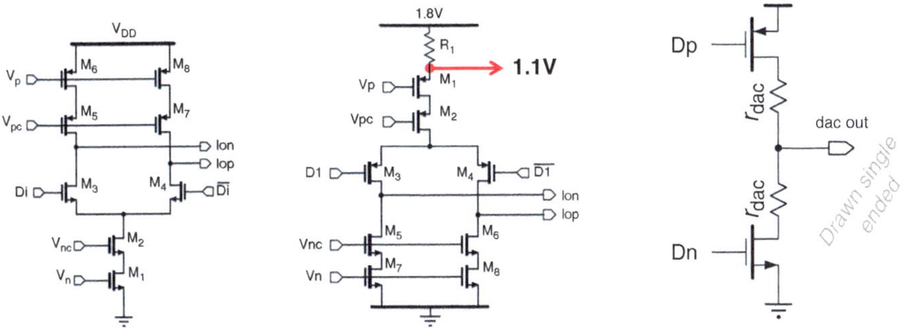


*(a)* *(b)* *(c)*

**Fig. 4.12** Current steering (**a**), resistively degenerated current steering DACs [26] (**b**), and resistive DAC [23] (**c**)

## 4.3.2 *Feedback DAC*

The feedback DAC defines the accuracy of a CTΔΣ modulator. The main types of architectures are resistive and current steering DACs as shown in Fig. 4.12. For converters targeting BW smaller than 60 MHz using a resistive DAC is advantageous because a resistor is often less noisy than an active current steering (CS) DAC for the same supply voltage [1, 24]. On the other hand, for BWs larger than 60 MHz, a current steering DAC brings speed advantage but it contributes more thermal noise than a resistive DAC [7, 9, 10, 12, 16, 22]. To reduce the thermal noise contribution of a CS DAC, designers often employ supply voltage higher than the nominal supply voltage of the process node and use resistively degenerated CS architecture [7].

A multi-bit DAC will introduce non-linearity due to mismatches of unit current sources. The static mismatch errors can be minimized either by scaling the unit current sources for matching or by calibrating the static mismatch sources. At GHz sampling rates, the dynamic error sources such as inter-symbol interference (ISI) and timing errors limit the accuracy of the converter even after calibrating the static mismatch sources.

#### Inter-Symbol Interference

As the sampling rate of a CTΔΣ modulator increases, the non-idealities related to timing and data-dependent switching artifacts limit the performance of the converter. Inter-symbol interference (ISI) which is caused by the non-equal rise/fall times of the DAC output currents is studied extensively in the literature [19, 23].

There are two main types of ISI compensation schemes:

- Switching schemes: return-to-zero [1, 43], differential quad switching (DQS) [43, 44], error-sampling scheme [26]
- Analog calibration: ISI error-amplifier [10], timing error calibration [9].

Return-to-zero switching scheme is the most effective ISI compensation scheme since for every sample, the DAC outputs the same amount of charge. The authors have demonstrated a CT$\Delta\Sigma$ modulator which achieved beyond 100dBc THD by employing a RTZ scheme [1]. The IM3 and HD3 performance of the other techniques are limited to around 80 dBc due to other circuit non-idealities.

#### Clock Jitter

A CT$\Delta\Sigma$ modulator is sensitive to clock jitter, which modifies the trigger moment of DAC unit and effectively acts as an additional noise source at the input of the modulator. To reduce the performance loss due to clock jitter, the clocking and DAC driver blocks must be co-optimized to achieve the target signal-to-jitter noise ratio (SJNR) which can be expressed as for a low pass CT$\Delta\Sigma$ [45]:

$$\mathrm{SJNR}_{\mathrm{LP}\Delta\Sigma} = 10\log_{10}\left(\frac{1}{4\cdot\sigma_j^2\cdot\sigma_Y^2\cdot f_s\cdot \mathrm{BW}}\right) \tag{4.8}$$

where $f_s$ is the sampling frequency, BW is the signal bandwidth, $\sigma_Y$ is the RMS output of a $\Delta\Sigma$, and $\sigma_j$ is the RMS jitter. Increasing sampling frequency and BW degrades SJNR.

A summary of techniques which improve the SJNR:

- Architecture level: Reduce $\sigma_Y$ by using a multi-bit or a finite impulse response (FIR) DAC architecture. Employ an NTF with lower out-of-band gain (OBG).
- At circuit level: Reduce $\sigma_j$, which basically means clock distribution network with lower noise. At GHz sampling rates this can drastically increase power dissipation of the modulator.

### *4.3.3 Loop Filter*

A CT$\Delta\Sigma$ modulator requires a continuous-time analog filter which provides gain and attenuates the quantization noise in the band of interest. At low frequencies, the filter can be approximated as a cascade of integrators, thus high gain. At high frequencies (fs/2), the filter should achieve a gain and phase response of a first order integrator for stability. The order of the filter is defined by the order of the NTF. The designs that target good linearity use RC based integrators, and the designs that are optimized to reduce power dissipation use $g_mC$ based integrators [7, 9, 12, 16, 24].

**Amplifier Architectures**

The basic building block of a loop filter is an amplifier. Historically, many CT$\Delta\Sigma$ designs have used traditional amplifiers such as telescopic, folded cascode, and two-stage amplifier. However, designing a multi-GHz CT$\Delta\Sigma$ and maintaining a good $FoM_W$ less than 100 fJ/cs requires an energy efficient amplifier architecture. As also indicated in Table 4.1, many $\Delta\Sigma$ modulators use an inverter based amplifier architecture. The supply regulation strategy is an important design aspect of an inverter based amplifier. Inverter based amplifiers with an integrated LDO can achieve similar or better power supply rejection ratio (PSRR), and linearity than traditional amplifier architectures. In [1], a CT$\Delta\Sigma$ modulator, which employs an inverter based biquad-integrator, has been presented with the state-of-the-art THD better than -102 dBc in 25 MHz BW.

Most of the filter stages of a GHz sampling CT$\Delta\Sigma$ use high-order amplifiers to achieve gain required to suppress the quantization error. However, these amplifiers should also achieve the target phase margin at fs/2 and their frequency compensation is crucial. Some examples from the literature are summarized as follows:

- Multi-path multi-stage amplifiers
  - feedforward frequency compensation [7, 11, 12, 17, 20, 22, 24]
  - feedforward and Miller frequency compensation [14, 18, 19]
- Inverter based multi-stage amplifiers
  - Miller compensation [1]
  - Nested feedforward compensation [8, 9]
- Single-stage stacked inverter [10]

## 4.4 Conclusions

Over the last decade, CT$\Delta\Sigma$ modulators have expanded their application space and now can address the needs of the wideband software defined radio platforms such as AM/FM/DAB car radio products and beyond. This is enabled by employing innovations at the architectural level, high-frequency circuit techniques, and ELD delay compensation techniques.

## References


1. L. Breems et al., A 2.2 GHz continuous-time $\Delta\Sigma$ ADC with -102dBc THD and 25MHz BW, in *2016 IEEE ISSCC*, (IEEE, Piscataway, 2016), pp. 272–273
2. B. Murmann, ADC Performance Survey 1997–2018. http://web.stanford.edu/~murmann/adcsurvey.html

3. C. Liu, M. Huang, A 0.46mW 5MHz-BW 79.7dB-SNDR noise-shaping SAR ADC with dynamic-amplifier-based FIR-IIR filter, in *2017 IEEE ISSCC*, (IEEE, San Francisco, 2017), pp. 466–467
4. S. Karmakar et al., A 280μW dynamic-zoom ADC with 120dB DR and 118dB SNDR in 1kHz BW, in *2018 IEEE ISSCC*, (IEEE, San Francisco, 2018), pp. 238–240
5. H. Chandrakumar, D. Marković, A 15.2-ENOB 5-kHz BW 4.5uW chopped CTΔΣ -ADC for artifact-tolerant neural recording front ends. IEEE JSSC **53**(12), 3470–3483 (2018)
6. S. Hsieh, C. Hsieh, A 0.4V 13b 270kS/S SAR-ISDM ADC with an opamp-less time-domain integrator, in *2018 IEEE ISSCC*, (IEEE, San Francisco, 2018), pp. 240–242
7. M. Bolatkale et al., A 4GHz CT ΔΣ ADC with 70dB DR and- 74dBFS THD in 125MHz BW, in *2011 ISSCC*, (IEEE, Piscataway, 2011), pp. 470–472
8. H. Shibata et al., A 9-GS/s 1.125-GHz BW oversampling continuous-time pipeline ADC achieving −164-dBFS/Hz NSD. IEEE J. Solid State Circuits **52**(12), 3219–3234 (2017)
9. Y. Dong et al., A 930mW 69dB-DR 465MHz-BW CT 1-2 MASH ADC in 28nm CMOS, in *2016 IEEE ISSCC*, (IEEE, Piscataway, 2016), pp. 278–279
10. S.-H. Wu et al., A 160MHz-BW 72dB-DR 40mW continuous-time ΔΣ modulator in 16nm CMOS with analog ISI-reduction technique, in *2016 IEEE ISSCC*, (IEEE, Piscataway, 2016), pp. 280–281
11. M.B. Dayanik et al., A 5GS/s 156MHz BW 70dB DR continuous-time sigma-delta modulator with time-interleaved reference data-weighted averaging, in *2017 Symposium on VLSI Circuits*, (Japan Society of Applied Physics, Kyoto, 2017), pp. C38–C39
12. H. Shibata et al., A DC-to-1GHz tunable RF ΔΣ ADC achieving DR=74dB and BW=150MHz at f 0=450MHz using 550mW, in *2012 IEEE International Solid-State Circuits Conference*, (IEEE, Piscataway, 2012), pp. 150–152
13. S. Huang et al., A 125MHz-BW 71.9dB-SNDR VCO-based CT ΔΣ ADC with segmented phase-domain ELD compensation in 16nm CMOS, in *2017 ISSCC*, (IEEE, San Francisco, 2017), pp. 470–471
14. S. Ho et al., A 23 mW, 73 dB dynamic range, 80 MHz BW continuous-time delta-sigma modulator in 20 nm CMOS. IEEE JSSC **50**(4), 908–919 (2015)
15. C. Briseno-Vidrios et al., A 75 MHz BW 68dB DR CT-ΣΔ modulator with single amplifier biquad filter and a broadband low-power common-gate summing technique, in *2015 Symposium on VLSI Circuits (VLSI Circuits)*, (IEEE, Piscataway, 2015), pp. C254–C255
16. V. Srinivasan et al., A 20mW 61dB SNDR (60MHz BW) 1b 3rd-order continuous-time delta-sigma modulator clocked at 6GHz in 45nm CMOS, in *2012 IEEE ISSCC*, (IEEE, Piscataway, 2012), pp. 158–160
17. A. Jain, S. Pavan, A 13.3mW 60 MHz bandwidth, 76 dB DR 6 GS/s CTΔΣM with time interleaved FIR feedback, in *2016 IEEE Symposium on VLSI Circuits*, (IEEE, Piscataway, 2016), pp. 1–2
18. D.-Y. Yoon, S. Ho, H.-S. Lee, An 85dB-DR 74.6 dB-SNDR 50MHZ-BW CT MASH ΔΣ modulator in 28nm CMOS, in *2015 IEEE ISSCC*, (IEEE, Piscataway, 2015), pp. 1–3
19. B. Young, K. Reddy, S. Rao, A. Elshazly, T. Anand, P.K. Hanumolu, A 75dB DR 50MHz BW 3rd order CT-ΔΣ modulator using VCO-based integrators, in *2014 Symposium on VLSI Circuits Digest of Technical Papers*, (IEEE, Piscataway, 2014), pp. 1–2
20. K. Reddy et al., A 54mw 1.2 GS/s 71.5 dB SNDR 50Mhz BW VCO-based CT ΔΣ ADC using dual phase/frequency feedback in 65nm CMOS, in *2015 Symposium on VLSI Circuits*, (IEEE, Piscataway, 2015), pp. C256–C257
21. T. He, M. Ashburn, S. Ho, Y. Zhang, G. Temes, A 50MHZ-BW continuous-time ΔΣ ADC with dynamic error correction achieving 79.8dB SNDR and 95.2dB SFDR, in *2018 IEEE International Solid - State Circuits Conference - (ISSCC), San Francisco, CA*, (IEEE, Piscataway, 2018), pp. 230–232
22. Y. Dong et al., A 235mW CT 0-3 MASH ADC achieving- 167dBFS/Hz NSD with 53MHz BW, in *2014 IEEE ISSCC*, (IEEE, Piscataway, 2014), pp. 480–481
23. S. Loeda, e. al, A 10/20/30/40 MHz feedforward FIR DAC continuous-time ADC with robust blocker performance for radio receivers. IEEE JSSC **51**(4), 860–870 (2016)

24. P. Shettigar, S. Pavan, A 15mW 3.6 GS/s CT-ΣΔ ADC with 36MHz bandwidth and 83dB DR in 90nm CMOS, in *2012 IEEE International Solid-State Circuits Conference*, (IEEE, Piscataway, 2012), pp. 156–158
25. S. Norsworthy, R. Schreier, G. Temes, *Delta-Sigma Data Converters (Theory, Design, and Simulation)* (Wiley, New York, 1996)
26. M. Bolatkale, L. Breems, K.A.A. Makinwa, *High Speed and Wide Bandwidth Delta-Sigma ADCs* (Springer, Cham, 2014)
27. M. Vadipour et al., A 2.1mW/3.2mW delay-compensated GSM/WCDMA ΣΔ analog-digital converter, in *2008 IEEE Symposium on VLSI Circuits*, (IEEE, Honolulu, 2008), pp. 180–181
28. C.-Y. Ho et al., A 4.5 mW CT self-coupled ΔΣ modulator with 2.2 MHz BW and 90.4 dB SNDR using residual ELD compensation. IEEE J. Solid State Circuits **50**(12), 2870–2879 (2015)
29. I. Ahmed et al., A low-power Gm-C-based CT-ΔΣ audio-band ADC in 1.1V 65nm CMOS, in *2015 Symposium on VLSI Circuits (VLSI Circuits), Kyoto*, (IEEE, Piscataway, 2015), pp. C294–C295
30. C. Briseno-Vidrios et al., A 75 MHz BW 68dB DR CT-ΣΔ modulator with single amplifier biquad filter and a broadband low-power common-gate summing technique, in *2015 Symposium on VLSI Circuits (VLSI Circuits), Kyoto*, (IEEE, Piscataway, 2015), pp. C254–C255
31. G. Mitteregger et al., A 20-mW 640-MHz CMOS continuous-time ΣΔ ADC with 20-MHz signal bandwidth, 80-dB dynamic range and 12-bit ENOB. IEEE J. Solid State Circuits **41**(12), 2641–2649 (2006)
32. G. Wei et al., A 13-ENOB, 5 MHz BW, 3.16 mW multi-bit continuous-time ΔΣ ADC in 28 nm CMOS with excess-loop-delay compensation embedded in SAR quantizer, in *2015 Symposium on VLSI Circuits (VLSI Circuits), Kyoto*, (IEEE, Piscataway, 2015), pp. C292–C293
33. P. Fontaine et al., A low-noise low-voltage CT ΔΣ modulator with digital compensation of excess loop delay, in *ISSCC. 2005 IEEE International Digest of Technical Papers. Solid-State Circuits Conference, 2005, San Francisco, CA*, vol. 1, (IEEE, Piscataway, 2005), pp. 498–613
34. W. Tze-Chien et al., A 0.022 $mm^2$ 98.5 dB SNDR hybrid audio ΔΣ modulator with digital ELD compensation in 28 nm CMOS. IEEE J. Solid State Circuits **50**(11), 2655–2664 (2015)
35. M. Neofytou et al., A 1.9 mW 250 MHz bandwidth continuous-time ΣΔ modulator for ultra-wideband applications, in *2018 IEEE ISCAS, Florence*, (IEEE, Piscataway, 2018), pp. 1–5
36. R. Schreier, G. Temes, *Understanding Delta-Sigma Data Converters* (Wiley, Piscataway, 2005)
37. R. Rutten, L.J. Breems, G. Wetzker, Digital calibration of a continuous-time cascaded ΣΔ modulator based on variance derivative estimation, in *2006 Proceedings of the 32nd ESSCIRC, Montreux*, (IEEE, Piscataway, 2006), pp. 199–202
38. Y. Dong et al., Adaptive digital noise-cancellation filtering using cross-correlators for continuous-time MASH ADC in 28nm CMOS, in *2017 IEEE CICC, Austin, TX*, (IEEE, Piscataway, 2017), pp. 1–4
39. M. Zhou et al., A 2 GHz 0.98 mW 4-bit SAR-based quantizer with ELD compensation in an UWB CT ΣΔ modulator, in *2018 IEEE ISCAS, Florence*, (IEEE, Piscataway, 2018), pp. 1–5
40. D. Gubbins et al., Continuous-time input pipeline ADCs. IEEE JSSC **45**(8), 1456–1468 (2010)
41. M. Bolatkale, F6: pushing the performance limit in data converters, in *2017 IEEE ISSCC, San Francisco, CA, USA*, (IEEE, Piscataway, 2017), pp. 515–517
42. B. Wu et al., A 24.7 mW 65 nm CMOS SAR-assisted CT ΔΣ modulator with second-order noise coupling achieving 45 MHz bandwidth and 75.3 dB SNDR. IEEE JSSC **51**(12), 2893–2905 (2016)
43. J. Harrison et al., An LC bandpass ΔΣ ADC with 70dB SNDR over 20MHz bandwidth using CMOS DACs, in *2012 IEEE ISSCC*, (IEEE, Piscataway, 2012), pp. 146–148
44. S. Park, G. Kim, S.-C. Park, W. Kim, A digital-to-analog converter based on differential-quad switching. IEEE J. Solid State Circuits **37**(10), 1335–1338 (2002)
45. E.C. Dijkmans et al., A 0.2-mW CMOS ΣΔ modulator for speech coding with 80 dB dynamic range. IEEE JSSC **31**(12), 1873–1880 (1996)

# Chapter 5
# Continuous-Time Delta-Sigma Converters with Finite-Impulse-Response (FIR) Feedback

**Shanthi Pavan**

## 5.1 Introduction

The $\Delta\Sigma$ architecture has become the de facto choice for the realization of analog-to-digital converters that achieve high resolution over low and medium bandwidths. Since the late 1990s, there has been a resurgence of interest in continuous-time $\Delta\Sigma$ modulators (CT$\Delta\Sigma$M), where the loop filter of the $\Delta\Sigma$ loop was implemented using continuous-time circuitry. There were many motivating facts that prompted this wave. Designing switch-capacitor circuits at high speeds in processes that supported low supply voltages was becoming increasingly difficult. Driving the switched-capacitor input impedance of the modulator, especially in high-speed high-resolution converters, was a challenge. Both these problems were avoided with a CT$\Delta\Sigma$M. Furthermore, a CT$\Delta\Sigma$M has the remarkable property of implicit anti-aliasing, since the input is sampled *after* being filtered by the loop filter.

Unfortunately, it turns out that clock jitter degrades a CT$\Delta\Sigma$M's in-band SNR. The effect of jitter on the modulator loop can be modeled by errors at the input and output of the ADC and DAC, respectively, [1, 2] as shown in Fig. 5.1. It is immediately apparent that $e_{adc}$ is shaped by the modulator's NTF, just like quantization noise, and has virtually no effect on the in-band spectrum of the CT$\Delta\Sigma$M.

The story is different, however, with the error induced by jitter at the DAC output. As seen in Fig. 5.1, $e\_dac$ adds to the input of the modulator and degrades the modulator's in-band SNR. With an NRZ DAC, it can be shown that $e\_dac$ consists of a sum of slivers, whose width and height at $nT_s$ are given by $\Delta t[n]$ and height $(v[n] - v[n-1])$, respectively. Jitter in a particular clock edge,

S. Pavan (✉)
Indian Institute of Technology Madras, Chennai, India
e-mail: shanthi@ee.iitm.ac.in

© Springer Nature Switzerland AG 2020

A. Baschirotto et al. (eds.), *Next-Generation ADCs, High-Performance Power Management, and Technology Considerations for Advanced Integrated Circuits*,
https://doi.org/10.1007/978-3-030-25267-0_5

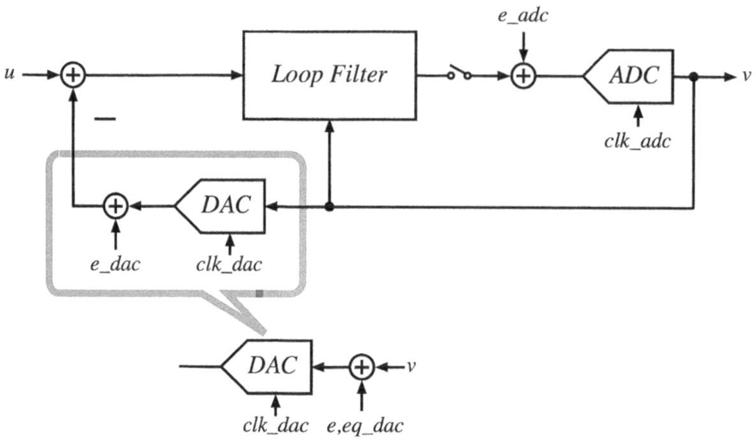


**Fig. 5.1** The clock-jitter problem in a CTΔΣM. $e_{adc}$ and $e_{dac}$ model jitter-induced errors. In-band SNR degradation is due to $e_{dac}$

therefore, introduces an error at that edge only if the modulator output changes in that cycle. Furthermore, the noise is dependent on the height of the transitions $(v[n] - v[n-1])$ in the output waveform. Using a multi-bit quantizer, therefore, is a very good way of reducing the sensitivity of a CTΔΣM to clock jitter.

From the discussion above, we see that clock jitter can be a potential show-stopper for a single-bit CTΔΣM. This motivated research into techniques that aim to address this problem. One such approach is the use of switched-capacitor (SC) feedback [3]. An SC-feedback DAC attempts to reduce jitter sensitivity by feeding a packet of charge into the loop filter in an impulsive fashion. Unfortunately, this greatly increases the peak-to-average ratio of the feedback waveform, and causes linearity problems in the loop filter. Furthermore, SC DACs severely compromise alias rejection of the modulator around multiples of the sampling frequency [4].

A particularly elegant and effective approach to address the jitter sensitivity of a single-bit CTΔΣM is the use of FIR feedback [5, 6]. We illustrate with the single-bit example shown in Fig. 5.2. The 2-level output sequence $v$ is filtered by an $N$-tap low pass FIR filter with transfer function $F(z)$, before exciting the main feedback DAC. The DAC has an NRZ pulse shape. For simplicity, the tap weights of $F(z)$ are assumed to be identical. Since $v = \pm 1$, the magnitude of transitions in $v$ is 2. With $F(z)$ in place, this is reduced to $2/N$. Thus, the magnitudes of the steps in the feedback DAC waveform (denoted by $v_1(t)$ in Fig. 5.2a) are $N$ times smaller than what they would have otherwise been. Since noise due to clock jitter is proportional to the height of the transitions in the DAC output, it follows that the in-band mean square noise due to jitter is reduced by $20\log(N)$ dB.

The FIR DAC has other important benefits. Since $F(z)$ is a low pass filter, the input component of $v$ is not affected, but the power of the shaped noise is reduced. The DAC's output $v_1(t)$, therefore, has a reduced high frequency content that closely follows the input $u$. The error processed by the loop filter, which is $u(t) - v_1(t)$, is

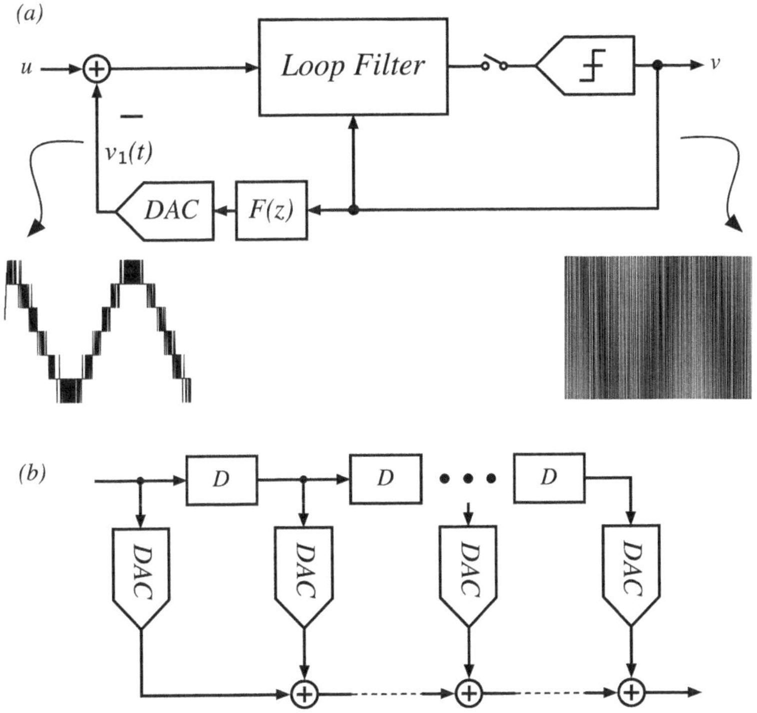


**Fig. 5.2** (**a**) Using an FIR feedback DAC to mitigate noise due to clock jitter in a single-bit modulator. (**b**) Semi-digital implementation of the FIR DAC

therefore much smaller. This relaxes the linearity requirements of the loop filter, just like in a CTΔΣM with a multi-bit DAC. Intuitively, since $v_1(t)$ "looks" like the output of a multi-bit DAC, one should expect similar benefits with respect to clock jitter and loop-filter linearity. For a detailed analysis of clock jitter with FIR feedback, see [5, 7, 8].

Implementing the FIR DAC as drawn in Fig. 5.2a is problematic when the DAC levels are not equally spaced (due to mismatched components). Recognizing that $v[n]$ is a two level sequence, a linear DAC-filter combination can be realized using the semi-digital approach [9], as shown in Fig. 5.2b. Here, the delays are implemented digitally, while the individual DAC outputs (which are assumed here to be currents) are weighted and summed in the analog domain. It is easy to see that DAC mismatch modifies the transfer function of the filter, but does not cause nonlinearity.

To summarize, using a plain 1-bit quantizer, while being a very efficient use of hardware, is severely impacted by jitter, and is problematic for the linearity of the CTΔΣM. Fortunately, using FIR feedback with a one-bit ADC not only addresses the jitter problem, but also greatly relaxes the linearity requirements of the loop filter, just like with a multi-bit quantizer. Unlike a multi-bit design, whose linearity is impacted by element mismatch in the feedback DAC, an FIR DAC implemented using a semi-digital approach is inherently linear. FIR feedback, therefore, combines the beneficial aspects of single-bit and multi-bit designs, without the key disadvantages of either approaches.

The conventional wisdom in the $\Delta\Sigma$ community is that most of the power in a CT$\Delta\Sigma$M is consumed in its input stage (to achieve low noise). At the high clock rates needed to achieve wide bandwidths, however, the stage driving the quantizer consumes significant power. The power dissipated in the clock generation and distribution circuitry also becomes problematic. Yet another issue is that DEM may not be feasible, as it adds an unacceptably large delay into the loop. These difficulties might have led to a renewed interest in CT$\Delta\Sigma$Ms with a single-bit quantizer and FIR feedback [8, 10, 11]. In such an approach, the quantizer design is greatly simplified and DEM is not necessary. There have been many designs reporting the use for FIR feedback in recent years (see, for instance, [7, 12–16]), targeting low and high bandwidths. While CT$\Delta\Sigma$Ms with single-bit quantizers and FIR feedback are particularly efficient, ADCs with multilevel quantizers and FIR feedback have also been reported [17].

This chapter is intended as a review of FIR feedback in CT$\Delta\Sigma$Ms [18]. The rest of this chapter is organized as follows. Section 5.2 addresses the issue of loop compensation when an FIR DAC is used. This is particularly important, since introducing a delay in the feedback path of a $\Delta\Sigma$ modulator can destabilize the loop. In Sect. 5.3, we discuss the issue of design centering a CT$\Delta\Sigma$M with FIR feedback, in the light of practical circuit nonidealities like finite OTA gain and bandwidth. Yet another recently discovered [14, 19] advantage of FIR feedback is that it enables the use of "chopping for free." The intuition behind this is discussed in Sect. 5.4. Section 5.5 concludes the chapter.

## 5.2 Compensation

Using FIR feedback introduces delay in the $\Delta\Sigma$ loop, and most likely renders the modulator unstable. One of the key design challenges with an FIR DAC, therefore, is to compensate the loop for the effect of the FIR filter. It turns out that the NTF of a $\Delta\Sigma$ loop with an FIR DAC can be restored *exactly*. This can be done by modifying the loop-filter coefficients and by adding a *direct path FIR filter* around the quantizer.[1] The method of moments allows us to quickly and simply determine the modified coefficients. A detailed analysis is given in [21, 22]—here we give the intuition behind the results.

[1]This is similar to compensating a CT$\Delta\Sigma$M for excess delay, where coefficient tuning and a direct path around the quantizer is usually needed [2, 20].

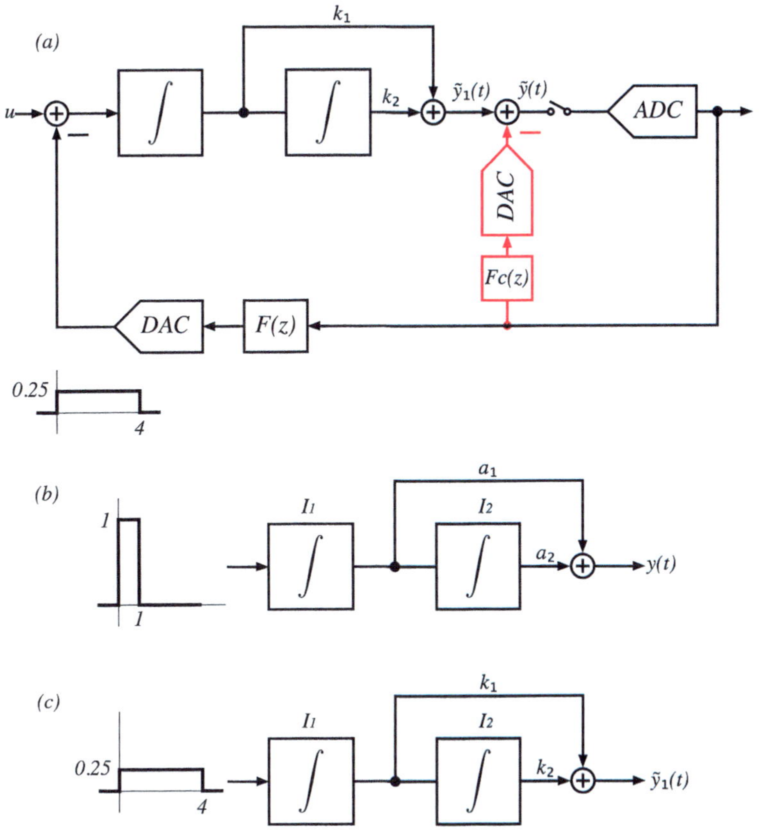


**Fig. 5.3** (**a**) A second-order CT$\Delta\Sigma$M with a 4-tap FIR DAC compensated for the delay of the FIR DAC. $F_c(z)$ is also a 4-tap FIR filter. (**b**) Determining the pulse response $y(t)$ of the loop filter without the FIR DAC. (**c**) $\tilde{y}_1(t)$ can be made to equal $y(t)$ for $t \geq 4$ by modifying coefficients. $F_c(z)$ is needed to compensate for the difference in the sampled pulse response for $t < 4$

To illustrate the process of compensation, assume that we have on hand a normalized[2] second-order cascade of integrators with feedforward (CIFF) CT$\Delta\Sigma$M prototype that achieves a desired NTF. Let the loop-filter transfer function be $L(s) = a_1/s + a_2/s^2$. The main feedback DAC of the prototype is modified to a 4-tap FIR DAC with equal tap weights (each being 0.25), as shown in Fig. 5.3a. If nothing else was done, the NTF of the loop will change. What should we do to restore the NTF? We need to ensure that the sampled pulse response of the prototype modulator and that of the loop filter with the FIR DAC are identical.

[2]Sampling rate of 1 Hz.

The prototype's pulse response is determined using the signal flow graph of Fig. 5.3b. After the DAC pulse has "died," the output of $I_1$ is 1. Thus, for $t \gg 1$, the output of $I_2$ will grow as $t$, and $y(t)$ will grow as $a_2 t$. Consider now the loop filter excited by the FIR DAC, as shown in Fig. 5.3c. The area of the pulse exciting the loop filter is unity. Thus, after $t \gg 4$, $I_2'$s output will grow as $t$, and $\tilde{y}_1(t)$ will grow as $k_2 t$. If the NTF has to be restored, it is immediately apparent that $k_2 = a_2$.

Choosing $k_2 = a_2$ ensures that $y(t)$ and $\tilde{y}_1(t)$ in Fig. 5.3b and c asymptotically grow at the same rate. However, $\tilde{y}_1(t)$ for large $t$ will be delayed with respect to $y(t)$, since the FIR-DAC pulse is delayed with respect to the NRZ-DAC pulse. To address this, $\tilde{y}_1(t)$ must be "advanced." This is possible by recognizing that for any waveform $g(t)$ with continuous derivative, $g(t + t_d) = g(t) + t_d \dot{g}(t) + (t_d^2/2)\ddot{g}(t) + \ldots$. In other words, by adding suitable amounts of the first and higher order derivatives of a waveform, the waveform can effectively be advanced.[3] In our example, the derivative of the output waveform can be generated by suitably weighting the output of $I_1$. Analysis, omitted here due to lack of space, shows that the $k$'s must be chosen as

$$k_2 = a_2, \quad k_1 = a_1 + 1.5a_2. \tag{5.1}$$

This will ensure that $\tilde{y}_1(t)$ and $y(t)$ are equal for $t \geq 4$. Figure 5.4 shows $\tilde{y}_1(t)$ and $y(t)$ in our 4-tap example after the $k$'s have been chosen as per the equations above. To achieve the desired pulse response for $t < 4$, a four-tap compensation filter ($F_c(z)$) must be used in a direct path around the quantizer, as shown in Fig. 5.3a. The taps of $F_c(z)$ can be computed using the following steps:

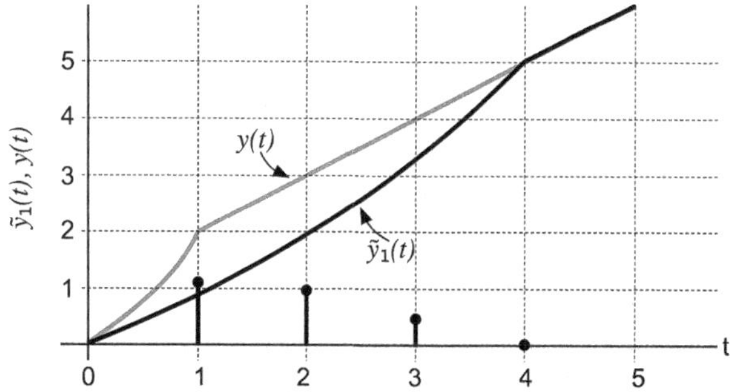


**Fig. 5.4** Loop-filter pulse responses of the NRZ prototype, the (coefficient tuned) loop filter with a 4-tap FIR DAC, and the response of the compensation filter. The main FIR DAC taps are all assumed to be equal. The direct path DAC with response $F_c(z)$ should make up for the difference $(\tilde{y}_1(t) - y(t))$

[3]The idea is the same as introducing zeros in a loop to stabilize a feedback system. In the time-domain, this corresponds to adding a derivative(s).

(a) Using $a_1, a_2$ from the prototype, and the coefficients of the main FIR DAC, use (5.1) to compute $k_1, k_2$.
(b) Determine the pulse response of the prototype loop filter and the coefficient tuned loop filter with the FIR feedback DAC. These responses will be identical beyond $t = M$, where $M$ denotes the number of FIR DAC taps.
(c) Determine the difference between the pulse responses in step (b) above that lasts for a duration $M$. The direct path filter taps are the samples of this difference at times $1, \ldots, (M - 1)$.

It is not necessary for the compensation path to appear directly around the quantizer. It can be moved to the input of the second integrator.

### *5.2.1 STF Peaking*

As seen from the discussion above, the process of compensating the $\Delta\Sigma$ loop for the delay introduced by the FIR DAC increased the gains of the $1/s$ and $1/s^2$ paths. This then means that the signal transfer function (STF) of the CIFF structure, which already has peaking, will increase further. This can be problematic in wireless applications and can be mitigated somewhat by adding input feed-ins into the integrators. Loeda et al. [7] is an industrial example of a CT$\Delta\Sigma$M that uses feed-ins to flatten the STF.

### *5.2.2 FIR Feedback Without Compensation*

Observing Fig. 5.3 shows that inserting the FIR filter into the outermost feedback path of the CT$\Delta\Sigma$M adds delay to *all* the feedback paths around the quantizer. It is therefore reasonable to conclude that the modulator would become unstable if it was not properly compensated. In some CT$\Delta\Sigma$M architectures (for example, the cascade of integrators with feedback (CIFB)), the outermost feedback DAC is not part of the "fast feedback" path around the quantizer. It is then conceivable that the CT$\Delta\Sigma$M will remain stable (though its NTF changes) when an FIR filter is inserted before DAC in the outermost path. This was employed in [6].

## 5.3 Design Centering

Design centering the modulator in the face of practical integrator nonidealities (like finite gain and parasitic poles) is perhaps the least appreciated aspect of CT$\Delta\Sigma$M design. We use the modulator of Fig. 5.3 as an example to illustrate the design-centering process. The loop filter (without the compensating FIR DAC) is shown in Fig. 5.5. In practice, $I_1$ and $I_2$ are nonideal—they will have finite dc gain and

Fig. 5.5 Loop filter with practical integrators

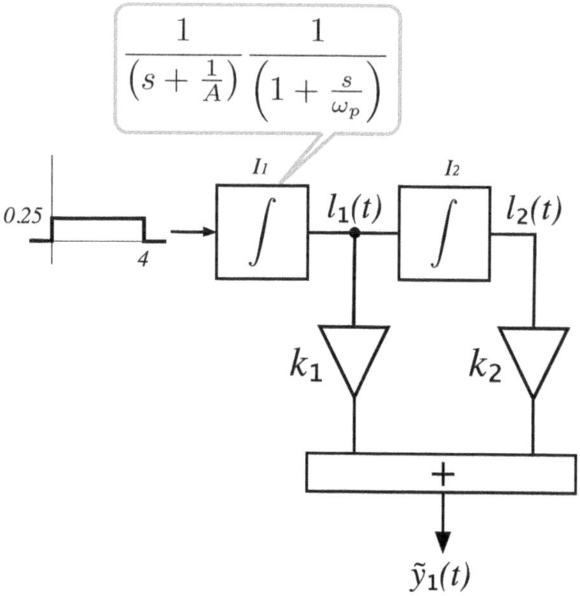


parasitic poles. Using such integrators in place of ideal ones will change the NTF that is realized, if one uses the coefficients corresponding to ideal integrators. The poles of the discretized loop filter are no longer at $z = 1$ (due to finite gain). Furthermore, the order of the loop filter has increased (in our example) to four. It is *not possible*, therefore to be able to achieve the second-order NTF of the ideal prototype. What we can hope for is that with an appropriate choice of the $k$'s the NTF actually realized approximates the desired NTF well.

A tempting, albeit numerically unstable, method of determining the $k$'s and the compensating FIR filter taps is to use "open-loop" fitting according to

$$\begin{bmatrix} l_1[1] & l_2[1] & 1 & 0 & 0 \\ l_1[2] & l_2[2] & 0 & 1 & 0 \\ l_1[3] & l_2[3] & 0 & 0 & 1 \\ l_1[4] & l_2[4] & 0 & 0 & 0 \\ \vdots & \vdots & \vdots & \vdots & \vdots \end{bmatrix} \begin{bmatrix} K \\ F_c \end{bmatrix} = \begin{bmatrix} l[1] \\ l[2] \\ l[3] \\ l[4] \\ \vdots \end{bmatrix} \tag{5.2}$$

where $l_1[n]$ and $l_2[n]$ denote the sampled responses at the outputs of $I_1$ and $I_2$ when the loop filter is driven by the FIR-DAC pulse as shown in Fig. 5.5. $K^T = \left[k_1\ k_2\right]^T$, $F_c^T = \left[f_{c1}\ f_{c2}\ f_{c3}\right]$, and $l[n]$ is the response of the ideal loop filter. $l_1$ and $l_2$ are easily obtained from simulations (of either schematic or layout). The set of equations above attempt to find that linear combination of $l_1$, $l_2$ and compensation filter taps that approximate the response of the ideal loop filter with the least square error. As shown in [23], the solutions to $K$, $F_c$ obtained above vary significantly with the number of equations used in the fitting procedure.

A robust way of determining the coefficients and taps of the compensation FIR DAC is to use "closed-loop fitting" [23] where we first form the sequences $h_1[n] = (l_1 * h)[n], h_2[n] = (l_2 * h)[n]$, where $h[n]$ denotes the impulse response corresponding to the desired NTF, and * represents convolution. The equations to be solved in the closed-loop fitting technique are given by

$$\begin{bmatrix} h_1[1] & h_2[1] & h[0] & 0 & 0 \\ h_1[2] & h_2[2] & h[1] & h[0] & 0 \\ h_1[3] & h_2[3] & h[2] & h[1] & h[0] \\ h_1[4] & h_2[4] & h[3] & h[2] & h[1] \\ \vdots & \vdots & \vdots & \vdots & \vdots \end{bmatrix} \begin{bmatrix} K \\ F_c \end{bmatrix} = \begin{bmatrix} -h[1] \\ -h[2] \\ -h[3] \\ -h[4] \\ \vdots \end{bmatrix}$$

where $K^T = \begin{bmatrix} k_1 & k_2 & k_3 \end{bmatrix}^T$ and $F_c^T = \begin{bmatrix} f_{c1} & f_{c2} & f_{c3} \end{bmatrix}$.

Closed-loop fitting has been used in several successful designs across industry and academia [16, 17, 24–30].

## 5.4 FIR Feedback and Chopping

Flicker noise degrades the in-band SNR of a CT$\Delta\Sigma$M. In ADCs intended for low-frequency applications, using small device sizes can lead to a $1/f$ noise corner that is larger than the signal bandwidth. The commonly used brute-force approach is to simply increase device dimensions until the $1/f$ noise corner is reduced to a sufficiently low frequency. This, however, is problematic on two counts. First, the area occupied by the input stage can become very large. Second, the flicker-noise corner of a device can vary significantly with process changes. It is not uncommon, especially in technologies with small geometries, for the $1/f$ noise spectral density to increase by a factor of 10 in the worst-case process corner. This implies that the area of the input stage should be made ten times larger to accommodate the worst case!

It is tempting to think of flicker noise as being merely a problem afflicting low-bandwidth converters. Unfortunately, high-speed CT$\Delta\Sigma$Ms are also affected by $1/f$ noise, albeit in a more subtle manner, as described below. In wideband ADCs, one is often forced to use the smallest feasible transistor geometry to achieve the desired speed of operation. The flicker noise due to this choice can occupy a non-negligible portion of the (large) signal bandwidth. Attempting to reduce the contribution of the flicker noise by device sizing significantly increases the parasitic capacitance at the input of the input OTA. This degrades the high-frequency loop-gain around the OTA, increasing the delay through the integrator and reducing its linearity. From the discussion above, it is seen that flicker noise is a problem with all flavors of CT$\Delta\Sigma$Ms, regardless of bandwidth. The problem is further exacerbated due to $1/f$-noise modeling issues in nanometer CMOS processes, which make it difficult to rely on simulations.

Chopping is a traditional way of eliminating flicker noise, and works by modulating it out of the signal band [31]. It can also be applied to the OTA in the input integrator of a CTΔΣM. The hope is that chopping allows the use of small input devices, without being concerned about flicker noise (since it will be eliminated anyway). Thus, handling inaccuracies in modeling and accommodating the worst-case $1/f$-noise corner are no longer problematic. Chopping in a CTΔΣM comes with its own challenges, as described below.

Figure 5.6a shows the input section of a CTΔΣM with an integrator employing a chopped OTA. For simplicity, we assume a single-stage OTA, and a resistive feedback DAC. If the input and feedback DAC resistors are each equal to $2R$, the integrator can be modeled as shown in Fig. 5.6b. $cp1$ and $cp2$ denote the parasitic capacitors at the input and output of the OTA, respectively. The chopping frequency is $fc$. The single-ended equivalent of the chopped-integrator is given in Fig. 5.6c. The effect of chopping the OTA can be modeled as follows [32]. The parasitic capacitances $cp1$ and $cp2$ are replaced by their chopped versions. The OTA's $1/f$ noise is modulated to around odd multiples of $f_c$, and will eventually

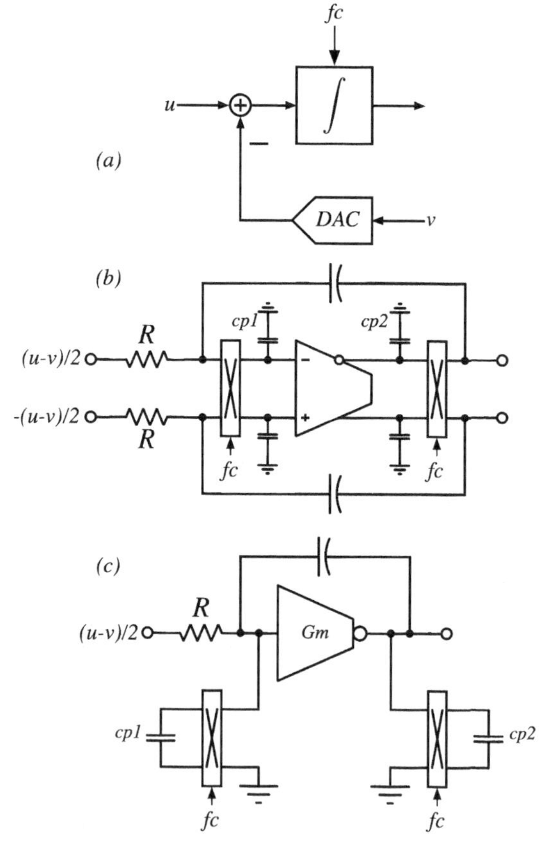


**Fig. 5.6** (**a**) Input stage of a CTΔΣM with a chopped integrator. (**b**) Simplified schematic, using a single-stage OTA-RC integrator. (**c**) Single-ended equivalent

be eliminated by the decimator. $(u - v)/2$ consists of shaped quantization noise, whose out-of-band noise spectral density is several *orders of magnitude* larger than that of the noise in the signal band. This is problematic due to the following. Due to the chopped capacitors, the integrator is a linear periodically time-varying (LPTV) system. It turns out that time-varying operation causes out-of-band noise from around multiples of *twice* the chopping frequency to fold into the signal band, dramatically degrading in-band SNDR. In the case of the single-stage OTA, most of the noise-folding is due to the inversion of the voltage across $cp1$ at every edge of the chopping clock. This is equivalent to injecting into the integrating capacitor $C$, a charge proportional to the virtual ground voltage at every chopping edge. It thus follows that chopping related artifacts are inversely proportional to $1/(g_m R)$. Another problematic aspect of chopping is the switched-capacitor resistor created by periodically switching $cp2$. This reduces the OTA's dc gain, thereby increasing the input referred $1/f$ noise due to the rest of the loop filter. Chopping artifacts can be reduced using an OTA with a higher transconductance, or increasing the number of quantizer levels, or both. None of these strategies is power efficient. An alternative solution to the problem is to chop the OTA at the modulator's sampling rate [33]. This way, noise folds from integer multiples of $f_s$, where the feedback DAC waveform has spectral nulls anyway. Unfortunately, chopping at such high speeds degrades the OTA's dc gain and linearity, and increases charge injection errors due to mismatches in the chopping switches.

The use of FIR feedback enables the use of a low chopping frequency, while greatly reducing the aliasing of shaped quantization noise. The basic idea behind this is illustrated using Fig. 5.7 [14, 19]. $F(z)$ is assumed to have $N$ equal taps. The $F(z)$-DAC combination is implemented using semi-digital techniques [9]. The DAC output waveform $v_1(t)$ has nulls at $f_s/N$. If the chopping frequency is chosen to be $f_s/(2N)$, the shaped noise that can alias into the signal band is greatly attenuated, as shown using a 12-tap example in Fig. 5.7b. The parts of the spectra in red show the spectral content that would fold back to low frequencies with and without FIR feedback. Calculations show that with OSR $= 128$, the shaped noise is attenuated by at least 35 dB before it aliases into the signal band. To achieve the same level of chopping artifacts by increasing the OTA's transconductance, $g_m$ should increase by a factor of 60. Alternatively, the number of quantizer levels should increase by a factor of 60. It is thus seen that FIR feedback, originally invented to the sensitivity of a CT$\Delta\Sigma$M to clock jitter, allows one to achieve "chopping for free." This is confirmed by the experimental results in [16] and [14]. The only difference between the two designs is the use of chopping in the first stage of the two-stage OTA used in the input OTA. The two designs achieve similar performance with the same power dissipation, except that [14] achieves a much lower $1/f$ noise corner even while using input devices that are 16$\times$ smaller than those used in [16].

A general technique to analyze and simulate the effect of chopping in a CTDSM is given in [34]. The technique is capable of addressing chopping artifacts with arbitrary OTA topologies. It was used to estimate the level of chopping artifacts in the high-performance design presented in [35].

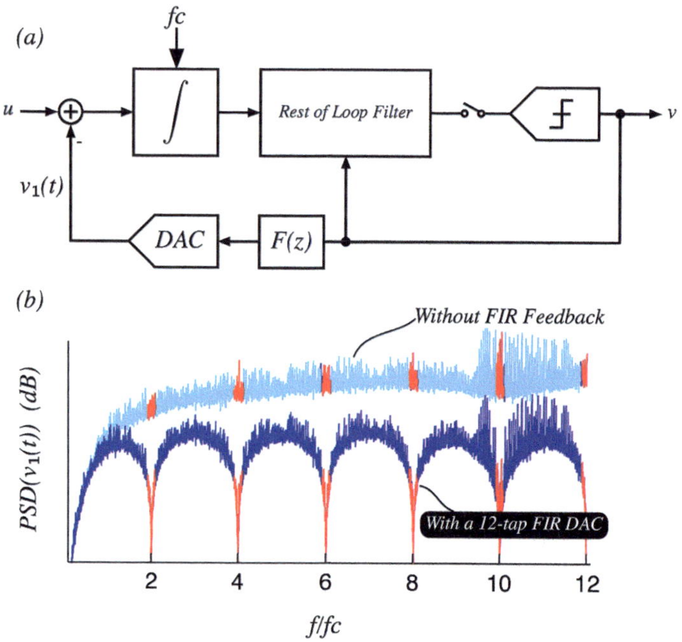


**Fig. 5.7** (**a**) CTΔΣM with FIR feedback. (**b**) PSD of the feedback waveform with and without FIR feedback. If the chopping frequency is chosen to be half the frequency of the lowest spectral null, chopping artifacts that alias into the signal band are small

## 5.5 Conclusions

This paper traced the motivation behind the use of FIR feedback in CTΔΣMs. The intuition behind compensation of the ΔΣ loop with FIR feedback was given, and design centering of a practical modulator was discussed. We described a recently discovered feature of FIR feedback, namely "chopping for free."

## References

1. J. Cherry, W.M. Snelgrove, Clock jitter and quantizer metastability in continuous-time delta-sigma modulators. IEEE Trans. Circuits Syst. II. Analog Digit. Signal Process **46**(6), 661–676 (1999)
2. J.A. Cherry, Theory, practice, and fundamental performance limits of high speed data conversion using continuous-time delta-sigma modulators, Ph.D. dissertation, Carleton University, 1998
3. M. Ortmanns, F. Gerfers, Y. Manoli, A continuous-time ΣΔ modulator with reduced sensitivity to clock jitter through SCR feedback. IEEE Trans. Circuits Syst. I Reg. Papers **52**(5), 875–884 (2005)

4. S. Pavan, Alias rejection of continuous-time delta-sigma modulators with switched-capacitor feedback DACs. IEEE Trans. Circuits Syst. I Reg. Papers **58**(2), 233–243 (2011)
5. O. Oliaei, Sigma-delta modulator with spectrally shaped feedback. IEEE Trans. Circuits Syst. II Analog Digit. Signal Process. **50**(9), 518–530 (2003)
6. B.M. Putter, $\Sigma\Delta$ ADC with finite impulse response feedback DAC, in *IEEE International Solid-State Circuits Conference Digest of Technical Papers (ISSCC)* (IEEE, Piscataway, 2004), pp. 76–77
7. S. Loeda, J. Harrison, F. Pourchet, A. Adams, A 10/20/30/40 MHz feedforward FIR DAC continuous-time $\Delta\Sigma$ ADC with robust blocker performance for radio receivers. IEEE J. Solid-State Circuits **51**(4), 860–870 (2016)
8. P. Shettigar, S. Pavan, Design techniques for wideband single-bit continuous-time $\Delta\Sigma$ modulators with FIR feedback DACs. IEEE J. Solid-State Circuits **47**(12), 2865–2879 (2012)
9. D.K. Su, B.A. Wooley, A CMOS oversampling D/A converter with a current-mode semidigital reconstruction filter. IEEE J. Solid-State Circuits **28**(12), 1224–1233 (1993)
10. P. Shettigar, S. Pavan, A 15 mW 3.6 GS/s CT delta-sigma ADC with 36 MHz bandwidth and 83 dB dynamic range in 90 nm CMOS , in *Digest of Technical Papers, International Solid State Circuits Conference* (2008), pp. 498–631
11. V. Srinivasan, V. Wang, P. Satarzadeh, B. Haroun, M. Corsi, A 20 mW 61 dB SNDR (60MHz BW) 1b 3rd-order continuous-time delta-sigma modulator clocked at 6 GHz in 45 nm CMOS, in *Digest of Technical Papers, International Solid-State Circuits Conference (ISSCC)* (2012), pp. 158–160
12. S. Zeller, C. Muenker, R. Weigel, U. Ussmueller, A 0.039 mm2 inverter-based 1.82 mW 68.6 dB-SNDR 10 MHz-BW CT-$\Delta\Sigma$-ADC in 65 nm CMOS using power-and area-efficient design techniques. IEEE J. Solid-State Circuits **49**(7), 1548–1560 (2014)
13. Y. Zhang, C.-H. Chen, T. He, G.C. Temes, A continuous-time delta-sigma modulator for biomedical ultrasound beamformer using digital ELD compensation and FIR feedback. IEEE Trans. Circuits Syst. I Reg. Papers **62**(7), 1689–1698 (2015)
14. S. Billa, A. Sukumaran, S. Pavan, Analysis and design of continuous-time delta–sigma converters incorporating chopping. IEEE J. Solid-State Circuits **52**(9), 2350–2361 (2017)
15. A. Jain, S. Pavan, A 13.3 mW 60 MHz bandwidth, 76 dB DR 6 GS/s CT$\Delta\Sigma$M with time-interleaved FIR feedback, in *Proceedings of the Symposium on VLSI Circuits* (IEEE, Piscataway, 2016), pp. C256–C257
16. A. Sukumaran, S. Pavan, Design techniques for audio single-bit continuous-time delta sigma modulators with FIR feedback. IEEE J. Solid State Circuits **47**(11), 1–15 (2014)
17. J. Gealow, M. Ashburn, C.-H. Lou, S. Ho, P. Riehl, A. Shabra, J. Silva, Q. Yu, A 2.8 mW $\Delta\Sigma$ ADC with 83 dB DR and 1.92 MHz BW using FIR outer feedback and TIA-based integrator, in *Proceedings of the Symposium on VLSI Circuits* (IEEE, Piscataway, 2011), pp. 42–43
18. S. Pavan, Finite-impulse-response (FIR) feedback in continuous-time delta-sigma converters, in *Proceedings of the IEEE Custom Integrated Circuits Conference (CICC)* (IEEE, Piscataway, 2018), pp. 1–8
19. S. Billa, A. Sukumaran, S. Pavan, A 280$\mu$W 24kHz-BW 98.5 dB-SNDR chopped single-bit CTDSM achieving $<$10 Hz $1/f$ noise corner without chopping artifacts, in *Proceedings of the IEEE International Solid-State Circuits Conference (ISSCC)* (IEEE, Piscataway, 2016), pp. 276–277
20. S. Pavan, Excess loop delay compensation in continuous-time delta-sigma modulators. IEEE Trans. Circuits Syst. II Express Briefs **55**(11), 1119–1123 (2008)
21. S. Pavan, Continuous-time delta-sigma modulator design using the method of moments. IEEE Trans. Circuits Syst. I Reg. Papers **61**(6), 1629–1637 (2014)
22. S. Pavan, R. Schreier, G. Temes, *Understanding Delta-Sigma Data Converters*, 2nd edn. (Wiley-IEEE Press, Piscataway, 2017)
23. S. Pavan, Systematic design centering of continuous time oversampling converters. IEEE Trans. Circuits Syst. II Express Briefs **57**(3), 158–162 (2010)

24. S. Ho, C.-L. Lo, J. Ru, J. Zhao, A 23 mW, 73 dB dynamic range, 80 MHz BW continuous-time delta-sigma modulator in 20 nm CMOS. IEEE J. Solid-State Circuits **50**(4), 908–919 (2015)
25. Y.-S. Shu, S. Ho, H.-H. Chen, B. Narasimhan, K.-H.L. Loh, Signal processing and analog/RF circuit design: cross-discipline interactions and technical challenges, in *APSIPA Transactions on Signal and Information Processing*, vol. 5 (Cambridge University Press, Cambridge, 2016)
26. J.-K. Cho, S. Woo, A 6-mW, 70.1-dB SNDR, and 20-MHz BW continuous-time sigma-delta modulator using low-noise high-linearity feedback DAC. IEEE Trans. Very Large Scale Integr. Syst. **25**(5), 1742–1755 (2017)
27. D.-Y. Yoon, S. Ho, H.-S. Lee, An 85dB-DR 74.6 dB-SNDR 50MHz-BW CT MASH $\Delta\Sigma$ modulator in 28nm CMOS, in *Proceedings of the International Solid-State Circuits Conference (ISSCC)* (IEEE, Piscataway, 2015), pp. 1–3.
28. D.-Y. Yoon, S. Ho, H.-S. Lee, A continuous-time Sturdy-MASH $\Delta\Sigma$ modulator in 28 nm CMOS. IEEE J. Solid-State Circuits **50**(12), 2880–2890 (2015)
29. R. Kaald, T. Eggen, T. Ytterdal, A 1 MHz BW 34.2 fJ/step continuous time delta sigma modulator with an integrated mixer for cardiac ultrasound. IEEE Trans. Biomed. Circuits Syst. **11**(1), 234–243 (2017)
30. A. Jain, S. Pavan, Continuous-time delta-sigma modulators with time-interleaved FIR feedback. IEEE Trans. Circuits Syst. I Reg. Papers **65**(2), 434–443 (2018)
31. C.C. Enz, G.C. Temes, Circuit techniques for reducing the effects of op-amp imperfections: autozeroing, correlated double sampling, and chopper stabilization. Proc. IEEE **84**(11), 1584–1614 (1996)
32. S. Pavan, Analysis of chopped integrators, and its application to continuous-time Delta-Sigma modulator design. IEEE Trans. Circuits Syst. I Reg. Papers **64**(9), 1953–65 (2017)
33. D.R. Welland, B.P. Del Signore, D.A. Kerth, Delta-sigma analog-to-digital converter with chopper stabilization at the sampling frequency, US Patent 5,039,989, 1991
34. R. Theertham, S. Pavan, Unified Analysis, Modeling, and Simulation of Chopping Artifacts in Continuous-Time Delta-Sigma Modulators. IEEE Trans. Circuits Syst. I. Reg. Papers **66**(8), 2831–2842 (2019)
35. R. Theertham, P. Kootala, S. Billa, S. Pavan, A 24mW Chopped CTDSM Achieving 103.5 dB SNDR and 107.5 dB DR in a 250kHz Bandwidth. Proceedings of the Symposium on VLSI Circuits, Kyoto, C226–C227 (2019)

# Chapter 6
# High-Speed ADCs for Wireless Base Stations

**Bob Verbruggen, Bruno Vaz, Christophe Erdmann, Diarmuid Collins, John Mcgrath, Ali Boumaalif, David Melinn, Edward Cullen, Darragh Walsh, Alonso Morgado, Conrado Mesadri, Brian Long, Rajitha Pathepuram, Ronnie De La Torre, Alvin Manlapat, Georgios Karyotis, Patrick Lynch, Peng Lim, Daire Breathnach, and Brendan Farley**

## 6.1 Introduction

The wireless revolution continues unabated with the impending roll-out of 5G networks. These networks aim to sate the seemingly limitless demand for wireless bandwidth, promising higher throughput, lower latency, and higher density to name but a few of the planned improvements. Clearly these promised improvements pose severe challenges for the notoriously tricky analog front-end (AFE) of the radio receiver.

Conventional radio receivers consisting of a direct-conversion or low-IF mixer stage, filtering, and relatively narrow-band ADCs [1] suffer from bandwidth limitations when instantaneous signal bandwidths exceed a few 100s of MHz. An inherently wideband approach, e.g., direct-RF sampling and digital down-conversion, avoids these bandwidth limitations and offers maximum flexibility. Because this approach quantizes unwanted spectrum and needs to digitally process this quantized spectrum there is an inherent penalty in power consumption. Nevertheless, with technology scaling and advances in ADC design, the direct-RF solution is now possible in an acceptable power envelope.

In this chapter we will discuss the implementation of such a direct-RF conversion system, part of the Xilinx RFSoC [2–4]. The rest of this chapter is organized as follows: Section 6.2 will describe the general ADC requirements and how they

B. Verbruggen (✉) · B. Vaz · C. Erdmann · D. Collins · J. Mcgrath · A. Boumaalif · D. Melinn
E. Cullen · D. Walsh · A. Morgado · C. Mesadri · B. Long · R. Pathepuram · R. De La Torre
A. Manlapat · G. Karyotis · P. Lynch · P. Lim · D. Breathnach · B. Farley
Xilinx, Dublin, Ireland
e-mail: bob.verbruggen@xilinx.com

© Springer Nature Switzerland AG 2020 
A. Baschirotto et al. (eds.), *Next-Generation ADCs, High-Performance Power Management, and Technology Considerations for Advanced Integrated Circuits*,
https://doi.org/10.1007/978-3-030-25267-0_6

are driven by the wireless application. Section 6.3 describes an ADC designed to meet these requirements, with measurements shown in Sect. 6.4. We will end with conclusions in Sect. 6.5.

## 6.2 ADC Requirements

For market success, direct-RF sampling ADCs should offer a compelling alternative to conventional receivers. To understand the ADC requirements for linearity and noise we can thus start by simply translating these receivers' specifications into equivalent ADC requirements. This will be done in the first part of this section for a few of the most important receiver specifications: noise figure, IIP2/3, and blocker noise figure. Next, we must consider any specific ADC circuit imperfections that could noticeably differentiate our RF-sampling ADC from the equivalent receiver, which is done in the second part of this section. The third part will discuss how the application drives the requirements on sampling rate and bandwidth.

### *6.2.1 Receiver Specifications Applied to an ADC*

The most obvious receiver specification is its noise figure. An equivalent ADC noise figure can be found by:

(1) measuring the ADC output noise spectral density (NSD) in dBFS/Hz,
(2) using the ADC full-scale input power to obtain input-referred NSD in dBm/Hz,
(3) subtracting the −174 dBm/Hz of an ideal 50 Ω resistor.

For example, an ADC with a 1 V peak-to-peak differential input range, 100 Ω differential input impedance, and NSD of −154 dBFS/Hz would have a −153 dBm/Hz input-referred NSD, and thus 21 dB noise figure. In general:

$$\mathrm{NF_{ADC}}\ (\mathrm{dB}) = \mathrm{NSD}_{\left(\frac{\mathrm{dBFS}}{\mathrm{Hz}}\right)} + 10\log_{10}\left(\frac{V^2_{\mathrm{FS,ppd(V)}}}{8 \times R_{\mathrm{in}(\Omega)} \times 1\mathrm{mW}}\right) + 174$$

An important caveat here is that the initial NSD should be measured with zero input, as NF is typically considered to be signal-independent and simply sets the smallest detectable signal. In a well-designed ADC the NSD should be limited by additive thermal noise.

In theory, the equivalent ADC IIP2 and IIP3 can be obtained by measuring IM2/3 with two tones at several input powers and doing the conventional extrapolation. In practice though, to get a true equivalent IIP2/3 great care must be taken that measured IM2 and IM3 are limited by "soft" non-linear effects that see expected 2 or 3 dB per dB scaling with input signal. In the presence of random INL effects, this

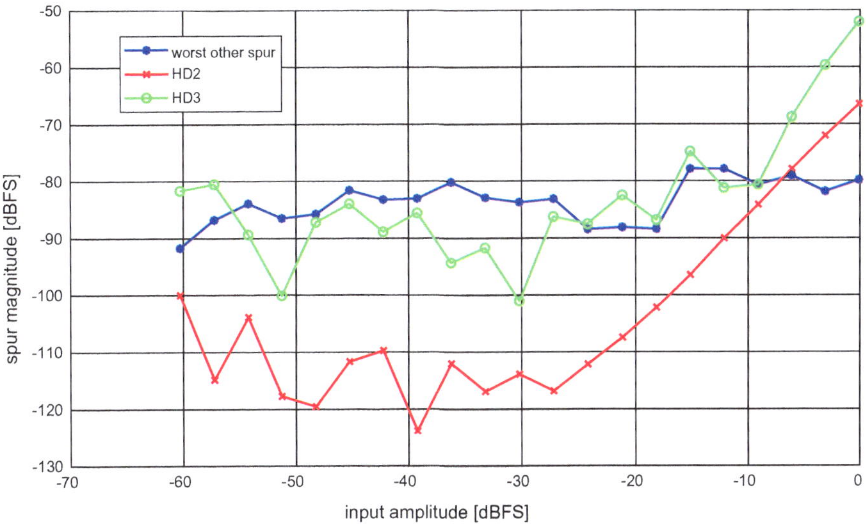


**Fig. 6.1** Worst case spur, HD2 and HD3 vs. input amplitude

is not a given, as these tend to set a "distortion floor" when input amplitude decreases as illustrated in Fig. 6.1 for a simple modeled ADC. As a result, it might be more reliable to instead measure IM2/3 with an input close to full-scale and specifying this, rather than the extrapolated IIP2/3. In most practical ADCs, these soft non-linear effects are pure analog effects such as input driver distortion or capacitance non-linearity.

Blocker noise figure is indicative of a receiver's noise floor degradation when large unwanted signals are applied. The equivalent ADC metric can be obtained by measuring the ADC noise figure with the above procedure when a full-scale signal is applied. Interestingly, just as in a conventional receiver, blocker noise figure is often limited by LO quality. In most RF-sampling ADCs clock jitter is the main contribution to noise floor degradation when a (high frequency) signal is applied.

### *6.2.2 ADC Specific Imperfections*

The above examples serve to illustrate the fact that conventional ADC testing, heavily focused on near full-scale single-tone signals, is insufficient when considering the ADC as a receiver. Moreover, commonly used "catch-all" metrics such as SNDR/ENOB are too crude, as they offer no information about where in the output spectrum errors will appear. This does not, however, imply that one can simply design an ADC for the abovementioned receiver specifications and assume the system will operate correctly. Indeed, many common ADC imperfections are unique to ADCs and have no equivalent receiver specification. Most significant among these are INL/DNL effects and interleaving impairments: these need to be dealt with separately.

As briefly mentioned above, INL/DNL effects tend to generate a "distortion floor." At high input amplitudes, soft non-linear and low-order effects tend to dominate, but when the signal is backed off, the random INL gains significance. Because the shape of the INL tends to be random, the order and magnitude of the generated distortion is very hard to predict. As a result, much more severe "small-signal" linearity requirements tend to be imposed. For example, while an RF-sampling ADC might be perfectly fine with a −60 dBFS second and third-order distortion near full-scale input, it might need <−90 dBFS SFDR at 20 dB back-off.

Interleaving is almost universally applied in RF-sampling ADCs, despite the well-known issues of channel offset, gain, and phase/time mismatch. These channel mismatch effects generate spurs and generally require calibration. While all of these spurs are undesirable and should be corrected to an acceptable level, offset mismatch is of specific note for wireless applications. Indeed, while gain and phase/timing mismatch generate spurs proportional to the input signal, offset mismatch is input independent. This means that when the ADC is operating with a backed-off input signal, offset spurs can easily become the most significant interleaving spur and therefore tends to have a more stringent requirement. For example, for a near full-scale input −80 dBFS might be an acceptable level for gain and phase/timing mismatch spurs, but offset spur requirements would be closer to −90 dBFS.

From the above description it should be clear that wireless applications impose very stringent requirements on spectral purity, especially at back-off. This focus on spectral purity means that absolute gain or offset requirements are comparatively relaxed, as they are linear effects and can easily be dealt with at the system level.

### 6.2.3 Sample Rate and Bandwidth

In theory, if we assume that the ADC can sub-sample the input signal, the sample rate required only depends only on the channel bandwidth. This seems to suggest that ADCs in conventional receivers and RF-sampling ADCs could operate with similar sample rates. This is not the case: RF-sampling ADCs need significantly higher sampling rates for reasons explained below. We will also touch on the input bandwidth requirements.

To illustrate how the application drives the sample rate requirement, let us consider the example of a 200 MHz channel around 3.5 GHz. Nyquist–Shannon correctly states that in theory, perfect reconstruction of this channel is possible with a 400 MS/s ADC, but there are several practical issues.

The first issue we run in to is that the 400 MS/s ADC would require a perfect anti-alias filter, in this case centered at 3.5 GHz. Without this filter the output of the ADC would get corrupted by adjacent spectral content, which is not acceptable. Such a filter is not practically realizable, so a certain amount of oversampling is required.

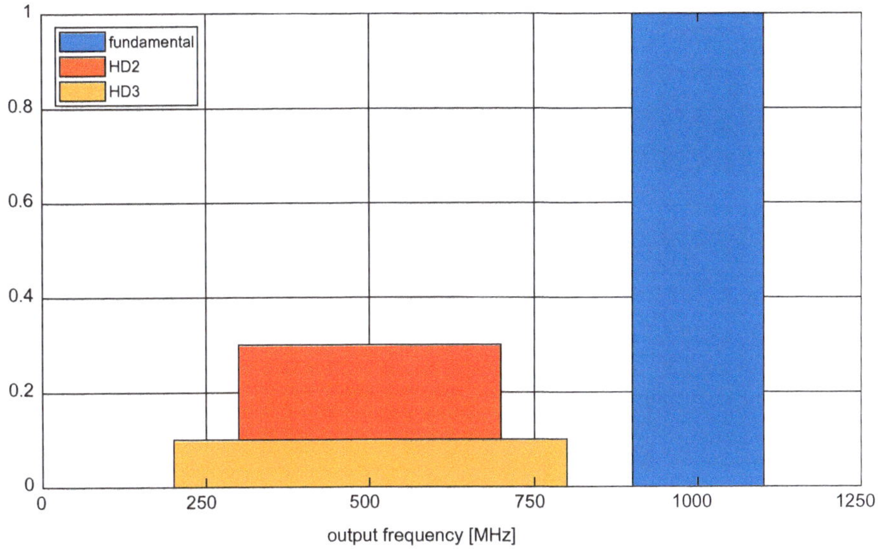


**Fig. 6.2** ADC output frequency plan for 2.5 GS/s sampling of 200 MHz band around 3.5 GHz

If we optimistically suggest 2× oversampling, our new ADC would operate at 800 MS/s as if it were quantizing a signal in the first Nyquist zone. In RF-sampling ADCs, however, the sampling clock also acts as a down-converting LO. This requires an increase of the sampling rate to 950 MHz, as with 800 MS/s clock the frequencies between 3.6 and 3.7 GHz fold on top of those between 3.6 GHz and 3.5 GHz, clearly still requiring ideal filtering to avoid corrupting the band of interest. At 950 MS/s there is some design space for the anti-alias filtering, now requiring a 200 MHz passband and 400 MHz stopband, still around 3.5 GHz. This filter would still be fairly tricky but might at least be possible.

The next issue we run into is harmonic folding. Indeed, our signal or neighboring channels could generate harmonic distortion components. With a 200 MHz channel, the second-order distortion products would be 400 MHz wide, and third-order distortion is 600 MHz wide. With only 400 MHz output bandwidth on the ADC, these harmonic components would clearly occupy the complete output spectrum, thus corrupting the desired channel.

This issue can be avoided by careful frequency-planning. Increasing sampling rate changes the Nyquist bandwidth and the folding locations of the harmonic components. For example, if our ADC sample rate is 2.5 GS/s we obtain the output spectrum shown in Fig. 6.2. Second and third-order distortion products are safely away from the desired band and can thus be filtered out digitally.

A final potential issue is the requirement for the same ADC design to be used in the transmitter feedback path. This feedback path is used to measure PA distortion and then apply digital pre-distortion (DPD) to avoid violating spectral masks. The issue here is that to correctly operate, this DPD algorithm needs to be able to read not just the desired spectrum, but also the spectral regrowth components. From the

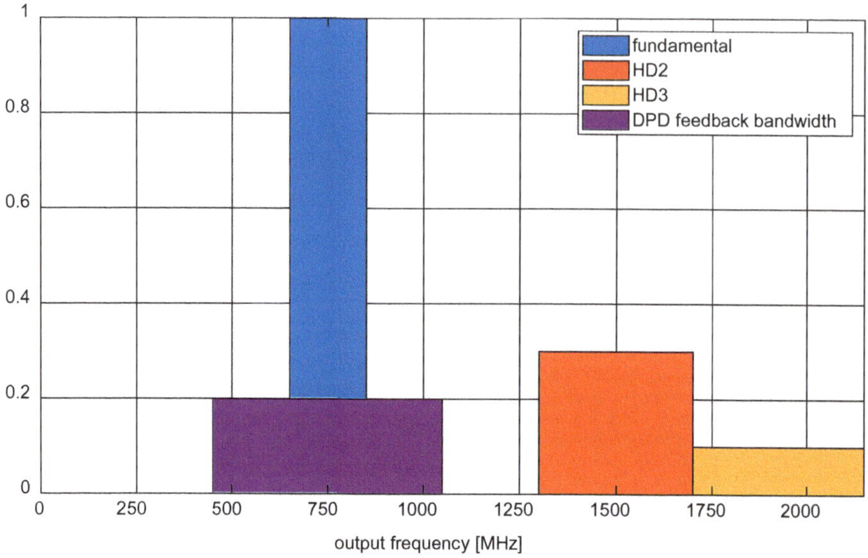


**Fig. 6.3** ADC output frequency plan for 4.25 GS/s sampling of 200 MHz band around 3.5 GHz with DPD feedback bandwidth

point of view of the ADC, this implies a $3\times$ or $5\times$ expansion on the bandwidth we need to quantize. In our example, instead of digitizing 200 MHz of spectrum, we now have to digitize 600 MHz or even 1 GHz bandwidth, all the while still ensuring no folded harmonics occupying the same desired spectrum. This is illustrated in Fig. 6.3 for an ADC running at 4.25 GS/s.

A final critical requirement for RF-sampling ADCs is their input bandwidth. Clearly, if the input signal is centered at 3.5 GHz, an input bandwidth of at least 4 GHz is required to maintain gain flatness and avoid phase distortion that would need to be corrected with digital equalization. Moreover, the ADC will be integrated with the rest of the receiver on board rather than on-die. Since this is a distributed 50-ohm environment, the ADC needs reasonably good input matching, e.g., S11 $<-15$ dB, to ensure the overall receiver response does not show ripple due to standing waves. While for both the input bandwidth and the input matching a bandpass response is in theory possible, the range of different RF frequencies in use today complicate this approach, and a low-pass approach is generally adopted.

## 6.3 Example ADC Implementation

In the previous section we discussed how the application drives the ADC requirements. We will now describe an ADC implementation targeted at this application and describe how the various key specifications are addressed. The simplified overall ADC architecture is shown in Fig. 6.4. It uses two input buffers sharing

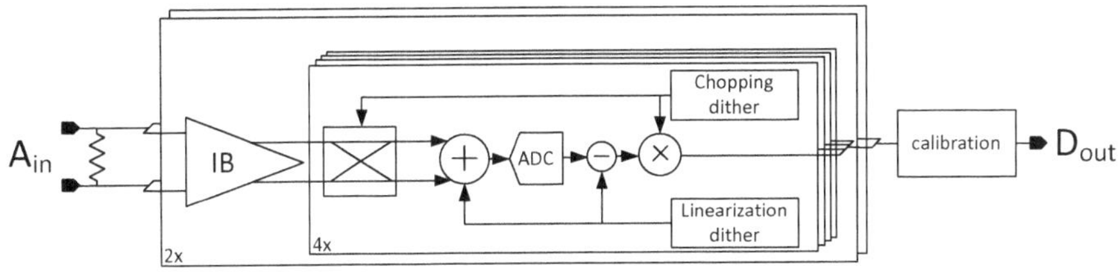


**Fig. 6.4** Simplified ADC architecture

**Fig. 6.5** Input buffer schematic

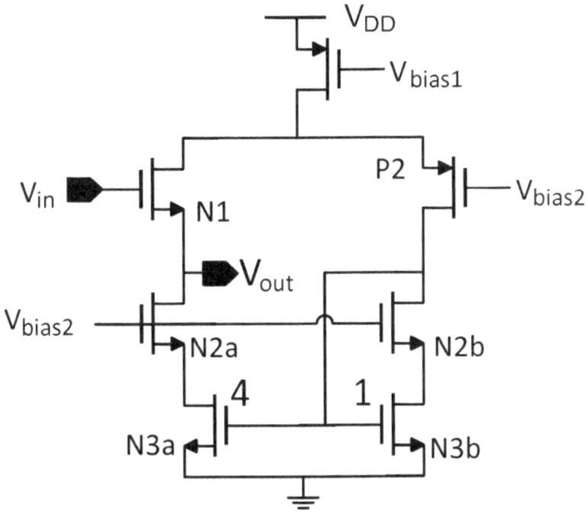


a single on-die termination to drive two sets of four ADC channels, for a total of 8× interleaving. Each ADC channel can operate up to 625 MS/s for an aggregate rate up to 5 GS/s. To achieve excellent spectral purity at back-off, two techniques are implemented. First, each ADC channel implements front-end chopping at the sampling stage. This randomizes the flicker noise of the ADC channels which would otherwise show up at offset spur locations with unacceptable magnitudes. Second, the overall INL/DNL of the ADC is scrambled using dither injection.

### *6.3.1 Input Buffer and Sampling Network*

The input buffer is based on an NMOS source follower as shown in Fig. 6.5. Current feedback is used to improve distortion: any signal-dependent increase in current drawn from N1 is sensed by P2 and fed back through cascaded mirror N2/3. A feedback ratio of 1–4 is used as a compromise between speed and linearity.

The sampling network and its clocking is shown in Fig. 6.6. The ADC implements top-plate sampling, which in this 16nm technology is possible with sufficiently low distortion using a standard bootstrap circuit [5]. The fairly poor flicker noise of this technology is suppressed by chopping the complete ADC channel. At the front-end this is done using two sets of differential sampling switches

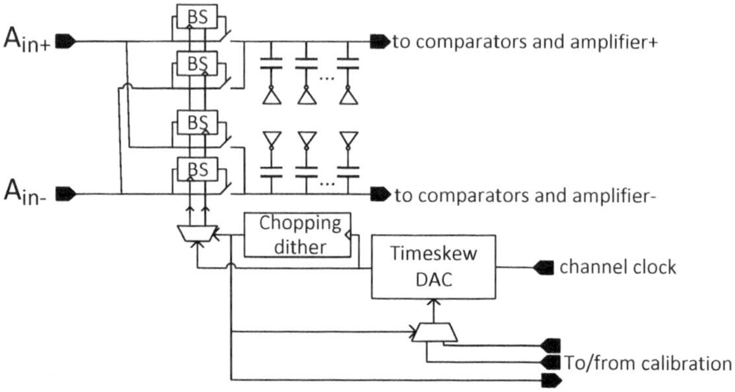


**Fig. 6.6** Sampling network and clocking

connected directly and with outputs swapped [6, 7]. Conveniently, this arrangement also cancels differential capacitive feedthrough. Unchopping is done digitally using a simple sign inversion as shown in Fig. 6.4.

These different physical sampling switches require different clocks which results in inevitable skew. Left uncorrected, this skew would translate into random jitter when these switches are pseudo-randomly selected. A chopping-aware time-skew calibration loop addresses this chopping switch time skew, as discussed in Sect. 6.3.3. To save space, both chopping switches share the same time-skew DAC, which does then need sufficiently fast settling to cope with fast-changing phase requirements.

### *6.3.2 ADC Channel Implementation*

The ADC channel is shown in Fig. 6.7 and consists of three asynchronous 5b SAR stages with two open-loop dynamic residue amplifiers [8]. The SAR stages use a split-capacitor DAC [9] for constant common-mode and are comparator-controlled. To reduce design and verification time, a single comparator and residue amplifier design is reused through the pipeline, albeit at a minor power penalty. Capacitor mismatch, comparator offsets, residue amplifier offsets, and residue amplifier gain are all calibrated on-chip as explained below.

Contrary to more conventional 1.5b or 3.5b pipeline stages, the 5b SAR stage implemented here exhibits a residue transfer discontinuity at zero differential input as illustrated in Fig. 6.8. The input signal shown in blue does not see any distortion from the 1.5b stage and would thus likely only be subject to soft distortion errors. For the 2b stage, the discontinuity is seen at any input level and gives rise to unacceptable INL/DNL errors even after calibration. These errors are randomized by injecting a 3.7b PRBS dither signal just after sampling. This dither is removed digitally after conversion and thus propagates through the full ADC channel, improving linearity at back-off to around the required −90 dBFS level.

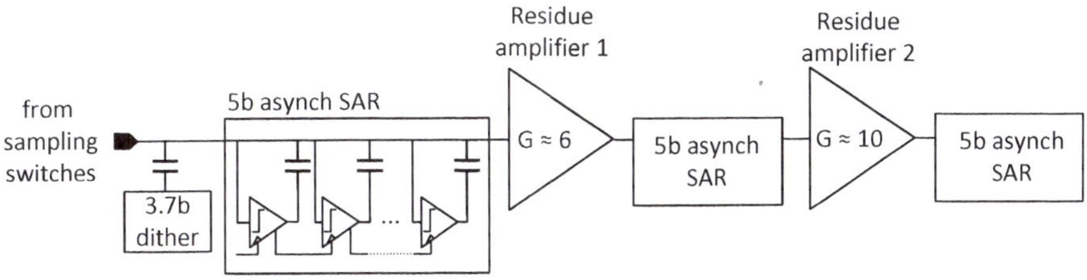


**Fig. 6.7** Simplified ADC channel

**Fig. 6.8** Residue transfer curves for conventional 1.5b and 2b stages

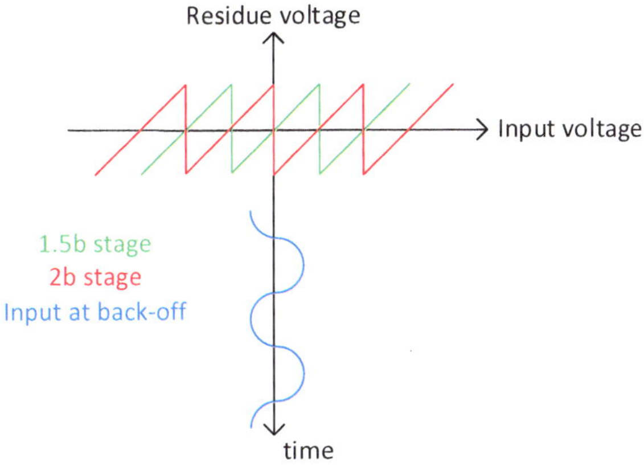


The DAC references are taken directly from the analog supply and ground. While this does impose a stringent requirement on external power and ground integrity, it simplifies the overall design by removing the need for on-die reference drivers, which will always be higher impedance than the off-chip power-delivery network. This lower impedance thus improves the speed and accuracy of the SAR DACs. A potential concern could be other signal-dependent current drawn from the power-delivery network corrupting the reference voltage, but this turns out not to be the case in our implementation. Absolute reference accuracy is also a potential concern, but in a wireless application the limited uncertainty on this external voltage is acceptable.

### *6.3.3 Digital Calibration*

As previously indicated this design leverages a significant amount of digital calibration. There are several loops which fall into two categories: single-channel calibration and inter-channel calibration.

The goal of the single-channel calibration loops is to linearize the ADC channels. This requires adjustments to DAC capacitor values, residue amplifier offset and gain, and comparator offsets. All of these are initially calibrated in foreground for fast start-up, and comparator offsets and residue amplifier gains have a supporting background calibration loop because they are sensitive to voltage and temperature.

Foreground calibration samples an internally generated common-mode voltage, which is used directly for comparator and residue amplifier offset calibration. For the other defects, the SAR DACs are used to generate either an error signal during capacitor calibration or a reference voltage during residue amplifier gain calibration. All these loops operate by essentially comparing the digital output of the ADC back-end with the desired result and adjusting analog trim controls to reduce this difference. Background calibration is done as in [10] for the comparators in the first two stages and the two residue amplifier gains. Third stage comparators are not calibrated in background, as due to the preceding residue amplifier gain their input-referred offset drift does not lead to significant DNL degradation.

Inter-channel calibration aims to adjust offset, gain, and time skews between the different channels. A standard gain calibration is implemented: the output power of each ADC channel is measured and adjusted digitally until they match. The offset and time-skew loops need to be modified to support the chopping architecture.

The standard offset calibration loop measures the average of each channel output and adjusts this average until they match. In the chopping ADC presented here this measurement and correction could be done either before or after the digital unchopping as illustrated in Fig. 6.9. If the offset calibration is placed after the unchopping, it will attempt to correct any signal at offset spur frequencies, but the

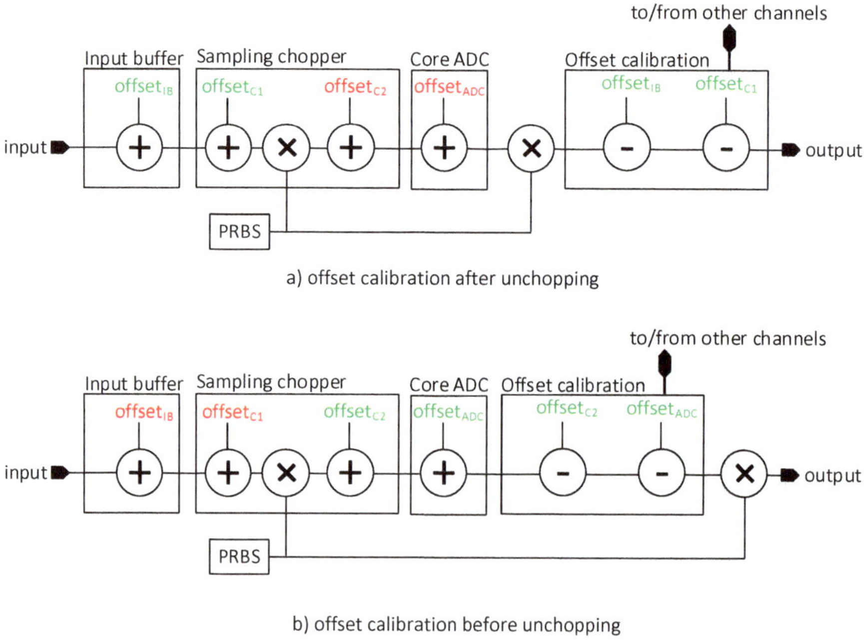


**Fig. 6.9** Offset sources in ADC and impact of where offset calibration is done

core ADC offset, which is modulated by the chopping signal, will be left uncorrected and turned into noise. If the offset calibration loop acts before the unchopping, it will correct the core ADC offset but offsets due to the chopping switches and input buffers can result in residual spurs appearing at the offset frequencies. To meet the desired linearity and noise targets, we thus use two loops, one acting inside the chopping and one outside.

The chopping architecture, which uses two different sets of switches for each channel, also requires a chopping-aware time-skew calibration block (TSCB) which now needs to adjust two sampling times for each channel. In addition to the increased number of corrections, the chopping switches are randomly selected and thus do not sample at regular intervals contrary to the standard implementation where each switch samples in turn. This non-uniform sampling creates potential interaction between the input signal and the chopping sequence, which is mitigated using the three improvements described below.

First, the chopping sequence and TSCB observation window are synchronized, ensuring that each update of the TSCB happens based on the same PRBS sequence. For a $2^N$ sample observation window, the chopping signal uses an $N$-bit LFSR with an added zero to balance the number of zeros and ones. Next, to ensure correct averaging between the two chopping switches when inputs coherent with the LFSR are used, the LFSR input is periodically inverted. This ensures that the ADC inputs that were sampled with an inverting switch on a first repetition of the LFSR are also sampled by the non-inverting switch on a next repetition and avoids any signal-dependent time skew being generated by the modified TSCB. Finally, because most real signals will not be coherent with the LFSR, the non-uniform sampling could require a larger averaging time for the same time-skew observation accuracy. This penalty is reduced by using independent loops for the time-skew average and difference of each pair of chopping switches. Because each pair of chopping switches samples uniformly, their average time skew can be adjusted fairly quickly, while their difference typically requires more averaging due to interaction between the chopping sequence and signal. A block diagram of the implemented arrangement is shown in Fig. 6.10.

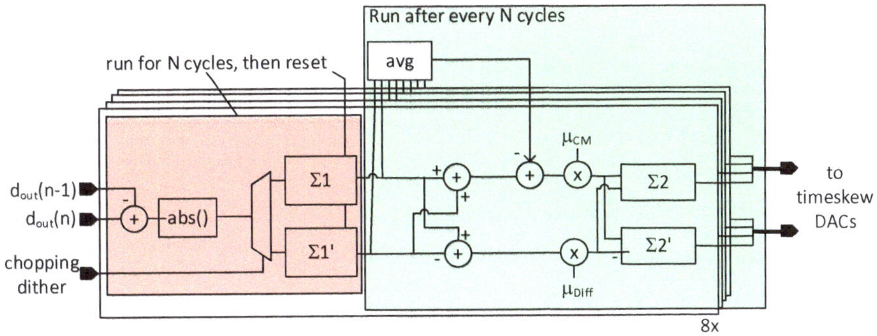


**Fig. 6.10** Block diagram for time-skew calibration engine

## 6.4 Measurements

The ADC is implemented in a 16 nm FinFET process and occupies a total area of 1.1 $mm^2$, including digital calibration and supply decoupling as shown in Fig. 6.11. Sampling at 5 GS/s, the ADC achieves a zero-input NSD of −154 dBFS/Hz with an input range of approximately 1 Vppd, for an equivalent NF of 21 dB. Figure 6.12 shows the single-tone spectrum with a −1 dBFS 2.4 GHz input sampled at 5 GS/s. SFDR and SNDR are 60.9 and 55.3 dB, respectively. Notably, HD2 of −70.4 dBFS and HD3 at −61.9 dBFS dominate distortion, with the largest remaining spur below −82 dBFS. HD2 and HD3 performance suggest equivalent IIP2 and IIP3 of approximately 63 and 26 dBm, respectively. The NSD with this high-frequency input is −152 dBFS/Hz, or a blocker NF of 23 dB with a 0 dBm continuous wave blocker at 2.4 GHz.

The ADC can operate up to 5.4 GS/s as shown in Fig. 6.13. The effectiveness of dither injection and chopping is illustrated in Fig. 6.14, which shows an input amplitude sweep: spur levels generally better than −90 dBFS are achieved at 20 dB back-off.

The performance of the ADC in a band 42 in-band blocking test for 5 MHz channels is shown in Fig. 6.15. The ADC input signal contains a near full-scale blocker channel and 52.4 dB smaller desired channel separated by 5 MHz, both near 3.5 GHz. These are sampled at 4423.68 MS/s and then down-converted and decimated using on-chip digital logic. The system achieves measured EVM and SNR of 11.6% and 18.7 dB for the first 5 MHz channel next to the blocker, which is dominated by ADC thermal noise. The ADC power consumption is 641 mW at 5 GS/s, which consists of 299 mW for input buffers, bandgap, and clock receiver, 191 mW for the ADC core and clocking, and 151 mW for digital calibration.

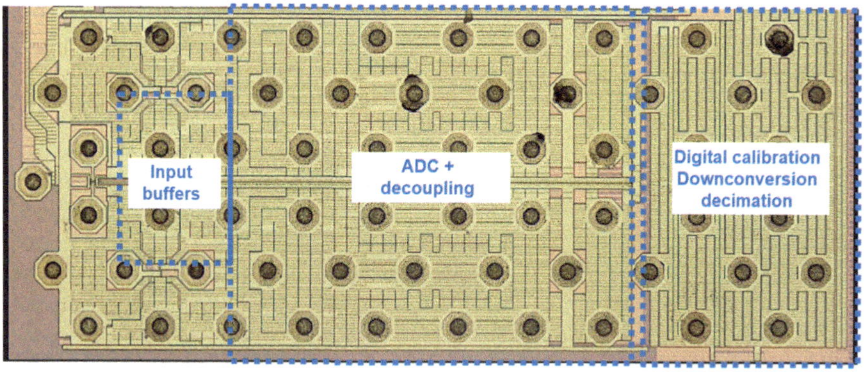


**Fig. 6.11** Die micrograph

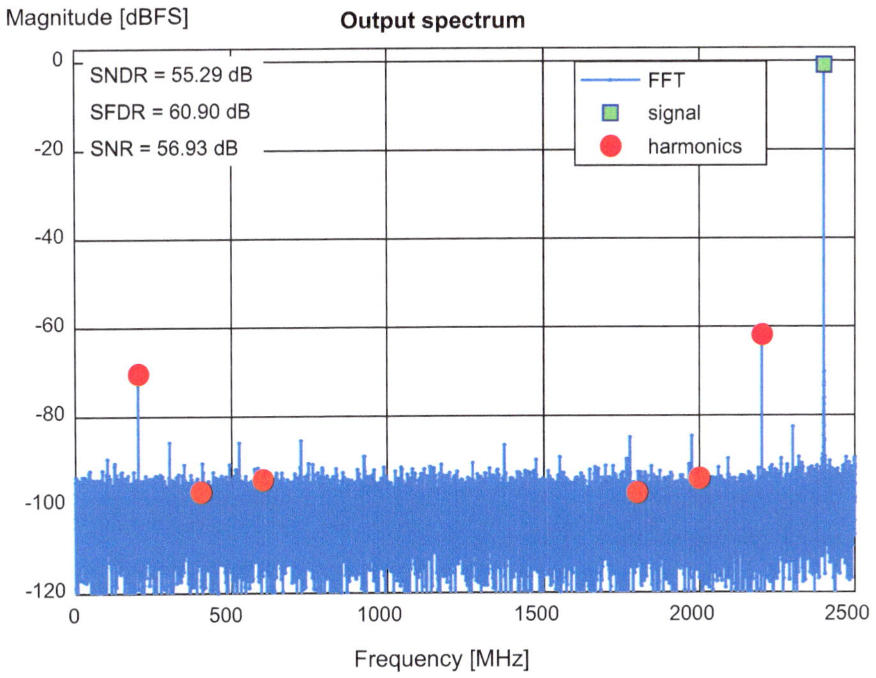


**Fig. 6.12** Single-tone output spectrum at 2.4 GHz, 5 GS/s

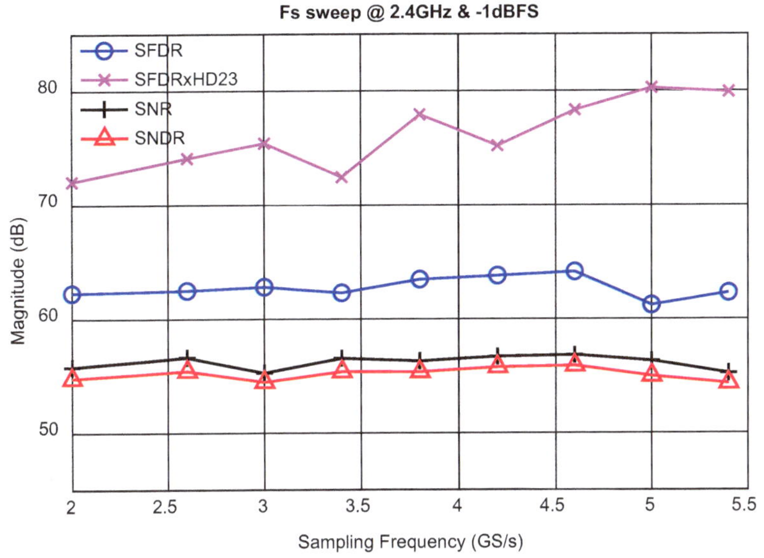


**Fig. 6.13** ADC performance vs. sampling frequency

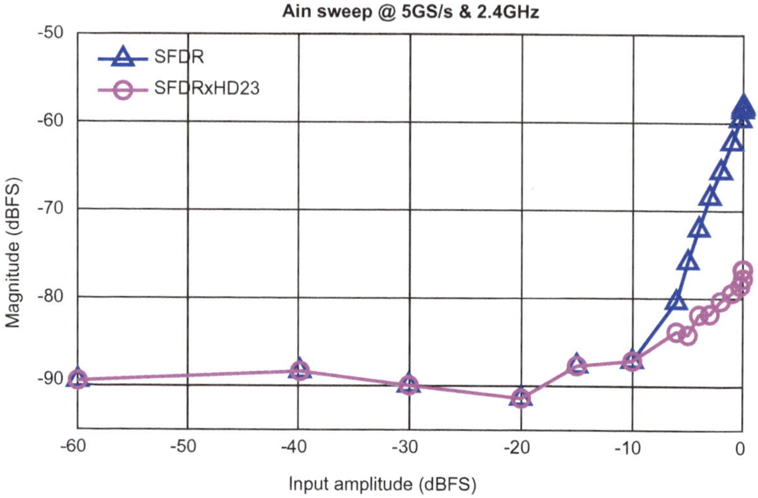


**Fig. 6.14** Spur magnitudes vs. input amplitude

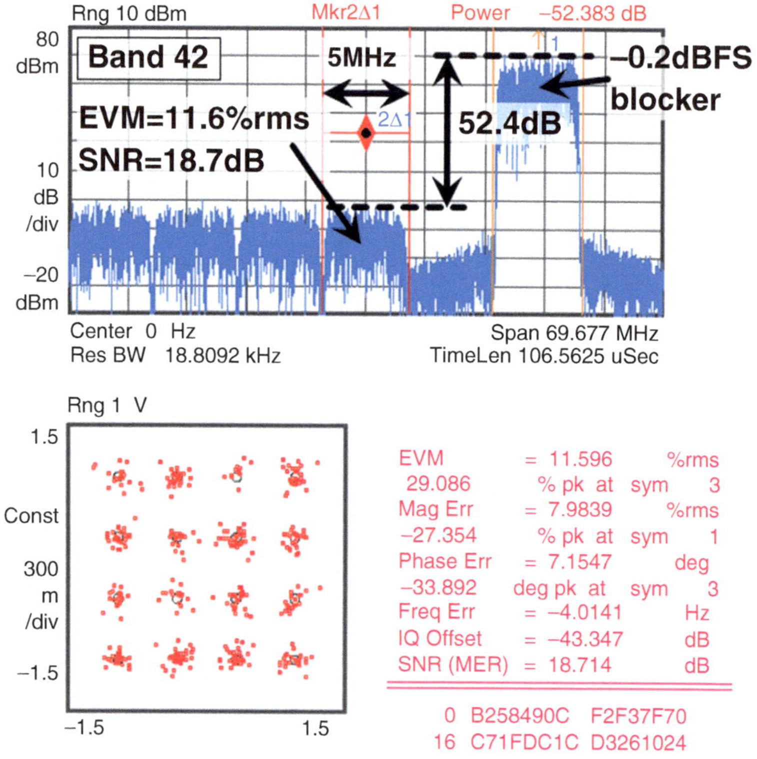


**Fig. 6.15** Measured performance in band 42 blocker test

## 6.5 Conclusions

RF-sampling ADCs are becoming a viable alternative for conventional down-conversion receivers. We have described how specifications for these ADCs are driven by the application, requiring mostly excellent noise floor and spectral purity at back-off. An example implementation uses dither and chopping to achieve excellent spectral purity at back-off, requiring only a few additional SAR DAC capacitors, an extra pair of sampling switches, and some modifications to standard digital calibration algorithms. The ADC achieves 11.6% EVM in a 5 MHz blocker test at 3.5 GHz.

## References


1. D. McLaurin et al., A highly reconfigurable 65nm CMOS RF-to-bits transceiver for full-band multicarrier TDD/FDD 2G/3G/4G/5G macro basestations, in *IEEE ISSCC 2018* (2018), pp. 162–163
2. B. Farley et al., A programmable RFSoC in 16nm FinFET technology for wideband communications, in *IEEE ASSCC 2017* (2017), pp. 1–4
3. C. Erdmann et al., A modular 16nm Direct-RF TX/RX embedding 9GS/s DAC and 4.5GS/s ADC with 90dB isolation and sub-80ps channel alignment for monolithic integration in 5G base-station SoC, in *IEEE Symp. VLSI Circuits* (2018), pp. 219–220
4. B. Vaz et al., A 13bit 5GS/s ADC with time-interleaved chopping calibration in 16nm FinFET, in *IEEE Symp. VLSI Circuits* (2018), pp. 99–100
5. A.M. Abo, P.R. Gray, A 1.5-V, 10-bit, 14.3-MS/s CMOS pipeline analog-to-digital converter. IEEE JSSC **34**(5), 599–606 (1999)
6. S. M. Jamal et al., A 10b 120M sample/s time_interleaved analog-to-digital converter with digital background calibration, in *IEEE ISSCC 2002* (2002), pp. 172–173
7. E. Janssen et al., An 11b 3.6GS/s Time-Interleaved SAR ADC in 65nm CMOS, in *IEEE ISSCC 2013* (2013), pp. 464–465
8. B. Vaz et al., A 13b 4GS/s digitally assisted dynamic 3-stage asynchronous pipelined-SAR ADC, in *IEEE ISSCC 2017* (2017), pp. 276–277
9. B.P. Ginsburg, A.P. Chandrakasan, 500-MS/s 5-bit ADC in 65-nm CMOS with split capacitor array DAC. IEEE JSSC **42**(4), 739–747 (2007)
10. B. Verbruggen, et al., A 2.1mW 11b 410MS/s dynamic pipelined SAR ADC with background calibration in 28nm digital CMOS, in *IEEE Symp. VLSI Circuits* (2013), pp. 268–269

# Part II
# High-Performance Power Management

The second part of this book is dedicated to developments in the field of high-performance power management. Chapters 7 and 8 discuss the advanced integrated power management systems for Internet-of-Things (IoT) applications and System-on-Chips (SoC). In Chap. 9, current sensing approaches are described, which are required in many power management tasks. Chapters 10 to 12 focus on the advances in technology development using non-conventional devices as well as high-voltage circuit designs.

Chapter 7, by Nicolas Butzen and Michiel Steyaert, discusses advanced multi-phasing concepts to push the limits of fully integrated switched-capacitor converters in terms of efficiency, chip density, and a scalable conversion ratio.

In Chap. 8, by Jürgen Wittmann, Francesco Cannillo, Dan Ciomaga, Mihail Jefremow, and Fabio Rigoni, the challenges of integrated power management ICs for IoT devices and wearables are discussed. Several circuit and architecture innovations are shown to minimize standby current and to maximize efficiency.

Chapter 9, by Mahdi Kashmiri, discusses current sensing techniques, as required, for instance, for monitoring battery and load conditions. Besides discussing the basic current sensing concepts, readout circuit solutions are also presented.

Chapter 10, by Carl-Mikael Zetterling, Saleh Kargarrazi, and Muhammad Shakir, introduces the use of wide-bandgap semiconductors (like GaN and SiC) as a solution for operating at higher voltages, higher temperatures, and high-radiation environments. Besides introducing the semiconductors, several example ICs are discussed as well.

In Chap. 11, Jef Thoné and Mike Wens describe the trade-offs and limits for the gate-driver design for wide-bandgap power transistors. Various devices are benchmarked, the main challenges and solutions for each technology are discussed, and various examples of gate drivers are presented.

Chapter 12, by Sergio Morini, Davide Respigo, and Martina Arosio, presents the challenges in driving superjunction power switches for motor drive applications. A circuit architecture is presented that can self-adjust the dV/dt slope to find an optimum balance between switching losses and electromagnetic interference compatibility.

# Chapter 7
# Advanced Multiphasing: Pushing the Limits of Fully Integrated Switched-Capacitor Converters

**Nicolas Butzen and Michiel Steyaert**

## 7.1 Introduction

Over the past decades, fully integrated power management has received a lot of attention in the literature [1]. For low-power internet-of-things (IoT), sensor nodes potentially combined with energy scavenging, the reduced PCB footprint, system height, and reduction of the number of external passives that monolithic power converters provide can have a tremendous effect on the total system size and cost. On the other end, the transportation of energy onto higher power systems-on-a-chip (SoCs) has become an effective bottleneck to these systems' performance. A bottleneck that could be solved using DC–DC converters that are integrated together with the load onto the same die, as illustrated in Fig. 7.1. Using an external point-of-load (POL) converter, the SoC has a certain intake current which induces substantial voltage droops in the power delivery network (PDN). In order to guarantee correct operation, the load consequently has to be designed with a certain voltage margin in mind [2]. However, this voltage margin does lead to a higher power consumption, even when it is not needed. Moreover, thanks to the continued scaling of technology's supply voltages [3], the intake current together with the voltage margins has increased making the PDN effects effectively large loss contributors [4]. In contrast, by shifting the POL converter on-chip, as shown in Fig. 7.1b, the intake current is reduced by the converter's voltage conversion ratio (VCR), causing the voltage margins and PDN losses to be reduced by the same factor as well.

In addition, fully integrated converters are a key enabler for extensive granularization of voltage domains in today's SoCs and processors [5, 6]. The central idea

N. Butzen (✉) · M. Steyaert
ESAT-MICAS KU Leuven, Heverlee, Belgium

© Springer Nature Switzerland AG 2020

A. Baschirotto et al. (eds.), *Next-Generation ADCs, High-Performance Power Management, and Technology Considerations for Advanced Integrated Circuits*,
https://doi.org/10.1007/978-3-030-25267-0_7

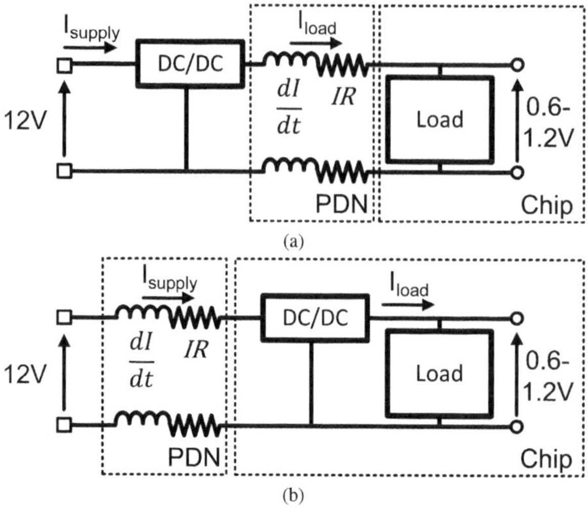


**Fig. 7.1** An example power delivery network (PDN) to an integrated load using (**a**) an external and (**b**) an on-chip DC/DC converter

is simple: if every core or functional block of a SoC would be supplied a voltage that is just high enough to fulfill its task in a given time, independent of the rest of the SoC, then a substantial amount of energy can be saved [7]. Depending on the workload, a processor's efficiency could be increased by as much as 21% if per-core dynamic voltage and frequency scaling (DVFS) is used [8]. Because this does mean that every granular block requires its own voltage domain and thus voltage regulator, realizing a high level of granularization is impractical with external converters.

By far the most popular type of POL converter is the buck converter. However, the lack of quality of integrated inductors has posed a serious problem for their full integration. To make things worse, this quality issue appears to be tied to the inductor's small size and is thus fundamental in nature [9]. As a result, most inductive converter designs in the literature have opted for a system-in-package (SiP) rather than a SoC approach where the inductor is not integrated on the die itself, but made out of bondwires [10], extra back end of the line (BEOL) metal and/or magnetic layers [11], PCB tracks [12], or a surface-mount device (SMD), possibly mounted directly on the silicon [13].

Switched-capacitor (SC) converters, in contrast, only need switches and capacitors, both of which are readily available in modern CMOS processes. Consequently, these converters have gained popularity as a promising candidate for full integration. Nevertheless, the monolithic context does pose its own challenges that inherently constrain the design space of fully integrated SC converters. This is portrayed in Fig. 7.2. At low-power densities, the converter's efficiency is limited by the relative size of the flying capacitor's parasitic substrate coupling, $C_{BP}/C_{par}$ and the flying capacitance, $C_{fly}$, itself due to a combination of charge-sharing losses in the flying capacitors and parasitic coupling, or bottom-plate (BP), losses [14, 15]. For increasingly large output powers, a clear efficiency-power-density trade-off becomes apparent due to power transistor losses gaining in importance. Here, it is

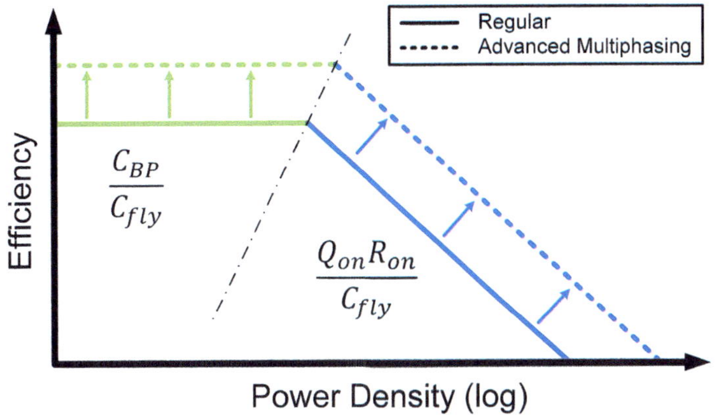


**Fig. 7.2** Effect of advanced multiphasing on the monolithic SC converter design space

the ratio of the transistor $Q_{on}R_{on}$ and $C_{fly}$ density that is the dominant factor [14]. Unfortunately, due to the planar nature of modern technology nodes, the parasitic coupling is relatively large while the flying capacitance density per unit area is small compared to external components. Thus, both at low- and high-power densities the design space is much more constrained compared with SC converters using external components.

Moreover, SC converters also have a topological issue: Due to the fact that charging capacitors with a voltage source or other capacitors inherently leads to charge-sharing losses, SC topologies have traditionally been designed to minimize the voltage swing across their capacitors' terminals [16]. With the capacitor voltages consequently being approximately constant over a full converter period, a SC converter topology has a fixed ideal voltage conversion ratio (iVCR) at which it could theoretically achieve 100% efficiency. At smaller VCRs, say when the output voltage is reduced, the efficiency drops rapidly. While multiple topologies can be combined together in a gearbox converter [17–23], this does require a lot of additional transistors to maintain a high efficiency over a wide voltage range, increases the system complexity, and ultimately reduces the efficiency and power-density for the full VCR range [24].

Multiphasing or interleaving is a popular technique in the literature where a converter is split into several converter cores that run in parallel but out of phase of each other to reduce the output voltage ripple [25, 26]. Especially in the monolithic context, where both capacitors and transistors can rather easily be split into smaller parts and where the achievable frequencies are much larger than required for a power converter, multiphasing can be implemented with very low overhead. With advanced multiphasing (AM), these out-of-phase cores are used as a resource: by having them interact with each other, the typical two-phase converter can be transformed to have many more distinct converter phases. These additional phases can then be used to boost the converter's performance or even to unlock fundamentally new types of SC converters.

This chapter is organized as follows. Section 7.2 introduces a first technique that focuses on the reduction of parasitic coupling losses in a SC converter. After, Sect. 7.3 focuses on improving the effective capacitance density on-chip, and Sect. 7.4 demonstrates a SC topology with a continuously scalable conversion ratio. Finally, Sect. 7.5 highlights the main conclusions of this chapter in a brief summary.

## 7.2 Scalable Parasitic Charge Redistribution

Figure 7.3a shows the main working principle of a two-phase SC converter that uses multiphasing from the point of view of its capacitor's bottom-plate node, $V_{BP}$.

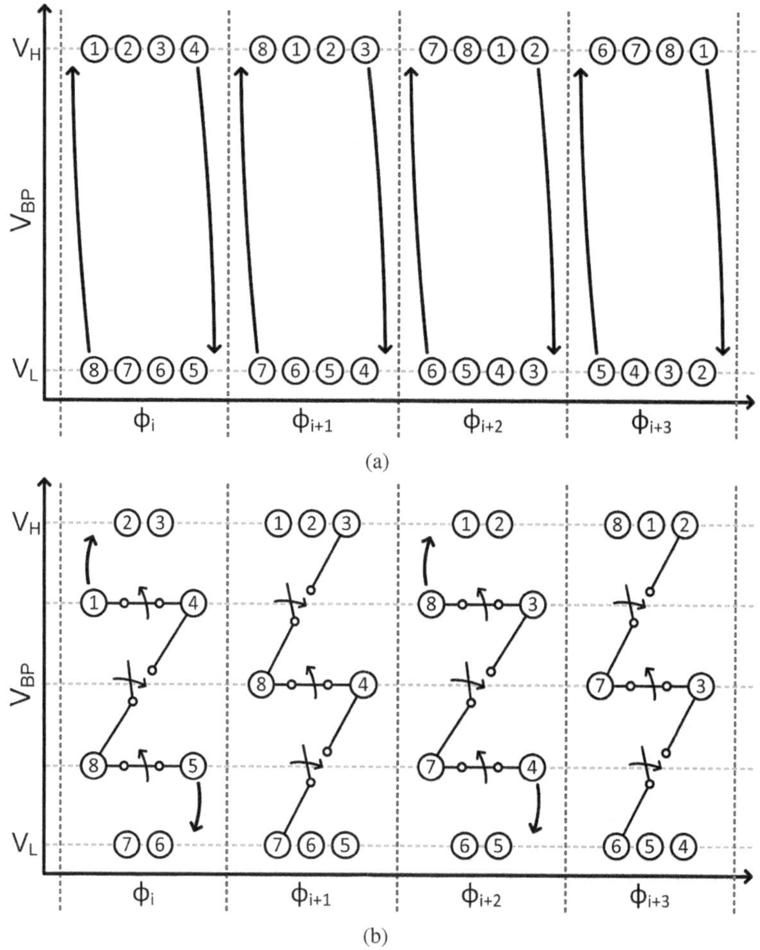


**Fig. 7.3** Bottom-plate voltage, $V_{BP}$, versus phase diagram of (**a**) a regular multiphasing converter with 8 cores, and (**b**) a converter using SPCR with 3 charge redistribution steps. Each labeled circle represents a different converter core. Arrows represent actions during phase transition

It can be appreciated that each core can either be in a high or low voltage state, corresponding to $V_H$ and $V_L$, respectively. At each switching event, the two cores that have been in the high/low state the longest transition to the next state by fully charging/discharging their $V_{\text{BP}}$ to $V_H/V_L$. In the charging case, $V_H$ needs to supply an amount of charge equal to the voltage difference between both states and the size of the parasitic coupling on the BP node.

Scalable parasitic charge redistribution (SPCR) introduces a dedicated BP charging and a dedicated BP discharging state [15, 27]. Instead of transitioning from a high state directly to a low state, cores will first enter the dedicated BP discharging state. Similarly, a core will go through the BP charging state when going from the low to the high state. Figure 7.3b illustrates an example SC converter using SPCR. Cores that are neither in the regular high nor in the regular low state are instead in the BP charging or discharging state. Here, all the regular power transistors are non-conducting and the core itself can only be at a set number of intermediate levels, chosen during design time. At every clock edge, each BP charging core is paired up with the BP discharging core which is in the closest, yet higher intermediate level. By shorting the BP nodes of each pair, their $V_{\text{BP}}$'s average out by transferring charge from the BP discharging to the BP charging core. This is called a charge redistribution step (CRS) and results in all paired BP charging cores going up, and all paired BP discharging cores going down one intermediate level. BP charging/discharging cores which are already at the highest/lowest intermediate level, and can consequently pair up no more, are instead pulled up/down to the high/low state. Furthermore, to keep this process going, every two phases, the two cores that have been in the high/low state for the longest time, are transferred to the BP discharging/charging state. The end result is that the low to high transition is now completed approximately adiabatically using a fixed number of CRS equal to the number of intermediate levels and that $V_H$ only needs to supply enough charge to pull the core up to the high state, which is in general (CRS + 1) times lower than the charge without SPCR. The BP losses are consequently also reduced by the same factor.

Ultimately, the losses associated with the parasitic coupling are thus reduced in a scalable manner. Considering the fact that these losses are one of the determining factors of the maximum obtainable converter efficiency on-chip [14], the efficiency of a SC converter can be substantially improved. For larger values of CRS, though, the extra transistors that enable the charge redistribution steps to take place do add significant additional losses such as leakage that cause a new efficiency ceiling to emerge [15]. Regardless, this new ceiling is generally substantially higher compared with the situation without SPCR, depending on the ratio of the parasitic coupling to the flying capacitance, and on how leaky the transistors are. To verify the obtainable efficiencies of the SPCR technique, a fully integrated 2:1 SC converter was designed in a 40 nm process using 16 cores and 9 CRS, thus reducing the parasitic coupling losses tenfold. A system overview of the converter is shown in Fig. 7.4. Rather than connecting the cores' BP nodes directly with each other, they connect through a charge redistribution bus (CRB). Furthermore, the bus they use depends on their resulting intermediate voltage level after their $V_{\text{BP}}$'s average out. The end result is

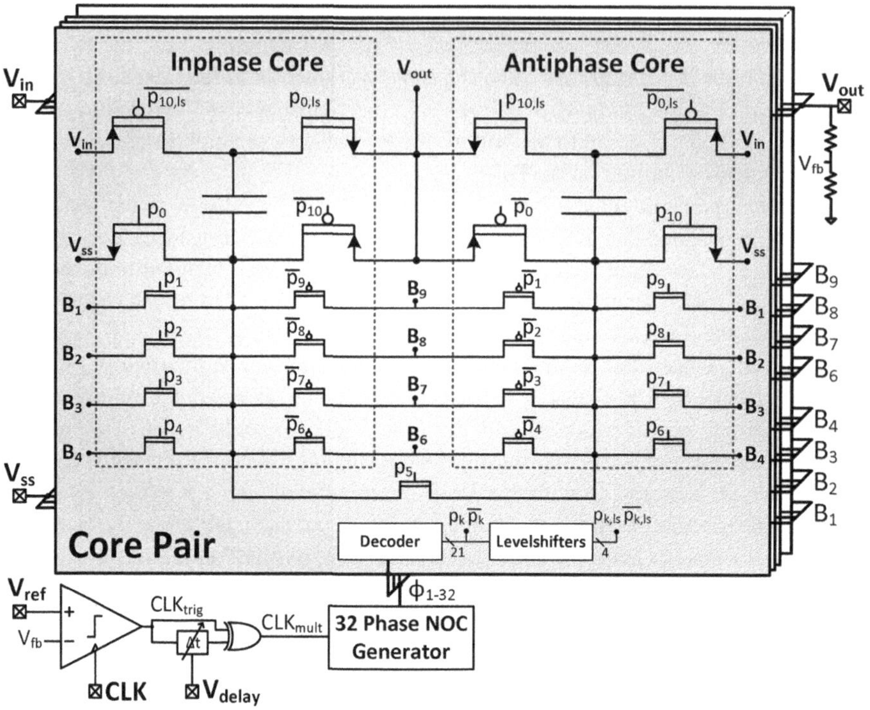


**Fig. 7.4** System overview of a converter implementing SPCR, showing the controller and transistor-level implementation of the converter cores

that significantly less area overhead is needed and that the voltage swing on each CRB is approximately zero, effectively making them DC voltage rails. The CRBs can also be used to power low-power circuitry within the converter itself [28, 29], and do not require any additional start-up circuitry to converge to their DC levels: Using the regular control signals, the CRBs naturally spread evenly over the full BP node swing.

From a conceptual point of view, the CRBs also greatly facilitate the design and implementation of this technique. A core simply needs to connect to these CRBs in a certain order when charging its BP node, and in the opposite order when discharging. The number of CRBs further determines the number of CRS and subsequently by how much the parasitic coupling losses are reduced. At the converter level, there need to be enough out-of-phase cores given the number of CRBs to make sure every time a core connects to a CRB, there is another core to exchange charge with.

The bottom line is that, thanks to the presented AM technique, the realized converter achieves a higher efficiency than any other fully integrated SC regulator, including those using deep-trench capacitors which have much smaller parasitic coupling but require extra masks [30, 31]. To this day, the achieved efficiency of 94.6% is still the highest converter efficiency demonstrated entirely on-chip.

## 7.3 Stage Outphasing and Multiphase Soft-Charging

Despite the success of SPCR in improving the efficiency of SC converters, its effect is reduced at higher power densities. This is simply because the parasitic coupling losses themselves are less important here. In contrast, stage outphasing (SO) and multiphase soft-charging (MSC) are two techniques that aim to improve the effective capacitance density by reducing the charge-sharing losses of the flying capacitors, thereby also having an impact at high-power densities. Although said techniques can be used for many topologies [32], they are demonstrated here using the Dickson converter [16].

In a regular $N$:1 Dickson converter, there are a total of N − 1 stages. If $V_{out}$ is close to the technology's supply voltage, the top-side switches are usually implemented as two stacked transistors to avoid using less-efficient I/O devices. This also leads to the creation of intermediate nodes ($k$), which are topologically speaking DC nodes and can be used as voltage rails. These intermediate nodes always connect a discharging flying capacitor of a stage to a charging flying capacitor of the next stage, also portrayed in Fig. 7.5.

Note that this situation is very similar to the discharging/charging of the parasitic coupling if SPCR is used with just one CRB. Recall that with SPCR, the number of CRBs can be scaled freely, and as long as the charging and discharging capacitors connect with the CRBs in opposite order, they will spread evenly over the full range, leading to the approximate adiabatic (dis)charging of the parasitic coupling. This is also the essence of the first technique, multiphase soft-charging, shown in Fig. 7.6: simply by splitting the intermediate node ($k$) up into multiple nodes ($k$,1) to ($k$, $M$) and having the charging and discharging flying capacitors connect to these in opposite order, they will spread evenly. Of course, this is assuming that the stages

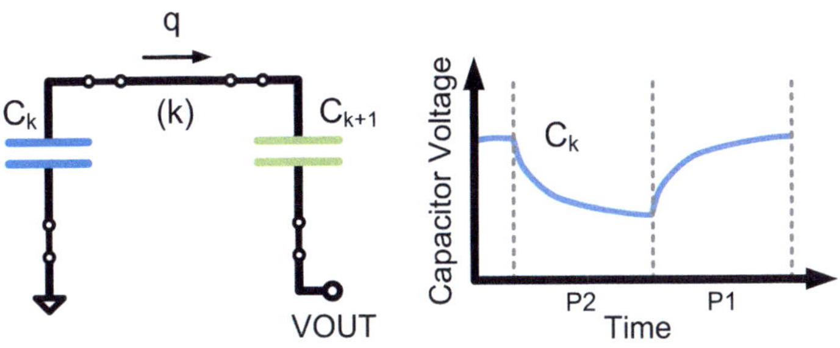


**Fig. 7.5** Charge transfer between two adjacent stages of a regular Dickson converter. $C_k$ and $C_{k+1}$ are the flying capacitances of stages $k$ and $k+1$, respectively, $q$ is the transferred charge

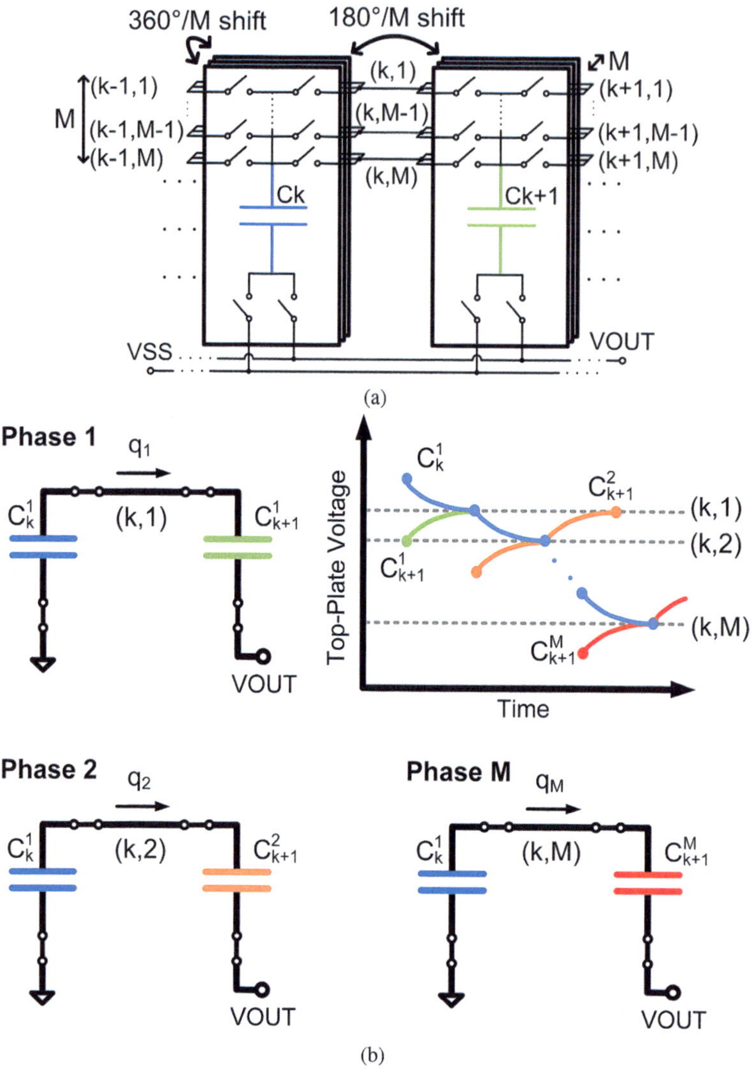


**Fig. 7.6** Application of multiphase soft-charging to a Dickson converter with factor $M$. (**a**) a schematic overview, (**b**) the charge transfers of two adjacent stages

themselves are split up into a sufficient number of out-of-phase cores. The resulting step-wise (dis)charging reduces the charge-sharing losses by the number of steps, $M$, which is fully equivalent to increasing the effective capacitance density by the same factor. In other words, the effective capacitance density can be scaled freely by spreading the charge transfers between the flying capacitors out over multiple steps in what we call soft-charging. To achieve this, however, extra power transistors need to be added that connect the capacitors to the many intermediate nodes, which

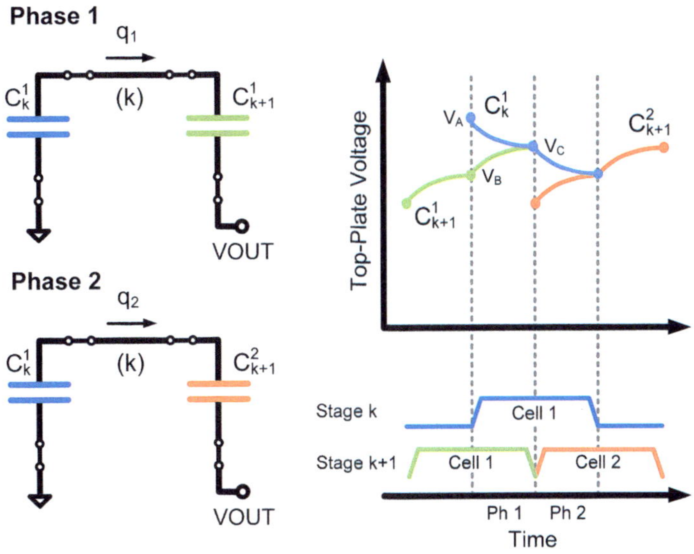


**Fig. 7.7** Application of stage outphasing to a Dickson converter

will increase the total transistor-related losses. In the end, the balancing between the increase of these losses with the reduction of the charge-sharing losses will determine the optimal number of $M$ and the overall efficiency improvement.

Stage outphasing is a second technique where adjacent stages are put out of phase with each other to, surprisingly, achieve the same effect as MSC with $M = 2$. Figure 7.7 illustrates this. Here, the discharging flying capacitor that was already connected to the intermediate node connects to the flying capacitor that has not been charged yet. In other words, the capacitor in its second discharging phase connects to the capacitor in its first charging phase. Similarly, when the converter switches again, the capacitor in the second charging phase will connect to another capacitor in its first discharging phase. In some way, the intermediate node ($k$) thus acts as two nodes through which the charging and discharging capacitors connect in opposite order. Unlike with MSC, though, no additional transistors have to be added, and the gain of SO consequently comes with no real cost.

Now, for a full converter there will be charge transfers that are between a capacitor and a combination of converter terminals. Because the voltage difference across these terminals is considered to be DC, these charge transfers cannot be soft-charged. Nevertheless, the Dickson converter only has two such transfers, regardless of the number of stages, which makes the use of MSC and SO more impactful as the number of stages and the VCR increases. Using a combination of SO and MSC, a fully integrated SC 3:1 DC–DC converter was made that has a 60% higher effective capacitance density [33, 34]. It is in part, thanks to these techniques, that said converter obtained 82% efficiency at a power-density of 1.1 W/mm$^2$

in measurements, which corresponds to 3× lower losses compared with similar power-density designs, or close to 30× higher power-density compared with similar efficiency designs in the literature.

## 7.4 Continuously Scalable Conversion Ratio

For the past decades, SC converters have been designed such that the voltage swing across their flying capacitors' terminals is minimized as shown in Fig. 7.8a, because this was considered the only way to minimize the charge-sharing losses and to obtain a high efficiency converter. Unfortunately, because the amount of charge that is transferred is proportionate to the capacitor voltage swing, this means that the output charge per cycle of these converters is minimized as well. If a higher output charge per cycle is required, the converter can only obey by increasing the voltage swing and thus lowering the efficiency, which also lowers the VCR. Consequently, a conventional SC converter's VCR, efficiency, and output charge per cycle are closely linked.

With the introduction of advanced multiphasing, however, we showed that charge-sharing losses can also be reduced by spreading the (dis)charging of flying capacitors out over multiple steps, over multiple phases. Consequently, rather than minimizing the voltage swing of the flying capacitors, and thus the output charge of the converter, the efficiency of the converter can instead be improved using soft-charging regardless of the capacitors' voltage swing. This offers the exciting opportunity to make switched-capacitor converters with large capacitor voltage swings that are soft-charged for high efficiency, as illustrated in Fig. 7.8b.

To arrive at such a converter, first a set of phases is chosen that maximizes the voltage swing on the flying capacitor while simultaneously taking into account that the transition from each phase to the next should lead to charge being transferred to and from the right converter terminals. In an efficient step-down converter, for example, charge should be transferred from the input and ground terminals to the output terminal. Figure 7.9 portrays such a set of phases, which we refer to as cornerstone phases. For a more detailed explanation on how to derive this set, the authors refer to [32, 35]. From these cornerstone phases, soft-charging can be implemented similarly to other AM techniques, by adding intermediate nodes through which charging and discharging capacitors can connect. Here, two sets of nodes are added for the top and bottom nodes of the flying capacitor, shown in Fig. 7.10. Because the charging and discharging capacitors connect to these nodes in opposite order, these nodes spread out between $V_{in}$ and $V_{out}$, and $V_{out}$ and $V_{ss}$, respectively.

Because the capacitor voltage does not stay approximately constant over a full clock cycle, this topology cannot be modeled using an ideal transformer with a finite

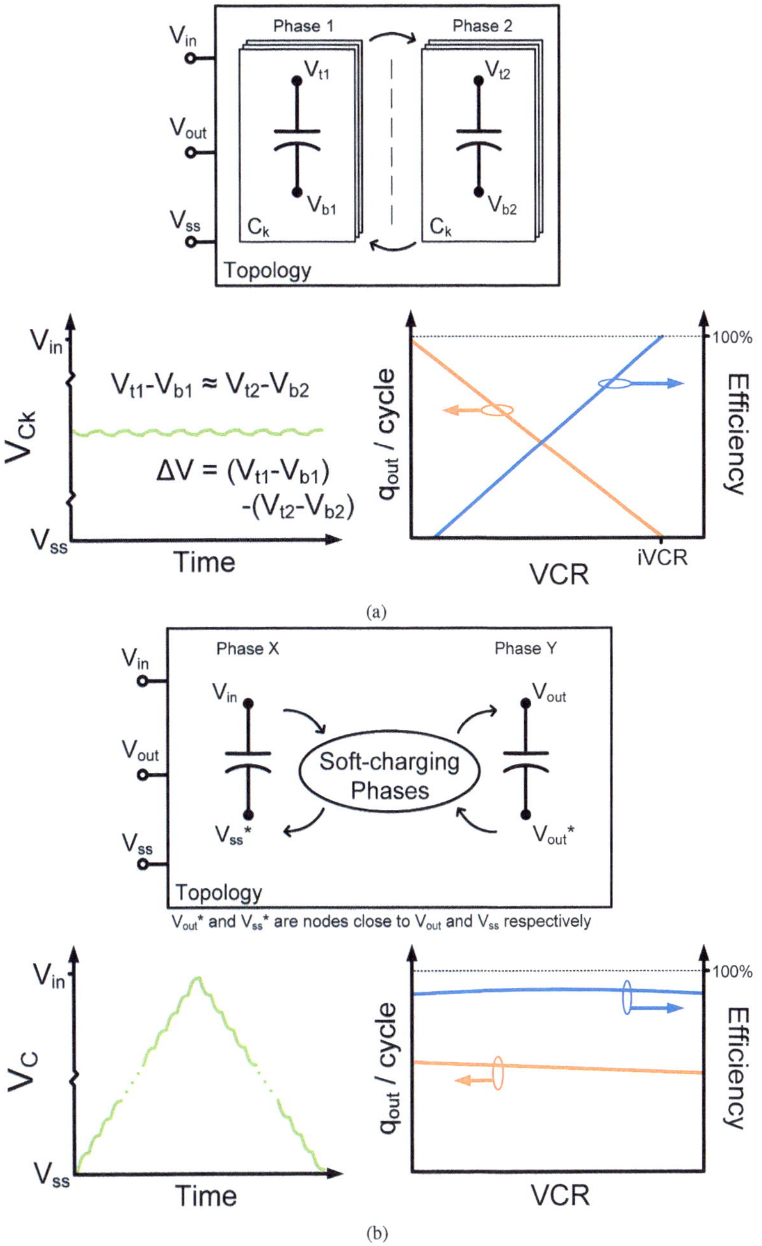


**Fig. 7.8** High level concept of (**a**) a conventional two-phase switched-capacitor converter, and (**b**) a large capacitor voltage-swing switched-capacitor converter

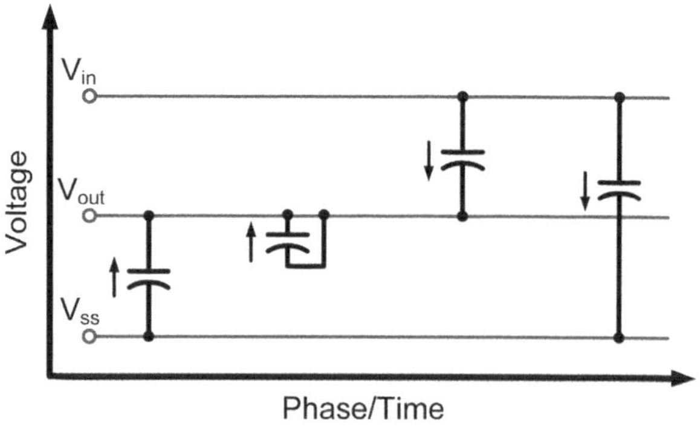


**Fig. 7.9** Cornerstone phases of large capacitor voltage-swing converter

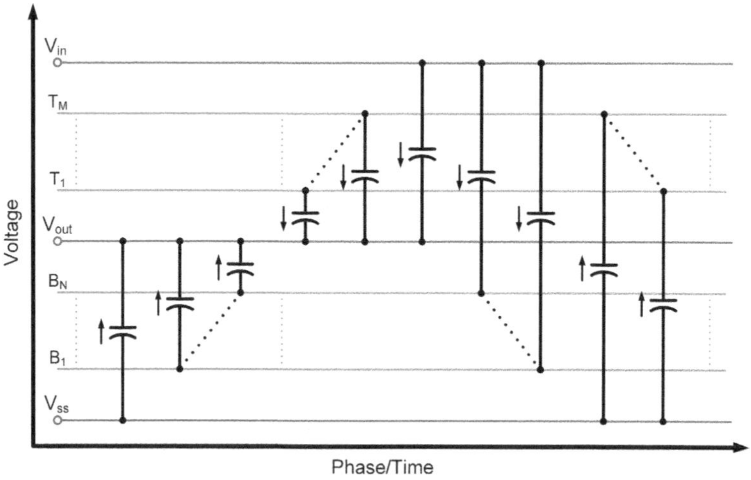


**Fig. 7.10** Voltage versus phase diagram of the presented topology

output impedance like regular SC converters can [36, 37]. Instead, it can be shown that this particular topology behaves like a gyrator [32]. That is, its output current is mostly proportionate to the input voltage, and not the output voltage. Thus, with constant input voltage, the output charge per cycle is mostly independent from the conversion ratio. At the same time, the input charge per cycle scales linearly with the output voltage. Figure 7.11 compares the theoretical efficiency of this topology to a regular SC converter, and demonstrates that, as the number of intermediate nodes increases, the converter tends towards an ideal gyrator which is lossless regardless of the VCR, both step-up and step-down.

The topology is realized in a 28 nm technology using 32 intermediate nodes at both the top and bottom side [38]. The micrograph of this design is shown in Fig. 7.12, while the measured efficiency is portrayed in Fig. 7.13 for an input voltage of 2 V. It can be appreciated that the latter demonstrates the gyrator behavior of the

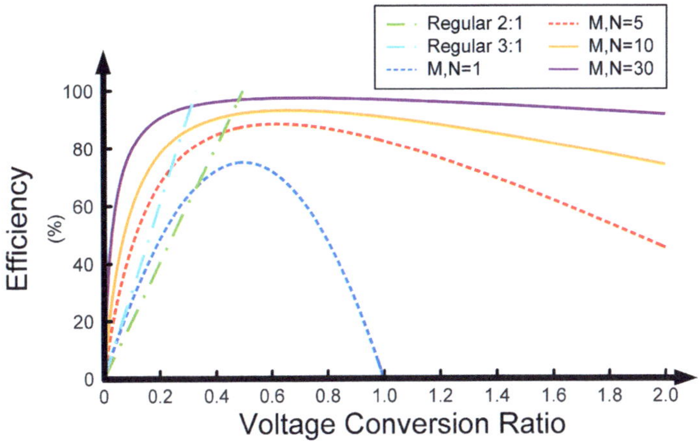


**Fig. 7.11** Efficiency of presented topology for different number of top- and bottom-side soft-charging phases, *M* and *N*, compared with a regular 2:1 and 3:1 SC converter

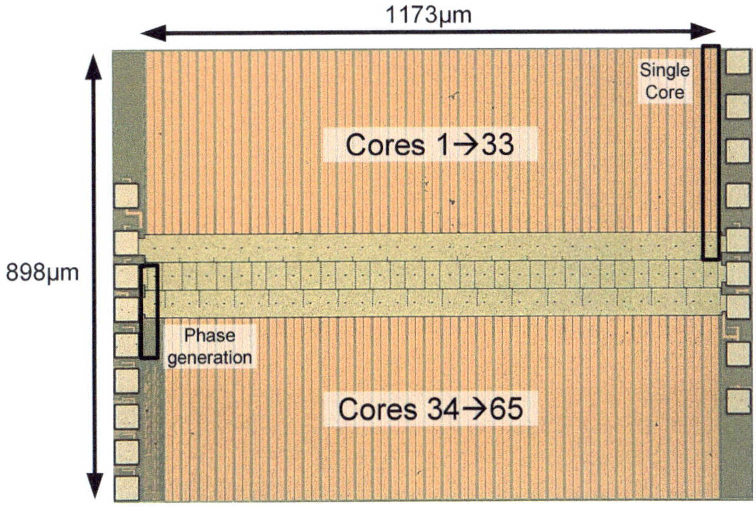


**Fig. 7.12** Micrograph of the monolithic SC converter using a continuously scalable-conversion-ratio topology

topology and that the converter maintains an efficiency of more than 80% over a continuous VCR range of 0.85, which is substantially higher than designs using regular SC gearbox converters in the literature [17, 18, 23]. In addition, this design achieves a peak efficiency of 93%.

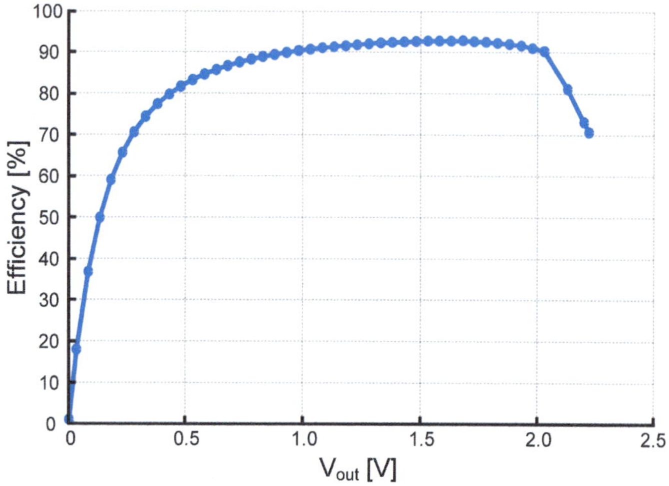


**Fig. 7.13** Measured efficiency versus output voltage of the presented converter at a fixed clock frequency

## 7.5 Conclusions

This chapter touched upon the main factors that have sparked the interest in monolithic power conversion and why switched-capacitor converters in particular make an excellent candidate. The large parasitic coupling to the substrate, the limited capacitor density on-chip, and the inherent constrained conversion ratio range were presented as challenges to the widespread adoption of this type of converter. With advanced multiphasing, multiple out-of-phase converter cores interact with each other to arrive at a switched-capacitor converter with more than the typical two phases that in the end has more capabilities and/or better performance. Several advanced multiphasing techniques were discussed that focus on both high- and low-power densities, and even allow for a new type of switched-capacitor converter which has a continuously scalable conversion ratio. Measurement results demonstrated the working principles of these techniques and showed the great potential of advanced multiphasing in pushing the limits of fully integrated power conversion further.

## References

1. M. Steyaert, N. Butzen, H. Meyvaert, A. Sarafianos, P. Callemeyn, T. Van Breussegem, M. Wens, DCDC Performance Survey, http://homes.esat.kuleuven.be/~steyaert/DCDC_Survey/DCDC_PS.html
2. L. Chang, D.J. Frank, R.K. Montoye, S.J. Koester, B.L. Ji, P.W. Coteus, R.H. Dennard, W. Haensch, Practical strategies for power-efficient computing technologies. Proc. IEEE **98**(2), 215–236 (2010)

3. ITRS, ITRS Reports (2015), http://www.itrs2.net/itrs-reports.html
4. F. Carobolante, Power supply on chip: from R&D to commercial products, in *International Workshop on Power Supply on Chip* (2014)
5. M. Steyaert, T. Van Breussegem, H. Meyvaert, P. Callemeyn, M. Wens, DC-DC converters: from discrete towards fully integrated CMOS, in *2011 Proceedings of the ESSCIRC (ESSCIRC)* (IEEE, Piscataway, 2011), pp. 42–49
6. T.M.V. Breussegem, M.S.J. Steyaert, Monolithic capacitive DC-DC converter with single boundary-multiphase control and voltage domain stacking in 90 nm CMOS. IEEE J. Solid-State Circuits **46**(7), 1715–1727 (2011)
7. T. Andersen, F. Krismer, J. Kolar, T. Toifl, C. Menolfi, L. Kull, T. Morf, M. Kossel, M. Brändli, P.A. Francese, A 10 W on-chip switched capacitor voltage regulator with feedforward regulation capability for granular microprocessor power delivery. IEEE Trans. Power Electron. **32**(1), 378–393 (2017)
8. W. Kim, M.S. Gupta, G.Y. Wei, D. Brooks, System level analysis of fast, per-core DVFS using on-chip switching regulators, in *IEEE 14th International Symposium on High Performance Computer Architecture, 2008. HPCA 2008* (2008, IEEE, Piscataway), pp. 123–134.
9. C.R. Sullivan, B.A. Reese, A.L.F. Stein, P.A. Kyaw, On size and magnetics: why small efficient power inductors are rare, in *2016 International Symposium on 3D Power Electronics Integration and Manufacturing (3D-PEIM)* (IEEE, Piscataway, 2016), pp. 1–23.
10. M. Wens, K. Cornelissens, M. Steyaert, A fully-integrated 0.18 um CMOS DC-DC step-up converter, using a bondwire spiral inductor, in *ESSCIRC 2007 - 33rd European Solid-State Circuits Conference* (IEEE, Piscataway, 2007), pp. 268–271
11. N. Kurd, M. Chowdhury, E. Burton, T.P. Thomas, C. Mozak, B. Boswell, P. Mosalikanti, M. Neidengard, A. Deval, A. Khanna, N. Chowdhury, R. Rajwar, T.M. Wilson, R. Kumar, Haswell: a family of IA 22 nm processors. IEEE J. Solid-State Circuits **50**(1), 49–58 (2015)
12. C. Schaef, J.T. Stauth, A 3-phase resonant switched capacitor converter delivering 7.7 W at 85% efficiency using 1.1 nH PCB trace inductors. IEEE J. Solid-State Circuits **50**(12), 2861–2869 (2015)
13. C. Schaef, E. Din, J.T. Stauth, A hybrid switched-capacitor battery management IC with embedded diagnostics for series-stacked Li-Ion arrays. IEEE J. Solid-State Circuits **52**(12), 3142–3154 (2017)
14. H.P. Le, S.R. Sanders, E. Alon, Design techniques for fully integrated switched-capacitor DC–DC converters. IEEE J. Solid-State Circuits **46**(9), 2120–2131 (2011)
15. N. Butzen, M.S.J. Steyaert, Scalable parasitic charge redistribution: design of high-efficiency fully integrated switched-capacitor DC-DC converters. IEEE J. Solid-State Circuits **51**(12), 2843–2853 (2016)
16. J.F. Dickson, On-chip high-voltage generation in mnos integrated circuits using an improved voltage multiplier technique. IEEE J. Solid-State Circuits **11**(3), 374–378 (1976)
17. D. El-Damak, S. Bandyopadhyay, A.P. Chandrakasan, A 93% efficiency reconfigurable switched-capacitor DC-DC converter using on-chip ferroelectric capacitors, in *2013 IEEE International Solid-State Circuits Conference Digest of Technical Papers (ISSCC)* (IEEE, Piscataway, 2013), pp. 374–375
18. W. Jung, D. Sylvester, D. Blaauw, A rational-conversion-ratio switched-capacitor DC-DC converter using negative-output feedback, in *2016 IEEE International Solid-State Circuits Conference (ISSCC)* (2016), pp. 218–219
19. L.G. Salem, P.P. Mercier, An 85%-efficiency fully integrated 15-ratio recursive switched-capacitor DC-DC converter with 0.1-to-2.2V output voltage range, in *2014 IEEE International, Solid-State Circuits Conference Digest of Technical Papers (ISSCC)* (IEEE, Piscataway, 2014), pp. 88–89
20. A. Sarafianos, M. Steyaert, Fully integrated wide input voltage range capacitive DC-DC converters: the folding Dickson converter. IEEE J. Solid-State Circuits **50**(7), 1560–1570 (2015)
21. S. Bang, A. Wang, B. Giridhar, D. Blaauw, D. Sylvester, A fully integrated successive-approximation switched-capacitor DC-DC converter with 31mV output voltage resolution, in

*2013 IEEE International Solid-State Circuits Conference Digest of Technical Papers* (IEEE, Piscataway, 2013), pp. 370–371
22. Y.T. Lin, W.H. Yang, Y.S. Ma, Y.J. Lai, H.W. Chen, K.H. Chen, C.L. Wey, Y.H. Lin, J.R. Lin, T.Y. Tsai, Unsymmetrical parallel switched-capacitor (UP-SC) regulator with fast searching optimum ratio technique, in *ESSCIRC 2017 - 43rd IEEE European Solid State Circuits Conference* (IEEE, Piscataway, 2017), pp. 287–290
23. Y. Jiang, M.K. Law, P.I. Mak, R.P. Martins, A 0.22-to-2.4V-input fine-grained fully integrated rational buck-boost SC DC-DC converter using algorithmic voltage-feed-in (AVFI) topology achieving 84.1% peak efficiency at 13.2mW/mm2, in *2018 IEEE International Solid - State Circuits Conference - (ISSCC)* (IEEE, Piscataway, 2018), pp. 422–424
24. C. Schaef, J.S. Rentmeister, J. Stauth, Multimode operation of resonant and hybrid switched-capacitor topologies. IEEE Trans. Power Electron. **33**(12), 10512–10523 (2018)
25. T. Van Breussegem, M. Steyaert, A fully integrated gearbox capacitive DC/DC-converter in 90nm CMOS: optimization, control and measurements, in *2010 IEEE 12th Workshop on Control and Modeling for Power Electronics (COMPEL)* (IEEE, Piscataway, 2010), pp. 1–5
26. G.V. Piqué, A 41-phase switched-capacitor power converter with 3.8mV output ripple and 81% efficiency in baseline 90nm CMOS, in *2012 IEEE International Solid-State Circuits Conference* (IEEE, Piscataway, 2012), pp. 98–100
27. N. Butzen, M. Steyaert, A 94.6%-efficiency fully integrated switched-capacitor DC-DC converter in baseline 40nm CMOS using scalable parasitic charge redistribution, in *2016 IEEE International Solid-State Circuits Conference (ISSCC)* (IEEE, Piscataway, 2016), pp. 220–221
28. N. Butzen, M. Steyaert, MIMO switched-capacitor converter using only parasitic capacitance with scalable parasitic charge redistribution, in *ESSCIRC Conference 2016: 42nd European Solid-State Circuits Conference* (2016), pp. 445–448
29. N. Butzen, M.S.J. Steyaert, MIMO switched-capacitor DC-DC converters using only parasitic capacitances through scalable parasitic charge redistribution. IEEE J. Solid-State Circuits **52**(7), 1814–1824 (2017)
30. L. Chang, R.K. Montoye, B.L. Ji, A.J. Weger, K.G. Stawiasz, R.H. Dennard, A fully-integrated switched-capacitor 2:1 voltage converter with regulation capability and 90%, in *2010 IEEE Symposium on VLSI Circuits (VLSIC)* (IEEE, Piscataway, 2010), pp. 55–56
31. T.M. Andersen, F. Krismer, J.W. Kolar, T. Toifl, C. Menolfi, L. Kull, T. Morf, M. Kossel, M. Bråndii, P.A. Francese, A feedforward controlled on-chip switched-capacitor voltage regulator delivering 10W in 32nm SOI CMOS, in *2015 IEEE International Solid-State Circuits Conference - (ISSCC) Digest of Technical Papers* (IEEE, Piscataway, 2015), pp. 1–3
32. N. Butzen, Fully integrated advanced multiphasing switched-capacitor DC-DC converters, Ph.D. dissertation, KU Leuven, 2018
33. N. Butzen, M. Steyaert, A 1.1W/mm2-power-density 82%-efficiency fully integrated 3:1 switched-capacitor DC-DC converter in baseline 28nm CMOS using stage outphasing and multiphase soft-charging, in *2017 IEEE International Solid-State Circuits Conference (ISSCC)* (IEEE, Piscataway, 2017), pp. 178–179
34. N. Butzen, M.S.J. Steyaert, Design of soft-charging switched-capacitor DC-DC converters using stage outphasing and multiphase soft-charging. IEEE J. Solid-State Circuits **52**(12), 3132–3141 (2017)
35. N. Butzen, M. Steyaert, Design of single-topology continuously scalable-conversion-ratio switched-capacitor dc-dc converters. IEEE J. Solid-State Circuits **54**(4), 1039–1047 (2019)
36. M.D. Seeman, A design methodology for switched-capacitor DC-DC converters, Ph.D. dissertation, EECS Department, University of California, Berkeley, 2009, http://www.eecs.berkeley.edu/Pubs/TechRpts/2009/EECS-2009-78.html
37. M.S. Makowski, D. Maksimovic, Performance limits of switched-capacitor DC-DC converters, in *Power Electronics Specialists Conference, 1995. PESC '95 Record, 26th Annual IEEE*, vol. 2 (IEEE, Piscataway, 1995), pp. 1215–1221
38. N. Butzen, M. Steyaert, A single-topology continuously-scalable-conversion-ratio fully integrated switched-capacitor DC-DC converter with 0–2.22V output and 93% peak-efficiency, in *2018 Symposium on VLSI Circuits* (IEEE, Piscataway, 2018), pp. 103–104

# Chapter 8
# Highly Efficient Power Management in Wearables and IoT Devices

**Juergen Wittmann, Francesco Cannillo, Dan Ciomaga, Mihail Jefremow, and Fabio Rigoni**

## 8.1 Introduction

The Internet of Things (IoT) market experienced a significant growth in terms of business and visibility in the last years. McKinsey Global Institute research estimates the impact of the Internet of Things on the global economy to be as high as $6.2 trillion by 2025 [1]. The purpose of IoT is to interconnect different applications in order to exchange information, e.g., sensor data or media, or to make this information available via the internet. IoT contains applications in a large range of different areas [2]:

- Smart buildings use sensors to identify the locations of people as well as the state of the building.
- Smart cities use sensors to monitor pedestrian and vehicular traffic and may integrate data from smart buildings.
- Medical systems connect a wide range of patient monitoring sensors that may be located at home, in emergency vehicles, the doctor's office, or the hospital.
- Wearables are mobile devices, wireless head phones, smart watches, Bluetooth head phones, performance measurements (heart rate, distance, and step counter, etc.).
- Industrial systems use sensors to monitor both the industrial processes themselves—the quality of the product—and the state of the equipment.
- Autonomous driving requires connected devices with power consumption due to high computational power mainly due to sensor and graphics evaluation.

J. Wittmann (✉) · F. Cannillo · D. Ciomaga · M. Jefremow · F. Rigoni
Dialog Semiconductor, Germering, Germany
e-mail: juergen.wittmann@diasemi.com

© Springer Nature Switzerland AG 2020

A. Baschirotto et al. (eds.), *Next-Generation ADCs, High-Performance Power Management, and Technology Considerations for Advanced Integrated Circuits*,
https://doi.org/10.1007/978-3-030-25267-0_8

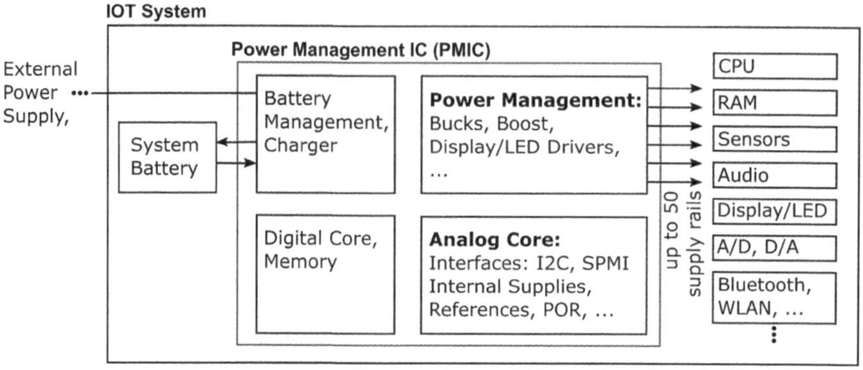


**Fig. 8.1** Overview of a typical IoT system

The challenge of IoT systems is that it involves different competences from software to hardware [3]. The hardware space includes a combination of different functions: sensors for external world data acquisition, connectivity with RF chips such as BLE (Bluetooth low energy), power management to supply the different part of the system, data conversion and computation to convert the data from the analog world to digital and compute the information as required, cybersecurity to ensure system security for applications involving the privacy or confidentiality of the users.

An overview of a typical IoT system is shown in Fig. 8.1. A main requirement of IoT devices is a high level of integration. Most of the functionalities are integrated on one or several different ICs to achieve a very small volume. This allows to reduce the application's size and cost (bill of materials, BOM), and at the same time to increase the number of functionalities in IoT devices, e.g., for standalone sensors, wearables as wireless ear plugs, mobile devices as smart watches, medical monitoring applications, and many others.

Most IoT or wearable systems contain batteries, to be functionally independent from a grid power supply connection. The battery is charged from an external power supply as a wall plug charger, a USB connection, a wireless charger, or even from energy harvesting sources. A long battery lifetime is essential and requires a highly efficient and low-loss energy management, which is provided by one or more power management ICs as shown in Fig. 8.1.

On the power management IC (PMIC), a battery management circuit charges the battery, and provides the main PMIC supply either from the battery or from the external supply, if available. An important circuit group in the PMIC is the analog core circuitry. This contains mainly all internal circuits required for the PMICs, which are the supply circuits of all internal circuits, interfaces (e.g., $I^2C$, SPMI, and general-purpose IOs), voltage and current references, sensing circuits, and several other supporting functions. The analog core circuitry is fundamental for the battery lifetime during standby mode of the IoT device. Sensors nodes, for example, can be in standby mode most of the time, in which the analog core circuit is still active to provide a communication and wake-up functionality. The power consumption (quiescent current, IQ) of the analog core in standby mode determines the overall battery lifetime. A digital core and memory support a high configurability of the PMIC.

Several power management circuits provide the power for different functions within the IoT device, which are accurately regulated and configurable voltage rails for other ICs or regulated power to control LEDs or displays with output voltages up to 20 V. The output power of some of these rails can be in the order of tens of Watts to do complex computation, e.g., in sensor devices, process media in mobile devices, or providing transmitting power from RF ICs. An increase of the power conversion efficiency is essential, as it directly impacts the functional battery lifetime.

A main challenge for PMICs in IoT devices is that a very large design parameter space needs to be covered. The supply voltage can be from a single Li-ion battery voltage up to 4.5 V, or two in series connected Li-ion batteries with supply voltage of up to 9 V, while the latest USB-PD (power delivery) standard allows to even provide up to 20 V charging voltage. e-Mobility or automotive applications, e.g., for autonomous driving, are supplied from batteries of 12 V, and observe a trend for battery voltages even up to 48 V.

Figure 8.2 shows an example of a typical PMIC used for wearable applications [4]. It has fully integrated battery management with overvoltage protection and a linear charger for the battery including the temperature and voltage supervision. It includes a digital part to control the main FSM (finite-state machine) and analog blocks. The buck converter supplies the host processor either from the battery (here VDD_SYS) or from the VDD_PWR rail. The boost converter can supply LEDs used for different applications. The three LDOs of the PMIC are usually supplying different sensors of the IOT system. The crucial spec for this kind of PMIC is the standby current, which is below 1 uA for the entire chip with buck enabled (no-load condition) and supplied from the battery.

This work describes advanced implementation techniques for PMICs used for different IoT applications. Section 8.2 focuses on the analog core implementation and demonstrates circuits and implementation techniques to achieve minimum IQ and reduce die area.

For lower input voltage, conventional buck converters provide a good ratio between high efficiency and die area as step-down converters, while the losses and converter size become more critical towards higher input voltages. Section 8.3 covers advanced architectures of voltage converters used for high-power delivery in the chargers, as well as for power management circuits, improving efficiency especially for higher input voltages and conversion ratios $V_{IN}/V_{OUT}$. It comprises the design of a switched-capacitor converter, and a 3-level buck converter. Conclusions are drawn in Sect. 8.4.

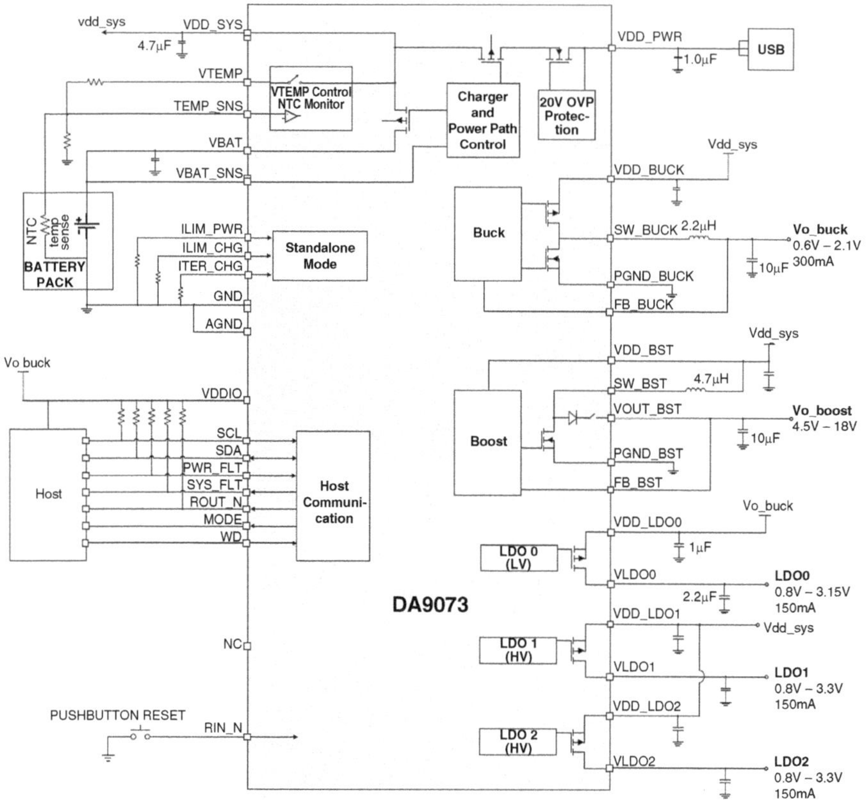


**Fig. 8.2** Example of a low-IQ PMIC used for wearables

## 8.2 Low-IQ Analog Core Design

Figure 8.3 gives an overview of the main functionalities of the analog core circuitry in a low-IQ PMIC. This section shows an implementation of the IQ critical circuits. As the main PMIC supply rail is typically connected to the battery, it is always supplying a voltage to some of the analog core circuits shown in Fig. 8.3. All circuits, which are required to power up the digital core of the PMIC, must be always ON, as they control not only the PMIC start-up, but also the start-up of the entire IoT device, as soon as a battery voltage is available. A pre-regulator is required to stabilize noisy external PMIC supply voltages. It supplies a bandgap voltage and current reference, which is required for a linear regulator to generate the internal PMIC supply (VDD). The IQ of these always ON circuits determines the overall battery lifetime in standby mode of the IoT system.

The analog core generates the initial power-on reset of the PMIC and the entire IoT system, as soon as the battery supply and the internal digital supply (VDD) are stable. Thus, the pre-regulator to supply a bandgap voltage and current reference and

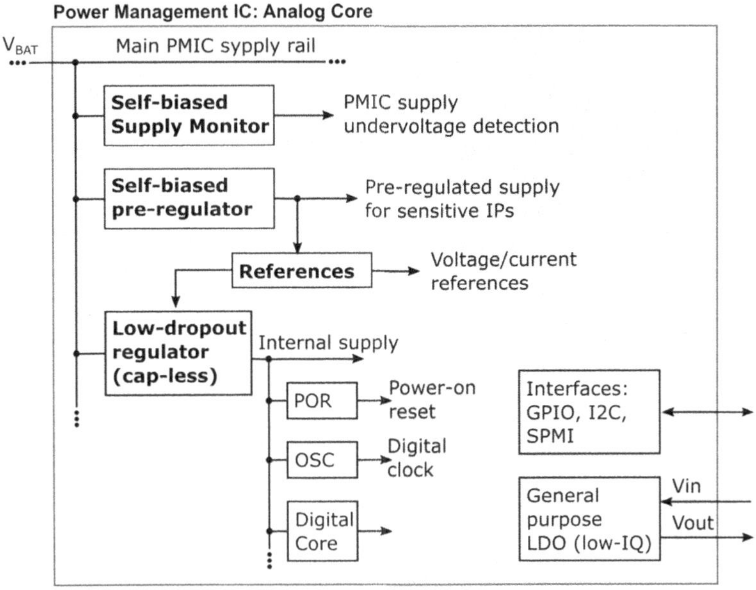


**Fig. 8.3** Typical analog core implementation of a low-IQ PMIC

the supply monitor circuit require to be self-biased, i.e., these circuits are operating autonomously, without relying on other PMIC functions to provide them with inputs and control signals, such as reference currents or voltages or digital control signals from the digital core. At the same time, they need to provide their designated functionality within acceptable margins of accuracy. An implementation example of a self-biased pre-regulator, supply monitor, and bandgap reference is shown in Sect. 8.2.1.

The oscillator requires either to have a low-IQ, slow clock mode, or it is completely turned off during standby mode by digital. This is possible if the digital core is activated by an interrupt received internally from the PMIC or externally from any of the communication interfaces. Therefore, some of the interfaces require to stay on in standby mode. The interfaces itself can be implemented without a significant current consumption, when they are not switching; however, for a larger amount of interfaces on the PMIC, the interface supply requires to be able to provide high power to supply all interfaces in case they are switching at the same time, but at the same time, this supply has to be very low-IQ in standby mode in case only few interfaces are to be supplied. In Sect. 8.2.2, an advanced IQ and area beneficial interface supply concept is presented.

The analog core often contains general-purpose low-dropout regulators (LDOs) supplying IQ critical functions on the IoT device, e.g., sensors, which might be required to be turned on in standby mode. Design examples of a low-IQ, general-purpose LDO are demonstrated in Sect. 8.2.3.

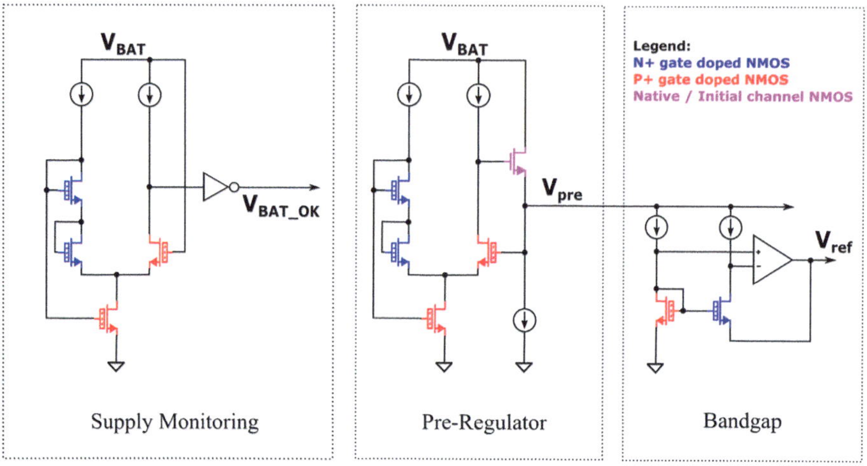


**Fig. 8.4** Proposed supply monitor, pre-regulator, and bandgap

## *8.2.1 Self-Biased, Self-Referenced PORs, Pre-regulators, Bandgaps*

The reference generator or bandgap in the analog core is typically supplied from a stabilized pre-regulator rail ($V_{pre}$). However, the bandgap is not available as a reference for the pre-regulator, as well as for the supply monitoring block (see Fig. 8.3). They require to be self-biased, i.e., generate their own internal reference with enough accuracy. In all circuits, a method is applied which generates the bandgap voltage through the polysilicon work function difference of a pair of MOS devices with controlled gate doping [5]. Several technologies provide an NMOS transistor with a P+ doped gate. The doping is performed to increase the threshold to higher voltages, such that the threshold difference compared to a standard NMOS transistor, which has an N+ doped gate, is shifted by one bandgap voltage (~1.2 V).

Figure 8.4 shows a possible implementation of a supply monitor, pre-regulator, and bandgap circuit with the help of a pair of differently gate-doped NMOS devices. The circuits are autonomous, and do not rely on inputs from other circuits and are able to provide their designated functionality as soon as the supply (VDD) is present.

In the configuration of Fig. 8.4, the supply monitor and pre-regulator provide a $V_{BAT\_OK}$ flag and $V_{pre}$ voltage at about two times one bandgap voltage. Alternative implementations are possible to generate the $V_{BAT\_OK}$ flag and $V_{pre}$ voltage at any other level higher than one bandgap voltage. The bandgap provides a reference equal to one bandgap voltage. The circuits can be designed to have a quiescent current less than 100 nA each.

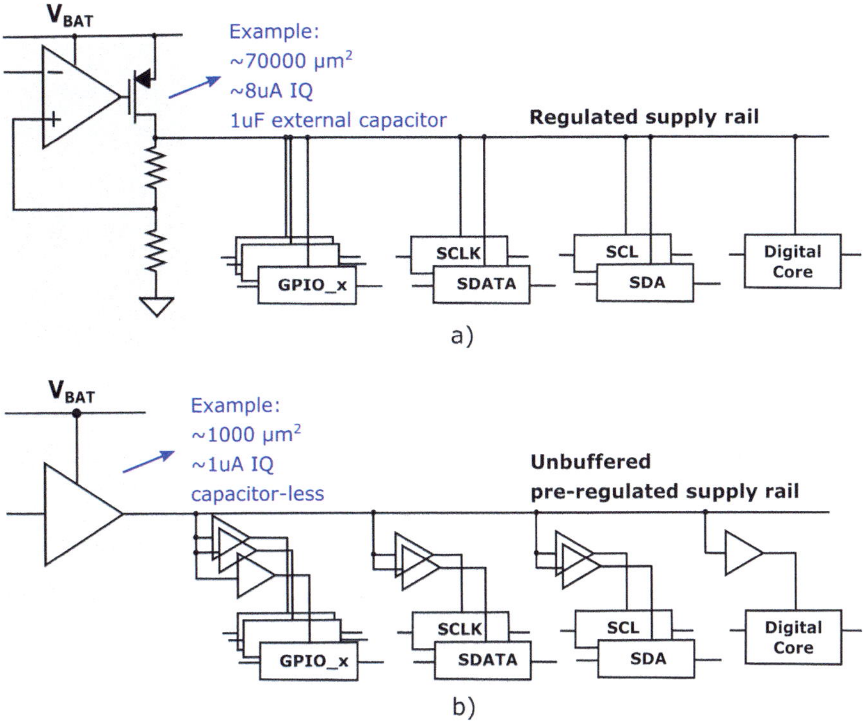


**Fig. 8.5** (**a**) Conventional supply concept for switching blocks; (**b**) IQ and area improved implementation of the switching blocks supply

### *8.2.2 Low-Power, Capacitor-Less Supplies for Switching Blocks*

A further IQ, area and BOM reduction is achieved with the power supply strategy of the switching blocks, which are mainly communication interfaces. A conventional implementation is shown in Fig. 8.5a. A main linear regulator is required to supply the worst case switching current, assuming all IOs are active and switching at the same time. This leads to an overdesign of this regulator, as most of the time none of the interfaces are switching. This leads to a large area of the regulator, as well as a high IQ in the regulator control circuit as it is designed for the maximum output current. Moreover, in standby or sleep mode, most of the interfaces could be turned off; however, the regulator IQ cannot be scaled down. Moreover, the regulator must fulfill the accuracy requirements of all analog blocks connected to the regulator. The large load often requires a large external output buffer capacitor, which increases the cost and size of the IoT device.

A more IQ beneficial implementation of the supply concept is shown in Fig. 8.5b. Instead of one large accurate regulator as shown in Fig. 8.5a, each interface

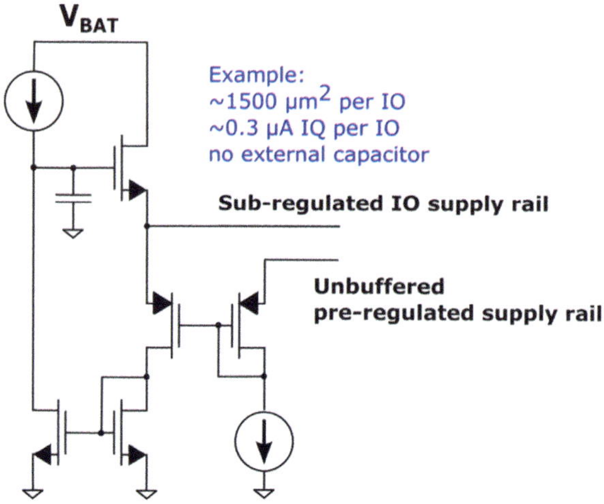


**Fig. 8.6** Implementation of a capacitor-less low-IQ sub-regulator

block and the digital core have their own small capacitor-less sub-regulator, just adjusted for the needs of each circuit block. The sub-regulator is supplied by the pre-regulator, which has less accuracy requirements, and can be implemented also as a capacitor-less low-IQ regulator as previously shown in Fig. 8.4. In standby, the sub-regulators are turned off together with the interface blocks, if it is not needed. This way, a significant IQ reduction can be achieved in standby, especially when only very few interfaces are active.

A typical implementation of such a capacitor-less low-IQ sub-regulator is shown in Fig. 8.6. The unbuffered pre-regulated supply rail is used as a voltage reference, which is then replicated by a native PMOS source follower. A simple and fast one-stage regulation loop controls an NMOS pass device. The sub-regulator design requires an area of ~1500 $\mu m^2$, consuming a total current of ~0.3 µA. This is about 25 times less IQ and nearly 50 times less area per sub-regulator, compared to the conventional approach shown in Fig. 8.5a. Thus, the total size and the IQ of the switching supply system are less, even if tens of interfaces are used.

### *8.2.3 Low-IQ General-Purpose LDOs for Power Delivery*

Linear regulators, especially low-dropout regulators (LDO) are often preferred, thanks to their better noise and PSRR performance. Moreover, they can be designed to have lower IQ over a wide range of load currents, but especially for small loads required in low-power or standby mode. The power efficiency of linear regulators improves for smaller voltage drops across the pass device. Therefore, linear regulators are preferred, if low load currents are required, or if the conversion ratio $V_{in}/V_{out}$ is reasonably small.

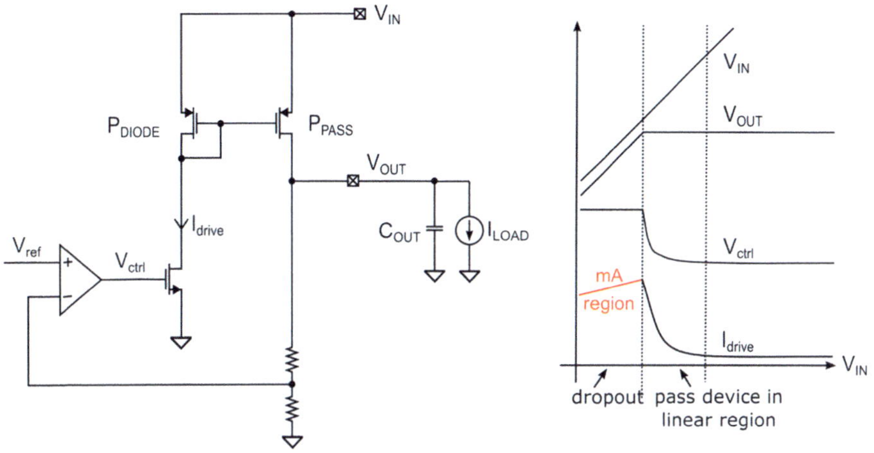


**Fig. 8.7** Conventional linear regulator and its behavior in dropout

One commonly used technique in linear regulators, which is depicted in Fig. 8.7, is to drive the pass device ($P_{\mathrm{PASS}}$) from a low impedance node, such as from a diode connected transistor $P_{\mathrm{DIODE}}$. This allows to achieve a good compromise between IQ, stability, and speed. A drawback of this approach is the behavior of the linear regulator close to dropout, i.e., for drain-source voltages lower than the saturation voltage of the pass device. In this region the loop needs to pull a significantly bigger current ($I_{\mathrm{drive}}$) from $P_{\mathrm{DIODE}}$ to maintain regulation at the output. Just before entering dropout, $I_{\mathrm{drive}}$ reaches a very large peak value in the milli-ampere range. This behavior is almost load current independent. Dropout mode is required, if the battery supply drops. Especially for low battery voltages, it is even less acceptable to have a high current consumption in the milli-ampere range, if the LDO is in dropout mode.

Figure 8.8 shows two possible implementations that effectively allow the $V_{\mathrm{IN}}$-$V_{\mathrm{OUT}}$ difference to control $I_{\mathrm{drive}}$ by a starved current buffer structure (SCB) to achieve a dropout behavior as depicted in the signal diagram in Fig. 8.8.

Another important benefit is a decreased dropout, compared to the implementation in Fig. 8.7, since a reduced $I_{\mathrm{drive}}$ current allows an increased swing of the gate of the pass device, thanks to less headroom being lost on the NMOS device controlled by $V_{\mathrm{ctrl}}$. The implementation of the SCB structures allows to significantly reduce the dropout current consumption from the milli-ampere range down to several micro-amperes only.

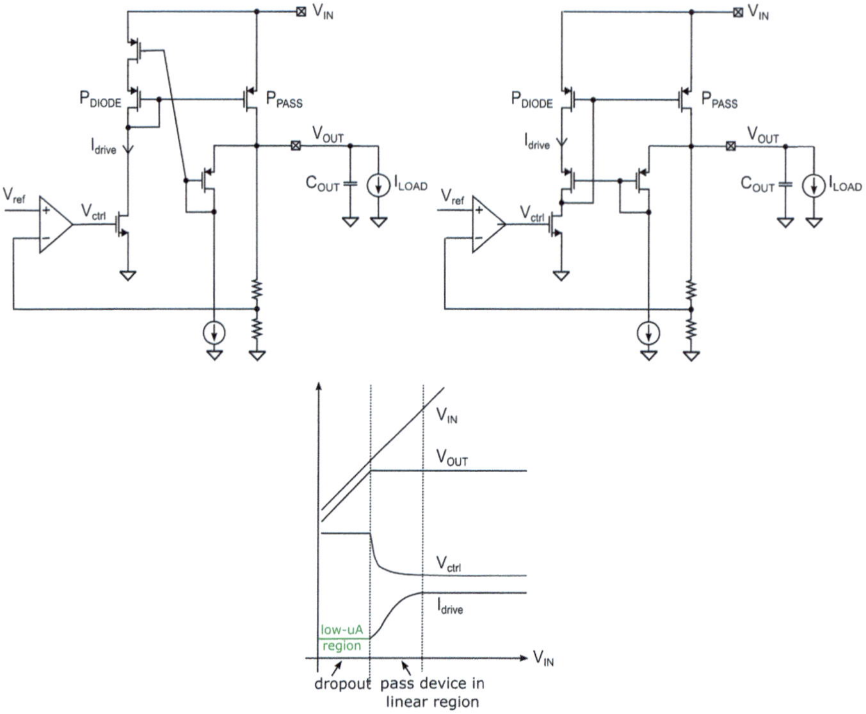


**Fig. 8.8** Linear regulators with starved current buffer (SCB)

## 8.3 Advanced Converter Design for Power Delivery

IoT and mobile computing devices require efficient power conversion with input to output voltage ratio $V_{IN}/V_{OUT} > 2$. For such purpose, key power converter topologies are the switched-capacitor (SC) divider [6, 7] and the multi-level switched-mode converter (MLC) [8–10].

For example, in the mobile computing space, the voltage developed by two series (2S) Li-ion batteries (typically 7.4 V) needs to be bucked down to a voltage of about 1.0 V used by the main core processor. The efficiency of the voltage down-conversion can be improved by introducing an unregulated intermediate supply between the input and the supply voltages of the different parts of the system. This is achieved with a 2:1 switched-capacitor (SC) power converter (Fig. 8.9).

The unregulated supply is then down-regulated by inductor-based switched-mode buck converters. An example for the inductor-based buck converter used in wearable applications is shown in Fig. 8.10. The buck converter architecture is based on current mode control allowing to use a reduced number of building blocks in the buck control, which can be switched off in the no load condition. Only the comp block stays alive together with VDAC to monitor the output voltage, this allows to

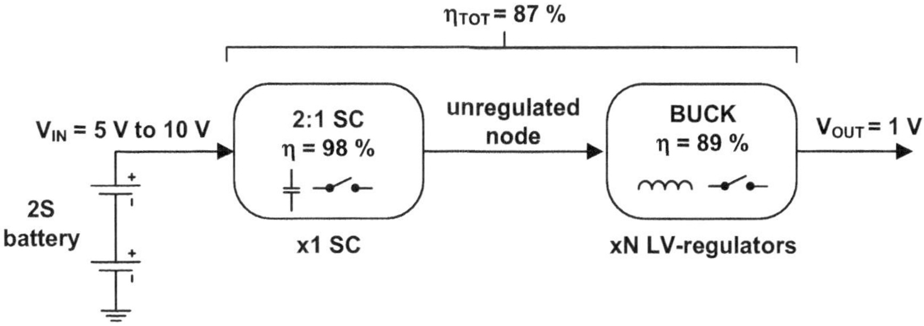


**Fig. 8.9** 2:1 SC converter generating an unregulated intermediate supply from a 2S battery

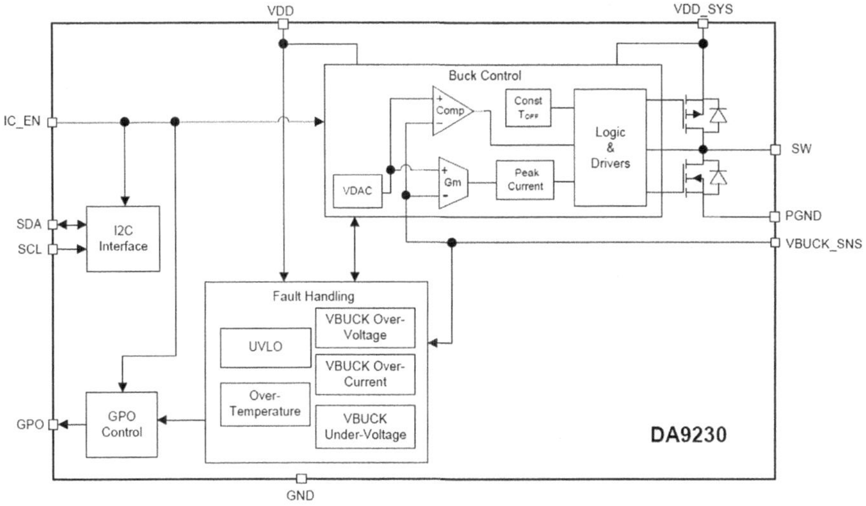


**Fig. 8.10** Example of a low-IQ, highly efficient buck converter [11]

reduce the entire buck input current to 750 nA at no load condition. The efficiency of this buck is above 90% for VDD_SYS of 3 V.

The overall system efficiency is improved because the SC divider generates the intermediate supply with much higher power efficiency (98%) than conventional buck converters. In addition, the buck converters performing the final regulation in Fig. 8.9 can use switches rated for lower voltages, and, therefore, can operate at higher switching frequency with a smaller inductor (with lower DC losses).

Another important use case is the charge of single cell (1S) Li-ion batteries, typically used in portable devices (Fig. 8.11). The user experience of portable devices demands longer usage time (higher battery capacity) and shorter battery re-charging time (higher charging current). The need for high charging currents (>1 A) causes higher power losses both in the charger IC (inside the portable device) and in the cable connecting the portable device to the AC/DC adaptor. Cable

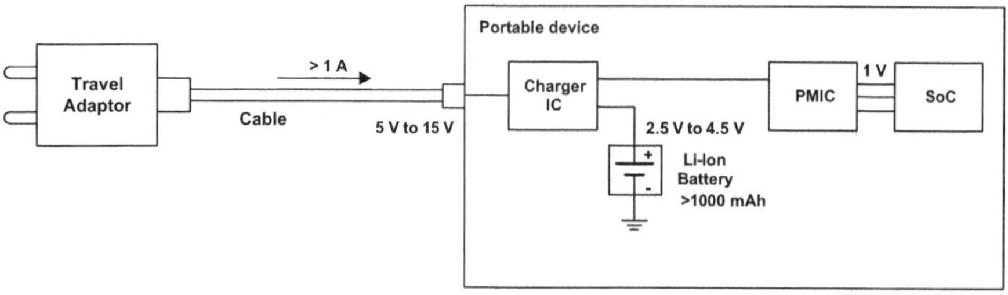


**Fig. 8.11** Charging solution in mobile computing devices with 1S battery

resistive losses can easily be reduced by increasing the input voltage of the charger IC provided by the AC/DC adaptor. The MLC buck topology is a good candidate to achieve voltage regulation having a conversion ratio $V_{IN}/V_{OUT} > 2$ with high efficiency, as it inherently generates an intermediate lower voltage. 3-level buck converters, for example, generate the $V_{IN}/2$ voltage over its flying capacitor. This level together with ground and $V_{IN}$ voltages is used for the power conversion.

### 8.3.1 Switched-Capacitor Converter with 2:1 Conversion Ratio

SC converters can achieve the highest theoretical power efficiency at a specific value of $V_{IN}/V_{OUT}$ ratio for the following two reasons:

- Switching losses can be minimized by reducing the time periods with a non-zero voltage on the switch and a non-zero current across it.
- The process of charging capacitors with a voltage source or with other capacitors approaches an efficiency of 100% when the initial condition on the capacitor approaches its final voltage value.

Figure 8.12 shows the schematic of a SC 2:1 voltage divider. It comprises four switches S1–S4 and one flying capacitors $C_F$.

Its basic operation consists of two phases alternating at frequency $f_{SW}$ with nominally 50% duty cycle: during one phase (Fig. 8.12a), both the flying capacitor $C_F$ and the output capacitor $C_{OUT}$ are connected in series and charged by $V_{IN}$; during the other phase (Fig. 8.12b), $C_{OUT}$ is charged by $C_F$ by placing $C_F$ in parallel with $C_{OUT}$. Therefore, the steady-state behavior of the SC converter can be modeled by a transformer with a ratio equal to the no-load conversion ratio ($V_{OUT}/V_{IN} = 1/2$) and an output resistance:

$$R_{OUT} = \frac{1}{4C_F f_{SW}} + \frac{R_S}{2} \tag{8.1}$$

with $R_S$ due to the ON resistance of the switches and PCB routing.

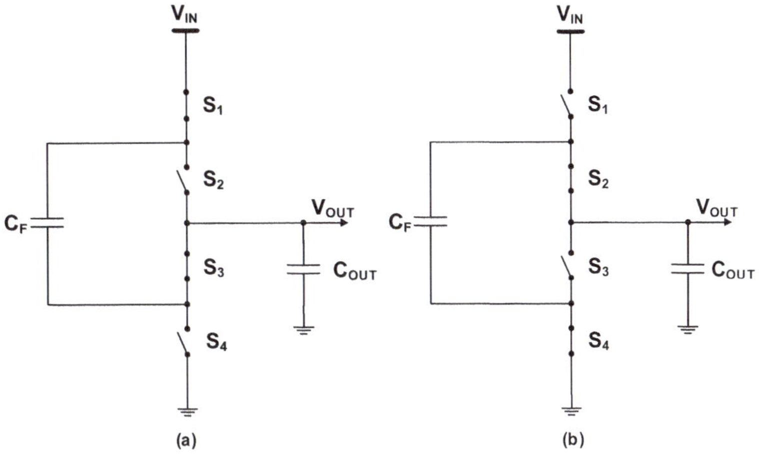


**Fig. 8.12** Switched-capacitor converter and its two operation phases: (**a**) $C_F$ connected in series to $C_{OUT}$; (**b**) $C_F$ connected in parallel to $C_{OUT}$

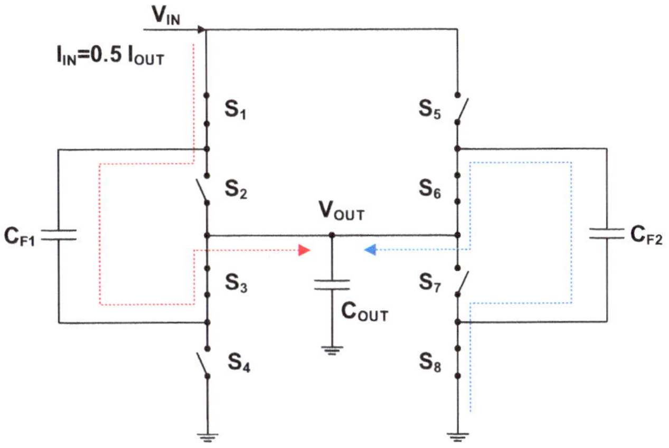


**Fig. 8.13** Interleaved switched-capacitor topology

In the topology of Fig. 8.12, the converter input current is equal to the output current for 50% of the time. This discontinuous input current causes unwanted voltage ripple at the input of the converter. The solution shown in Fig. 8.13 adopts an interleaved topology [7]. The two interleaved circuit parts operate in alternating phase, guaranteeing a constant input current of 0.5 $I_{OUT}$.

Figure 8.14 shows the efficiency measurement of a 2:1 SC converter [7] suitable for applications supplied by a dual (2S) Li-ion or Li-polymer stacked cell battery pack. The input voltage range goes from 5.5 to 10.5 V and the maximum output current is 10 A. The SC power converter operates with 98% efficiency for loads up to 3 A.

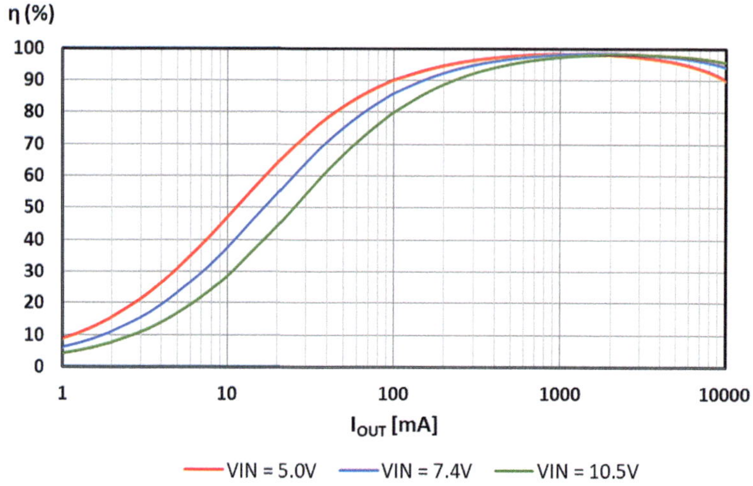


**Fig. 8.14** 2:1 SC converter efficiency in fixed frequency mode

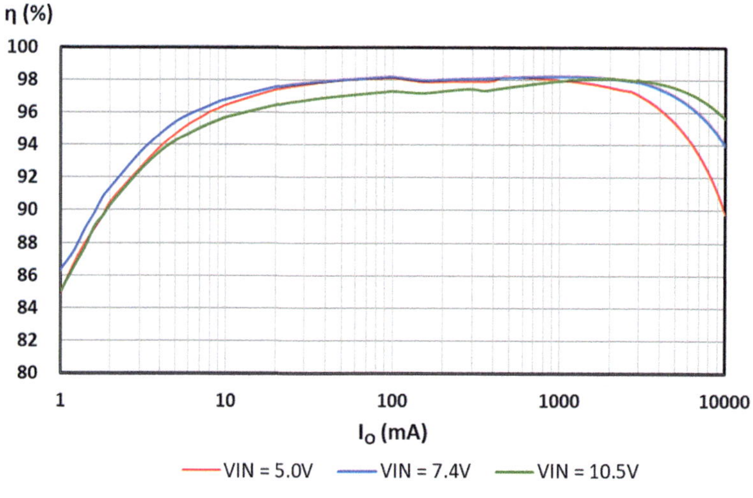


**Fig. 8.15** 2:1 SC converter efficiency in automatic frequency mode

The SC converter can operate at a fixed frequency of 500 kHz (Fig. 8.14) or in automatic frequency mode (Fig. 8.15). In automatic frequency mode, the switching frequency does not stay constant; the converter operates in discontinuous conduction mode in order to improve efficiency in light load conditions.

### 8.3.2 3-Level Buck Converter

Figure 8.16 shows a 3-level buck topology. It requires four switches S1–S4 and a flying capacitor $C_F$.

The switching node $V_X$ swings between $V_{IN}$, $V_{IN}/2$, and 0. Therefore, the required $V_{DS}$ voltage rating of the MOSFET devices is $V_{IN}/2$ instead of $V_{IN}$ as in conventional 2-level buck converters. The converter goes through 4 operation states while developing a voltage $V_{IN}/2$ over $C_F$. For output voltages below $V_{IN}/2$ (above $V_{IN}/2$) the converter goes through the sequence of states (a) → (b) → (c) → (b) ((d) → (a) → (d) → (c)). Since during each switching period the inductor is magnetized twice (by $V_{IN}$ if $V_{OUT}>V_{IN}/2$ or by the flying capacitor if $V_{OUT}<V_{IN}/2$), the frequency of the inductor current ripple is twice the switching frequency. Hence, the required inductance is reduced for a given target current ripple.

Figure 8.17 shows the measured efficiency of a 3-level buck converter switching at a frequency of 500 kHz and using an inductor of 0.47 μH and a flying capacitor $C_F$ of 2 × 22 μF. With an input voltage of 9 V and an output voltage of 4.4 V, the converter achieves a peak efficiency of ~96%. The efficiency advantage of the multi-level converter is evident by comparing it to a conventional buck converter operating at the same input and output voltages and with the same inductor. The conventional buck converter achieves a peak efficiency of ~92% while switching at 1.5 MHz. The two converters have been implemented in the same technology but use devices with different voltage ratings: while the conventional buck requires high voltage devices, the 3-level buck converter relies on low voltage devices. Consequently, even if the 3-level converter uses four devices (with two devices in series during each phase), the power FETs' area increases only by 80% compared to the 2-level converter.

## 8.4 Conclusions

The large parameter design space in wearables and IoT devices demands a power management, which addresses both a very low standby power consumption in, e.g., sensor applications, and a highly efficient high-power delivery for data processing, computational power, and transceivers, which can provide tens of Watts with negligible power losses. A low standby current consumption in the sub-microampere region is achieved by advanced design techniques for the analog core circuitry of power management ICs. This work showed power efficient implementations of self-biased always ON circuits. A reference circuit, pre-regulator, and a battery monitor utilize the work function difference of a pair of MOS devices as reference generators, which are suitable to be operating with a current consumption as low as 100 nA per block. A supply concept for switching blocks, like interfaces, is based on sub-regulators. They are designed for the current demand of each circuit block, avoid external buffer capacitors, and significantly reduce IQ and implementation area. Low-dropout regulator concepts, used for power delivery to IoT devices, are

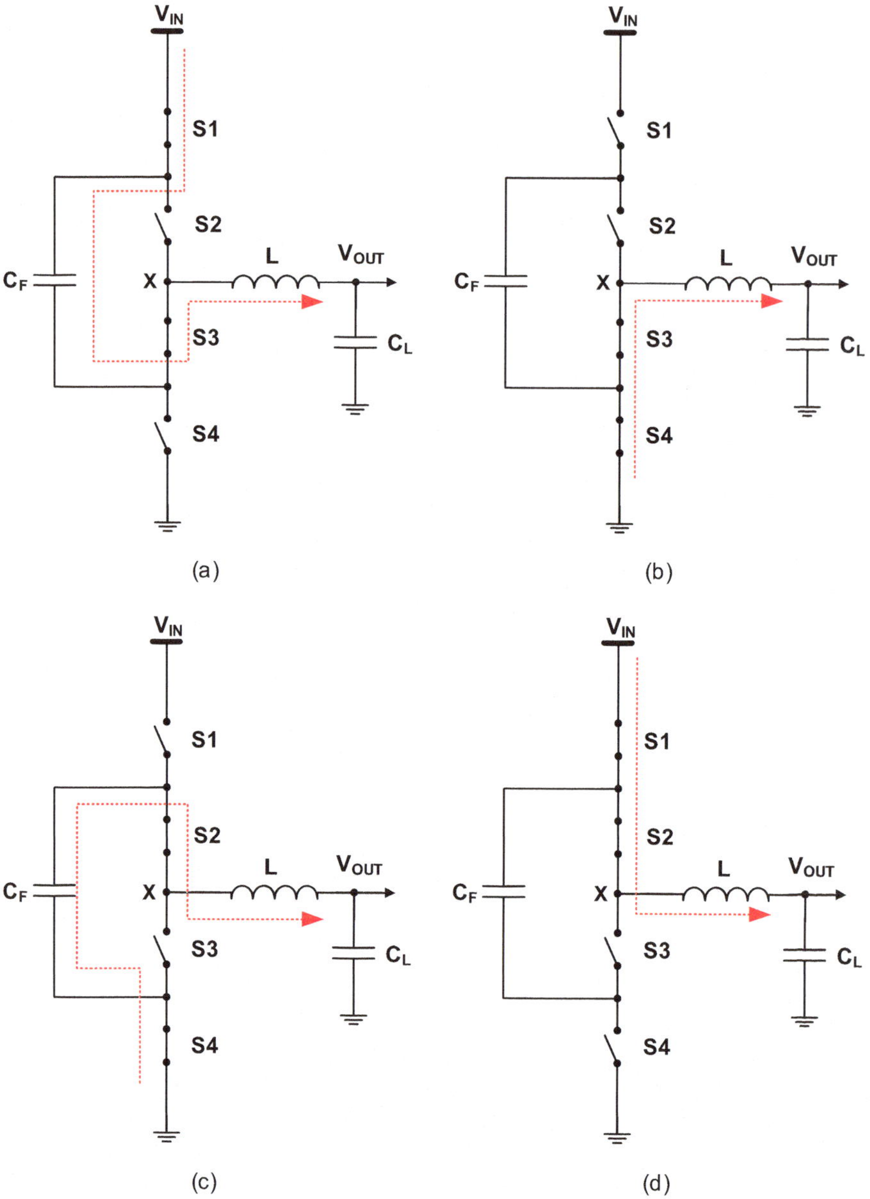


**Fig. 8.16** 3-level buck converter and its four operation phases

proposed, which reduce the dropout current consumption from milli-ampere to a few micro-amperes by utilizing starved current buffers.

This paper presented converter topologies applied to two typical power applications, which are utilized for high-power conversion, especially suitable for higher

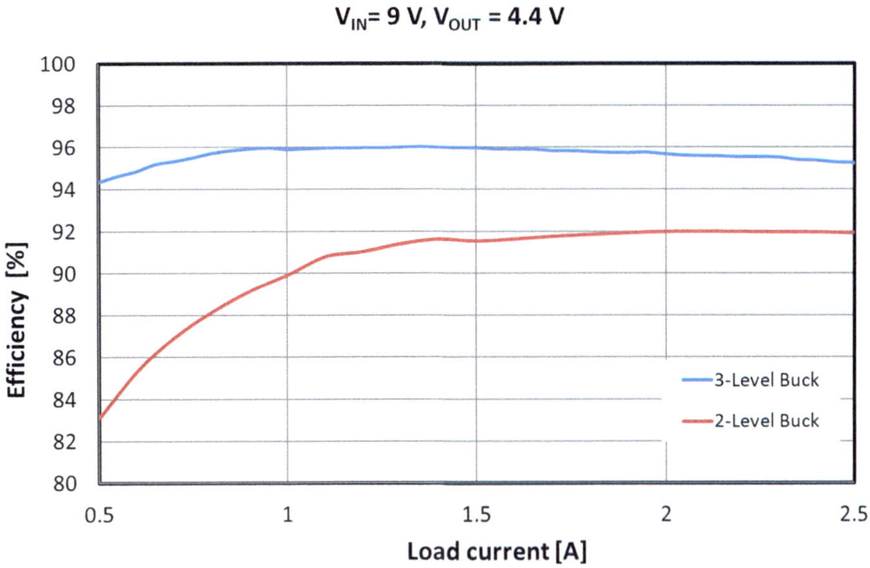


**Fig. 8.17** Efficiency measurement of a 3-level converter and a conventional 2-level buck converter operating at $V_{IN} = 9$ V and $V_{OUT} = 4.4$ V

supply voltage range up to 10 V: a 2:1 switched-capacitor converter used to efficiently (~98%) generate an intermediate voltage in power delivery trees, a low-power buck converter to generate the final low-power voltage rail achieving a peak efficiency above 90%, and a 3-level buck converter (~96% efficiency) used for efficient charging of Li-ion batteries.

## References


1. H. Bauer, M. Patel, J. Veira, The internet of things: sizing up the opportunity (2014)
2. D. Serpanos, M. Wolf, *Internet-of-Things (IoT) Systems* (Springer, Cham, 2018)
3. M.D. Saulles, *The Internet of Things and Business* (Routledge, Columbia, 2016)
4. Dialog datasheet, DA9313, high efficiency dual cell switched capacitor divider, in *Dialog Semiconductor* (2018). Available https://www.dialog-semiconductor.com/sites/default/files/da9313_datasheet.pdf
5. M. Al-Shyoukh, A. Kalnitsky, A 500nA quiescent current, trim-free, ±1.75% absolute accuracy, CMOS-only voltage reference based on anti-doped N-channel MOSFETs, in *Proceedings of the IEEE 2014 Custom Integrated Circuits Conference, San Jose, CA* (2014), pp. 1–4
6. M. Seeman, A design methodology for switched-capacitor DC-DC converters, University of California, Berkeley, Technical Report No. UCB/EECS-2009-78, 2009
7. L. Chang et al., A fully-integrated switched-capacitor 2:1 voltage converter with regulation capability and 90% efficiency at 2.3A/mm$^2$, in *IEEE Symposium on VLSI Circuits* (2010)
8. T.A. Meynard, H. Foch, Multi-level conversion: high voltage choppers and voltage-source inverters, in *IEEE Power Electronics Specialists Conference* (1992), pp. 397–403

9. V. Yousefzadeh et al., Three-level buck converter for envelope tracking applications. IEEE Trans. Power Electron. **21**(2), 549–552 (2006)
10. W. Kim et al., A fully integrated 3-level DC-DC converter for nanosecond-scale DVFS. IEEE J. Solid-State Circuits **47**(1), 206–219 (2012)
11. Dialog datasheet, DA9230, ultra-low quiescent current buck, in *Dialog Semiconductor* (2018). Available https://www.dialog-semiconductor.com/sites/default/files/da9230_datasheet_2v4.pdf

# Chapter 9
# Current Sensing Techniques: Principles and Readouts

**Mahdi Kashmiri**

## 9.1 Introduction

Current is the electrical parameter associated with the flow of charge, which runs through electrical circuits. These usually involve a supply, i.e., a battery and a load such as an electronic device or an electromechanical actuator such as an electric motor. In most cases, current measurement has many uses in monitoring, control, protection, and prediction. Current sensors vary in regard with their underlying physical principle of operation, dynamic range, accuracy, size, complexity, and isolation.

The effect of current is usually sensed through a transduction mechanism. This transduction is enabled by either of the following principles: Ohm's law, Faraday's law of induction, or Ampere's law through magnetic sensing (Fig. 9.1). In the case of Ohm's law a resistor is placed in series with the current and the voltage drop across the resistor is measured. This is an invasive method as the resistor contacts the conductor. The other two methods are non-invasive as they provide contactless current sensing. In case of Faraday's law of induction a transformer effect is used such that the main current carrying conductor acts as primary and a secondary coil wound around it picks up AC induction due to the AC current running in the primary. In case of magnetic based current sensors, Ampere's law denotes that the current running through a conductor results in a magnetic flux at a distance from the conductor that is proportional to the current and inversely proportional to the distance. This magnetic flux can then be measured by means of a magnetometer [1–3].

M. Kashmiri (✉)
Robert Bosch Research and Technology Center, Sunnyvale, CA, USA
e-mail: Mahdi.Kashmiri@us.Bosch.com

© Springer Nature Switzerland AG 2020

A. Baschirotto et al. (eds.), *Next-Generation ADCs, High-Performance Power Management, and Technology Considerations for Advanced Integrated Circuits*,
https://doi.org/10.1007/978-3-030-25267-0_9

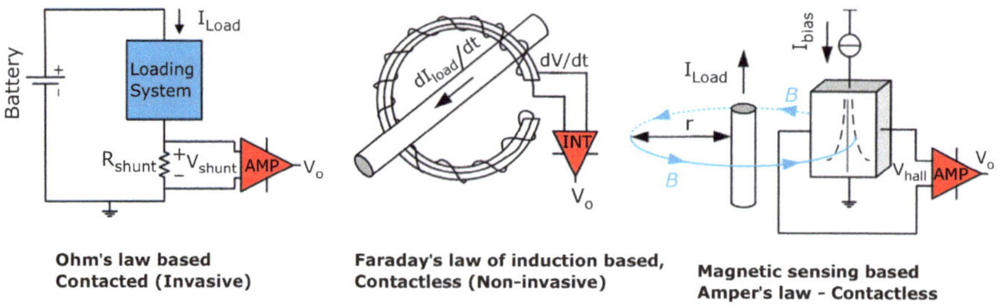


**Fig. 9.1** Physical underlying principles of current sensing

After this introduction section, an overview of some of the current sensing applications will be provided in Sect. 9.2. Section 9.3 discusses the principles of the two mainstream current sensing techniques: contacted and contactless. Section 9.4 provides an overview of some of the readout techniques for contacted current sensors (current sense amplifiers). Section 9.5 introduces some of the contactless current sensors (magnetic based such as Hall effect, fluxgate, etc.) The paper ends with conclusions in Sect. 9.6.

## 9.2 Overview of Applications

Some of the main applications of current sensing are in battery monitoring systems (BMS), which can be found in automotive and consumer spaces. Furthermore, automotive converters, motor control, and switching power supplies are some of the other key applications for current sensors [4–16].

### *9.2.1 Battery Monitoring Systems (BMS)*

Battery operated systems require a control of the charge and discharge scheme of the battery and the interaction with the charger. This is done by monitoring the state of charge in the battery through voltage, temperature, and current measurements. A current sensor measures the charge entering or exiting the battery (Fig. 9.2). A charging profile ensures safe and fast charging as well as battery life maximization through operating it within a limited range of the capacity (e.g., 20% and 80%).

An automotive battery monitor system also deals with operations such as plugin and regenerative charging, implementing accurate fuel gauge function, and providing over current protection (Fig. 9.3). Dynamic range, isolation from high voltage lines, and accuracy over the automotive temperature range and reliability are some of the challenges for these current sensors (Table 9.1).

Consumer battery monitor systems used in handheld devices or power tools usually deal with Li-Ion batteries at lower voltage and power ranges, e.g.,

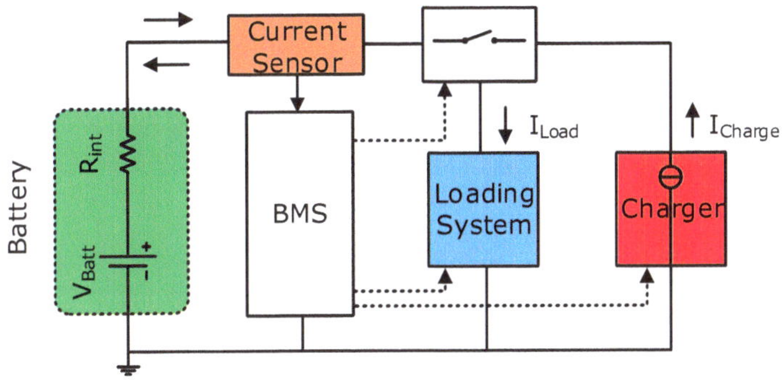


**Fig. 9.2** Battery monitoring systems

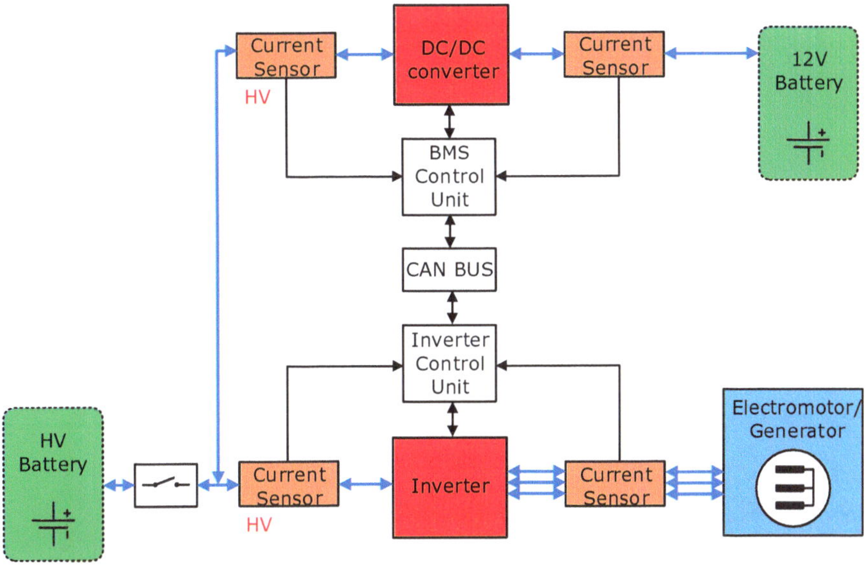


**Fig. 9.3** Current sensors in automotive battery monitor systems

**Table 9.1** Electro-mobility typical voltage and power ranges

| Application | Voltage (V) | Power (kW) |
|---|---|---|
| Electric vehicles | 400800 (near future) | 100 |
| Hybrid vehicles | 90 | 20 |
| Mild hybrid vehicles | 48 | 15 |
| Regular auto battery | 12 | 2–3 |
| e-Bikes | 36 | 0.5 |

3–30 V and few mA–few A. The charge phases should be controlled for thermal management, life time extension, and fast charging. The typical charging phase (Fig. 9.4) involves a pre-charge, then a constant current and fast charge phase followed by a constant voltage tapering [13].

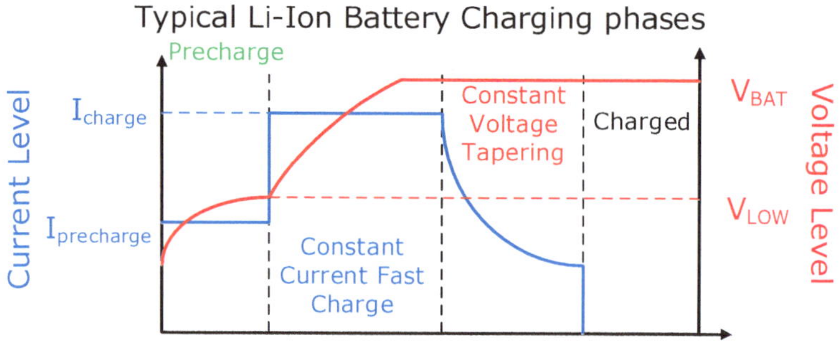


**Fig. 9.4** Typical Li-Ion battery charging phases

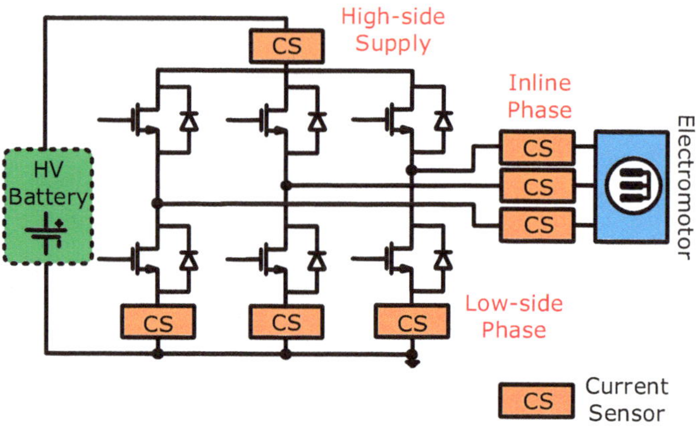


**Fig. 9.5** Current sensors in electric motor drive

## 9.2.2 *Motor Control Applications*

The other application for current sensors is in motor control. Electric motors are typically controlled electronically sometimes by means of a three-phase bridge circuit providing modulated drive to the three-phase lines of the motor. This usually involves Pulse-Width Modulated (PWM) type signals (Fig. 9.5). The current monitoring at various nodes of such system is essential: at the supply line, at the bridge branches as well as the phase lines for the control algorithm. The latter is challenging as the current sensor has to measure current while exposed to the PWM line voltages [14–16].

## 9.2.3 *Converters (Automotive and Switching Supplies)*

Finally, current sensors have their application in converters used in automotive and switching power supply domains. Automotive converters involve: DC/DC converters between HV and LV batteries, AC/DC converters in plugin charging or regeneration, and DC/AC converters for motor drive. Usually current is sensed for the purpose of load regulation through current feedback mechanisms. Efficiency of the converters is a key metric that should not be affected by the losses due to the current sensor's operation. Furthermore, high-voltage isolation as well as dynamic range and bandwidth are some of the challenges for the current sensors in converters.

## 9.3 Overview of Current Sensing Methods

The two mainstream methods for current sensing are called: Contacted, where the sensor is connected in series with the load and thus contacting the conductor and Contactless, where the effect of current is measured indirectly without physically touching the conductor (Fig. 9.6). The former method is simple and low cost but provides no isolation. The latter provides galvanic isolation and hence desired for application with high voltage lines; however, it entails cost and complexity.

### *9.3.1 Contacted Current Sensing*

This method relies on measuring the voltage drop across a resistor placed in series with the current. An amplifier connected in parallel (shunt) amplifies this voltage, hence these sensors are also called "shunt current sensors". Some of their benefits include low cost and simplicity as well as capability to measure from DC to relatively high bandwidths. Their drawbacks are usually related to their lack of isolation in high voltage applications and the cost and power losses of the sensing resistor (shunt resistor).

Depending on the location of the shunt resistor, these sensors are divided into two categories: when the resistor is at ground side: Low-side sensors; when the resistor is located at supply side: High-side sensors [17–21].

In the case of low-side sensors (Fig. 9.7) the shunt amplifier senses a low input voltage, hence more simple architectures such as single-ended, non-inverting amplifiers could be adopted. Some of the considerations for low-side sensors involve

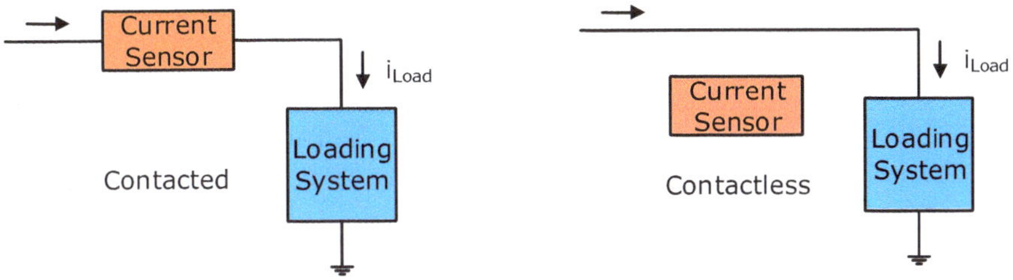


**Fig. 9.6** Contacted and contactless current sensors

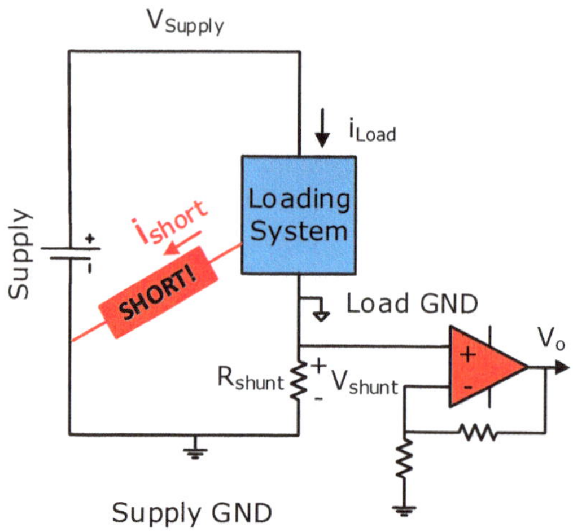


**Fig. 9.7** Low-side shunt current sensor

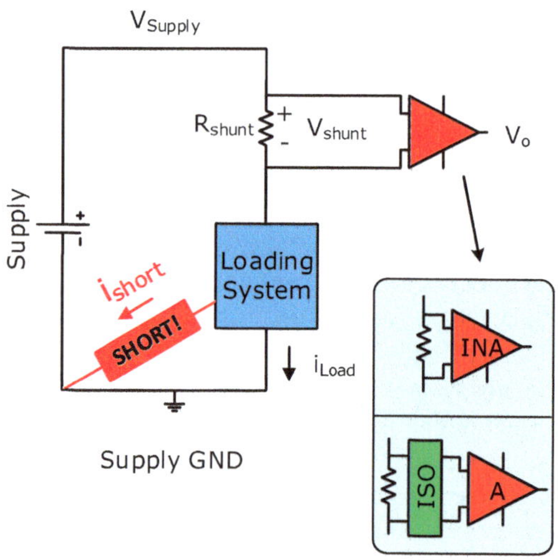


**Fig. 9.8** High-side shunt current sensor

the fact that the load ground is not at zero Volt due to the voltage drop across the shunt. Furthermore, short circuit currents from the load to supply ground (could be the system chassis) could bypass the shunt resistor and will not be sensed.

In the case of high-side sensing (Fig. 9.8), the shunt resistor is placed between the supply and the supply terminal of the load. The load and supply grounds are now at the same potential and any shorts from load to system chassis (supply ground) can be monitored as the short circuit current passes the shunt resistor. Usually the considerations with these sensors involve the readout complexity due to the requirements on the shunt amplifier that it has to access a small differential voltage across the resistor at a high voltage common-mode.

The shunt resistor itself has a few requirements to take into account. This involves some key electrical specifications such as resistance, max current rating

**Table 9.2** Some shunt resistor ranges

| Type | Current (A) | Temp. coeff. (ppm/K) | Power | Res | Notes |
|---|---|---|---|---|---|
| PCB trace | <10 | ~1000 | 100's mW | mΩ | Low cost |
| Surface mount/ integrated | <10 | ~1000 | 10's W | μΩ to mΩ | Low cost |
| Exposed element/ wire | Few 100 | <100 | 10's W | μΩ | |
| High power | Few 1000 | | Few kW | mΩ | Heatsink or liquid cooling |

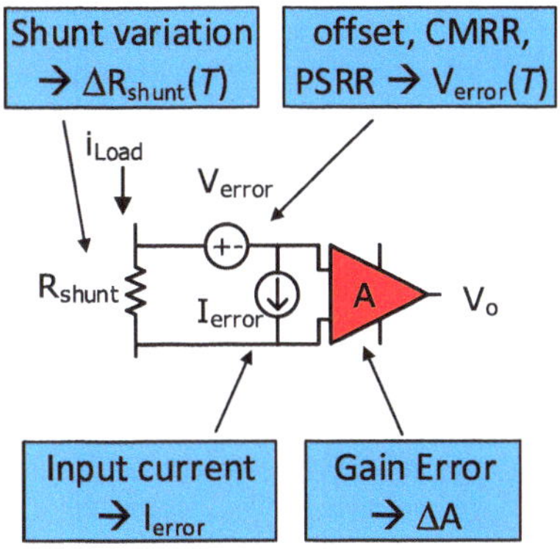


**Fig. 9.9** Shunt current sensor's error sources

(fusing current), continuous and duty cycled power ratings, temperature coefficient, inductance (for high speed applications), etc. Based on the current specifications, cost, and complexity there are several types of resistors that can be chosen (Table 9.2). Furthermore, it is essential to access the shunt resistor through a four-wire setup where the force and sense terminals are separated (Kelvin sensing) [22].

The accuracy of a shunt current sensor is affected by many parameters, among others the shunt resistor tolerance itself $\Delta R_{\text{shunt}}(T)$, which can be a function of temperature $T$ and lifetime. Furthermore, the current sense amplifier's input referred voltage errors $V_{\text{error}}(T)$ caused by offset, drift, Common-Mode Rejection Ratio (CMRR), and Power Supply Rejection Ratio (PSRR). Other error sources from the amplifier include input bias current, $I_{\text{error}}$ as well as gain error, $\Delta A$ (Fig. 9.9).

### *9.3.2 Contactless Current Sensing*

Contactless current sensing enables indirect sensing with no contact to the conductor, hence providing isolation. This is important especially if the conductor is at hundreds or even thousands of volts and the current sensor is in the low voltage domain.

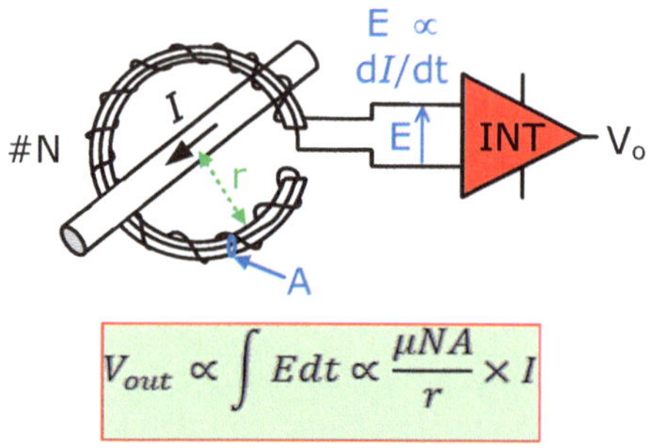


**Fig. 9.10** A Rogowski coil current sensing setup

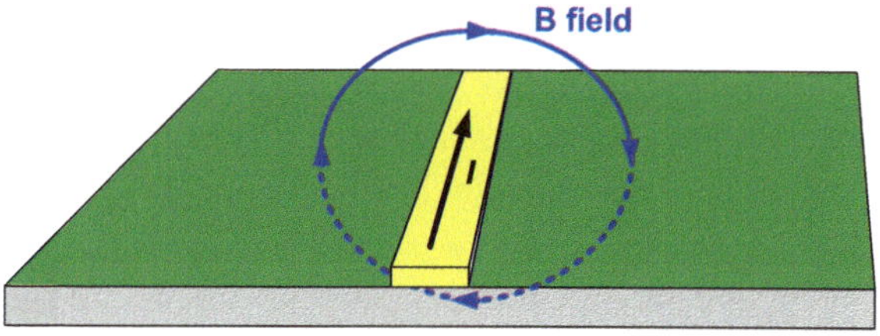


**Fig. 9.11** A PCB trace creating a magnetic field

The physical principles enabling contactless sensing include Faraday's law of induction such as Rogowski coils, which are also becoming integrated in CMOS chips, as well as magnetic sensing methods. A Rogowski coil is a helical coil of a wire with no magnetic core (Fig. 9.10). It surrounds the conductor carrying current. When AC current passes the primary, it creates a varying flux in the coil, which then induces a voltage across it. This voltage is proportional to the derivative of the current in the primary, which can be integrated to produce a voltage that is proportional to the current [2].

These sensors are relatively linear and due to the lack of magnetic core are not prone to saturation making relatively low cost and high bandwidth current sensors specifically for clamp on applications. Their drawback is that they are not suitable for DC current measurement and their integrator readout usually limits speed and can introduce errors such as drift.

For the case of magnetic based contactless current sensor, Ampere's law defines their physical underlying principle. A current carrying conductor with current $I$ creates a magnetic field $B$ at distance $r$. The conductor could be a wire or a PCB trace (Fig. 9.11). This field is proportional to the current and can be sensed with a magnetometer that senses "Strength" and "Direction" of the field. Magnetic based current sensors can measure both DC and AC currents at a range of accuracy and speed determined by the magnetometer. Some of their drawbacks include sensitivity to interference due to stray fields, cost and complexity of both sensors and their readouts.

Magnetometers are also available as integrated sensors in CMOS technology and have wide application in current sensing (Fig. 9.12). Some of the magnetometers used in current sensing include: Hall-effect magnetometers, which could be made of a semiconductor plate biased with a current. A magnetic field applied perpendicular to the plate causes a voltage induced at opposite sides, hence behaving like a

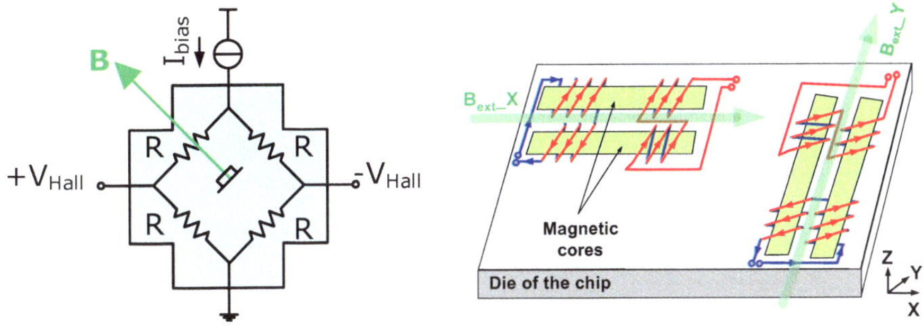


**Fig. 9.12** Hall and fluxgate magnetometers

**Fig. 9.13** Open-loop magnetic current sensors

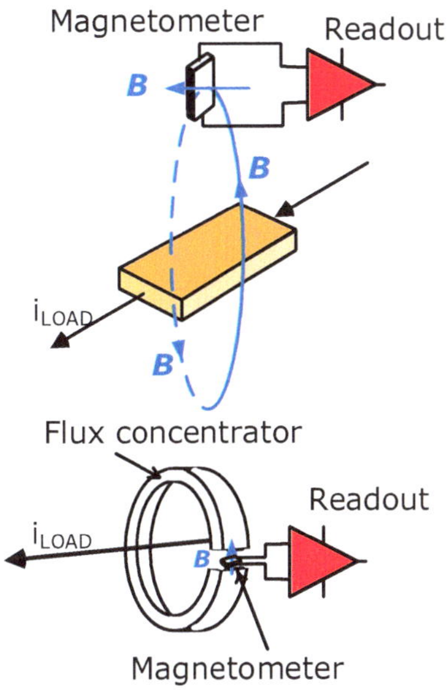


Wheatstone bridge. Also magneto-resistive sensors are used in current sensing. These are resistors whose resistance changes as a function of the applied magnetic field. Some of their variants include AMR, GMR, and TMR magnetometers. Finally fluxgate magnetometers are also available as integrated sensors, including magnetic cores with excitation and sense coils. AC excitation in the excitation coil couples to the sense coil, which gets imbalanced in presence of an external field resulting in an AC signal appearing across the sense coil. The magnetic sensors will be discussed further in the following section [23–43].

One important consideration when using magnetometers as current sensors is their configuration in terms of open-loop and closed-loop operation.

In an open-loop sensing configuration (Fig. 9.13), the sensor output is a direct function of the total *B* field induced by the current. These are attractive due to their

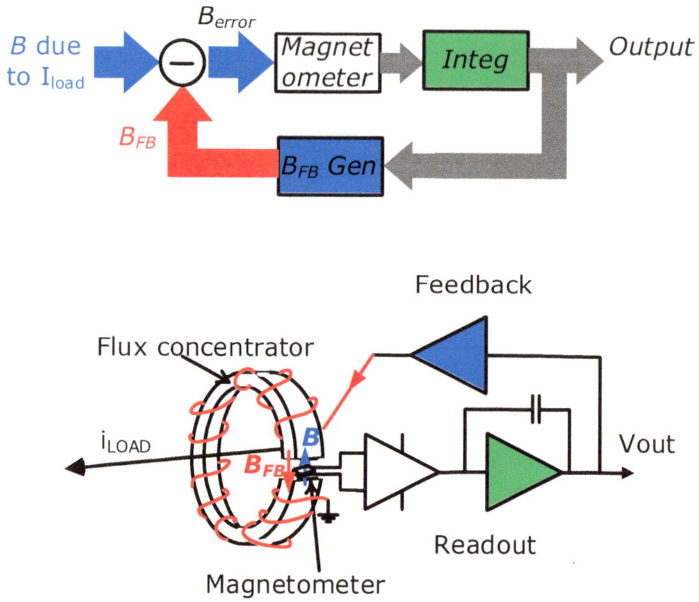


**Fig. 9.14** Closed-loop magnetic current sensors

simplicity and lower level of cost. The drawback of open-loop sensing is that the current sensor's transfer characteristic will be directly affected by the characteristic of the magnetometer, which in many cases can suffer from non-linearity and/or lack of dynamic range due to saturation effects. This could limit the performance of the current sensor significantly.

In contrast to that, a closed-loop sensing configuration can be adopted (Fig. 9.14), where a feedback loop in the magnetic domain ensures that the magnetic field created by the current of interest is compensated by a feedback magnetic field. When the loop is closed, the feedback forces the loop error signal, being the difference between the field due to the current and the feedback field ($B_{FB}$) to be driven to zero. This error field can then be sensed by a magnetometer, translated to an electrical signal, filtered and used to drive the feedback path. As a result of this configuration, the magnetometer senses a near zero signal and hence the non-linearity and lack of dynamic range in its transfer characteristic will be suppressed by the loop. This however comes at more cost, complexity, size, and power consumption.

## 9.4 Current Sense Amplifiers: Readout for Contacted Sensors

A shunt current sensor's readout is made of a current sense amplifier that has to access and amplify the shunt resistor's voltage to a level suitable for digitization by an analog to digital converter (ADC). The system dynamic range determines the accuracy requirements on the amplifier and specifically the need for reduction of losses requires ever smaller shunt resistors that translates to smaller voltage drop across the resistor. This will result in even more stringent requirements on the performance of the sense amplifier [18–31].

Some of the key performance metrics of the sense amplifier include: low offset, low dependence to input common-mode voltage (common-mode rejection), input common-mode range (sometimes beyond the amplifier's supply), gain accuracy, and finally noise performance that can determine the lower bound of sensitivity (Fig. 9.15).

Key input referred errors of the sense amplifier are shown as series error voltage sources and parallel error current sources as shown in the figure. These errors cannot be distinguished from the shunt voltage or the load current. One of the contributors is the offset (both voltage and current). Offset can be calibrated; however, it drifts as a function of temperature and lifetime, which might not be acceptable for some applications. For instance a 1 mV input referred offset is a respectable number; however, in comparison with a 5 mV shunt voltage drop this is equivalent to 20% error.

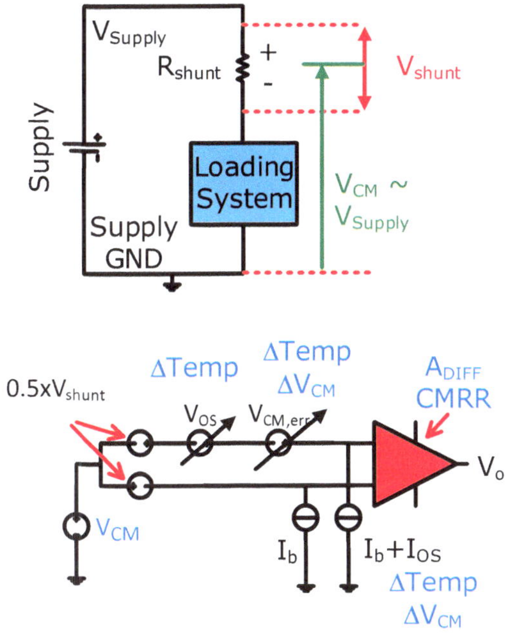


**Fig. 9.15** Current sense amplifier input referred errors

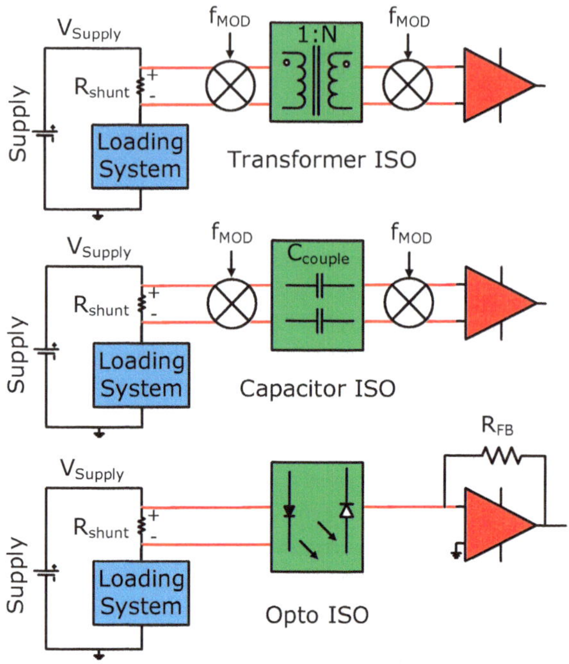


**Fig. 9.16** Isolation amplifiers

Depending on their input stage structure, some amplifiers have input bias current, which in case of mismatch between the two input terminals shows as an input offset current. Furthermore, an amplifier's CMRR (Common-Mode Rejection Ratio) determines how much of the input common-mode variation translates into a differential input referred error voltage. For instance a −100 dB CMRR (10 μV/V) means that a variation of 100 V in the input common-mode level results in 1 mV of input referred error.

Other performance metrics of the sense amplifier include Power Supply Rejection Ratio (PSRR), which is the rate at which power supply variations translate into input referred errors, and gain error. Another important metric is the input common-mode voltage range or CMVR. This has an operational range, within which the amplifier meets its performance specifications and an absolute maximum beyond which safe operation is not guaranteed as break down might occur.

Current sense amplifiers have various configurations. These are isolation amplifiers, difference amplifiers, and instrumentation amplifiers (Fig. 9.16). Isolation amplifiers provide a floating difference amplification and extend the CMVR beyond the amplifier's supply range by means of an isolation barrier that could be a transformer, a capacitive coupling, or even optical coupling. The first two require a modulation scheme to pass the signal through the transformer or capacitive coupling. The optical amplifiers do not provide accuracy and are less common.

The next configuration involves difference amplifiers [18] where a pair of resistors access the shunt voltage differentially from one side and the amplifier's

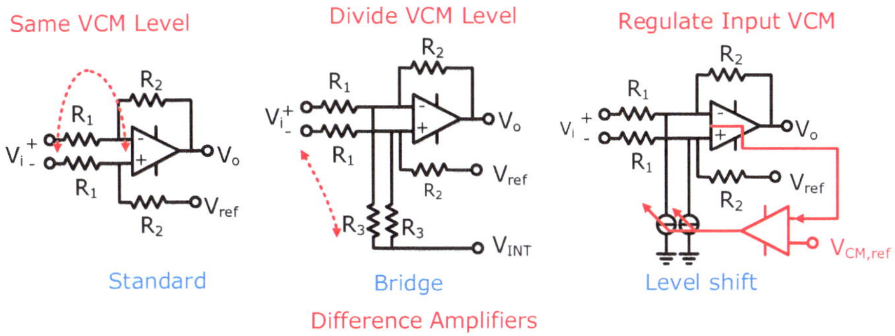


**Fig. 9.17** Difference amplifiers as current sense amplifiers

differential input (virtual ground) from the other side (Fig. 9.17). The shunt voltage is hence translated into a differential current and amplified to an output voltage by the amplifier. These amplifiers include the standard difference amplifier, where the shunt common-mode and amplifier common-mode are at the same level, the bridge amplifier, where the shunt common-mode voltage is reduced through a divider to a lower level tolerated by the amplifier, and finally level shift amplifier, where a feedback controls current sources that regulate the amplifier input common-mode level to a desired reference voltage [18].

The last category of current sense amplifiers includes instrumentation amplifiers (INA's). These combine accurate voltage gain, low offset, and high CMRR (Fig. 9.18). The classic INA is the three op-amp INA, where the differential input is reflected through two buffers across a resistor and amplified by a difference amplifier at the output. Its CMVR is limited by that of the input buffers and its CMRR is determined by matching. The next type of INA used in current sensing is the current feedback instrumentation amplifier (CFIA). These rely on their feedback being closed in the current-domain through dedicated input and output (feedback) transconductors. This configuration allows different common-mode levels at the input and output of the amplifier. The other type of INAs used specifically for high-side sensing is capacitive coupled instrumentation amplifier (CCIA). Inputs are capacitively coupled to the high side and a chopper (passive modulator) turns the shunt voltage into AC. The CMVR is defined through the isolation level provided by the capacitors.

## 9.5 Magnetic Sensors as Contactless Current Sensors

As discussed earlier, contactless current sensors mainly rely on two physical principles: the Faraday's law of induction or inductive based current sensors and magnetic based current sensors. Inductive based sensors can only sense AC currents; however, magnetic sensors are capable of both DC and AC at various dynamic

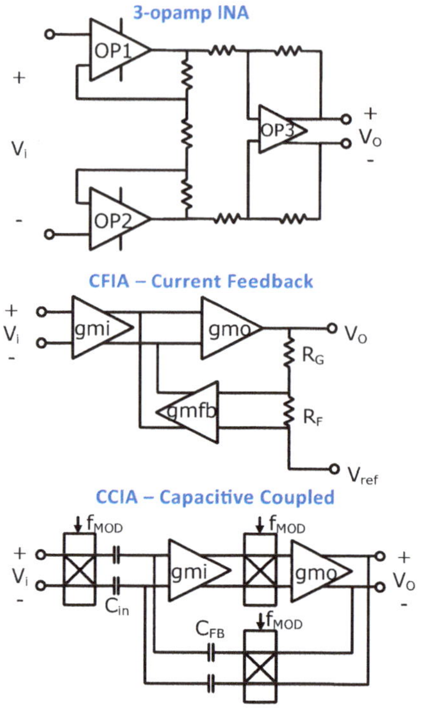


**Fig. 9.18** Instrumentation amplifiers used for current sensing

ranges and accuracies determined by the magnetometer. Inductive sensors have been shown to be combined with magnetic sensors in the form of hybrid sensors, in order to have very wide-band current sensors where low frequency response is determined by the magnetometer and the high frequency response by the inductor. In this section of the paper we will only discuss some of the magnetic sensors with some references provided to the hybrid sensors [32–63].

## 9.5.1 Magnetic Current Sense Modules

Magnetic based current sensors rely on Ampere's Law. This assumes the current conductor to be an infinitely long wire with negligible diameter. In real life we design current sensing modules, which means we deal with conductors and magnetometers with finite geometrical dimensions, positions, and given axis of sensitivity. This means that magnetic based current sensing boils down to the design of a module, which houses the conductor and the magnetometer at a defined position with given dimensions and geometries. This requires 3D vector field calculations to calculate the effect of geometrical parameters to determine sensitivity and dynamic range of the sensor.

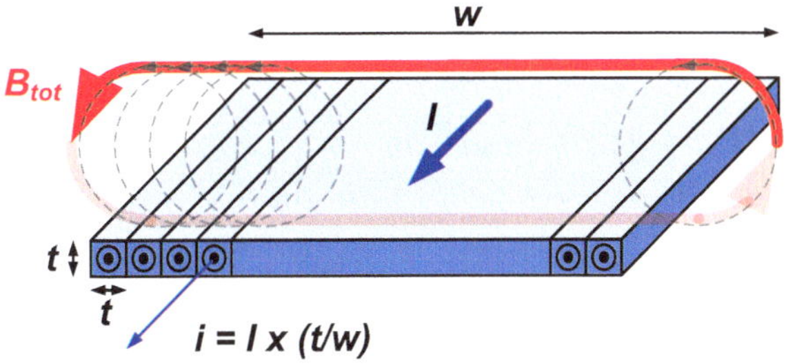


**Fig. 9.19** Real life geometries of conductors require finite element analysis to calculate magnetic field

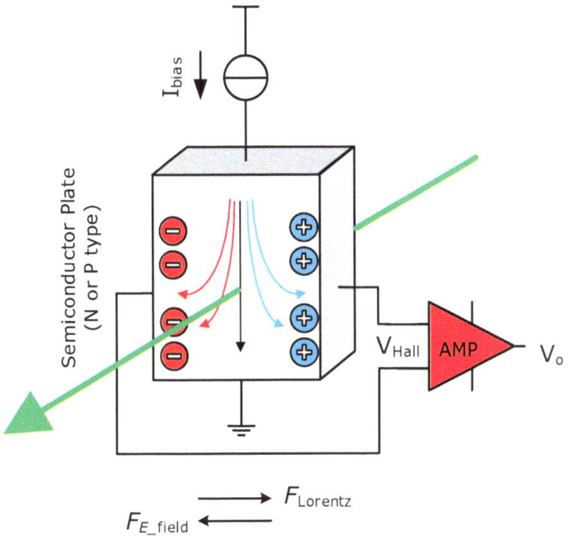


**Fig. 9.20** Hall effect

A current conductor can be broken into many smaller wires and the magnetometer can also be broken into many smaller point sensors (Fig. 9.19). The 3D vector effects of all the fields created by all small wires to the direction of sensitivity of each of those small magnetometers are then calculated, including magnetic field from any neighboring conductors creating interference. This allows the dynamic range of the current to be mapped to the magnetic dynamic range of the magnetometer through the 3D design of the structure. The misalignments in manufacturing will then result in a gain error that should be calibrated at the system level.

### *9.5.2 Hall Sensors*

Hall sensors are one of the popular magnetometers used in contactless current sensing. They operate based on the effect discovered by Edwin Hall. A Hall sensor is a plate, which could also be of a semiconductor material, where a current bias is applied across one side of it (Fig. 9.20). When a magnetic field is applied orthogonal

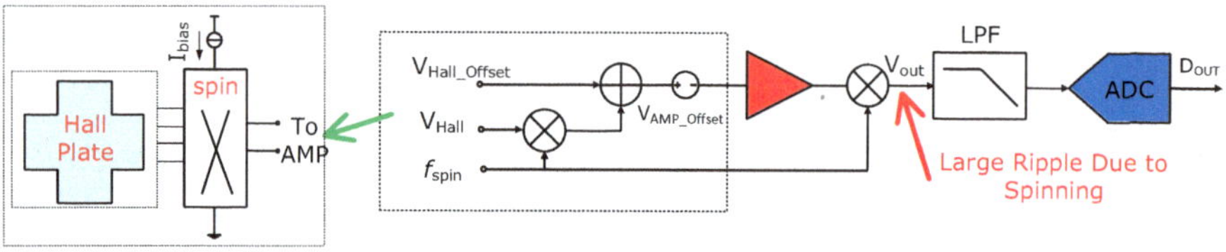


**Fig. 9.21** Current spinning hall sensor readout

**Fig. 9.22** Differential sensing readout with two hall sensors for over-the-trace current sensing

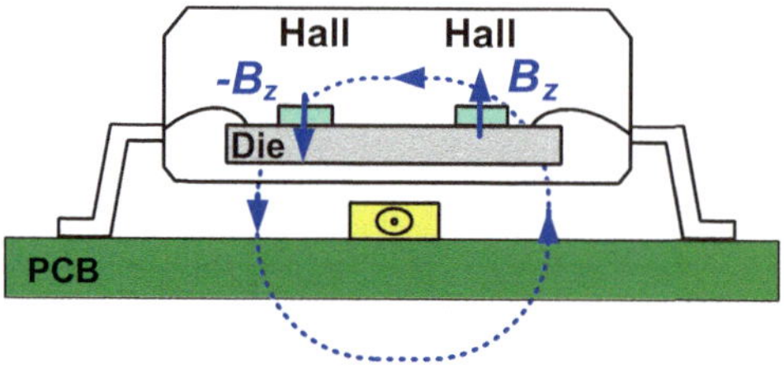


to the plate, a Lorentz force inside deflects the path of charge resulting in charge buildup that eventually results in an electric field that counter balances the Lorentz force and can be measured across the opposite sides of the plate.

Hall sensors are CMOS compatible and hence provide a low-cost solution for integrated contactless current sensors. One consideration with their use is due to their offset, which is a result of the limited homogeneity of the N-well plates. This results in a structure that resembles an imbalanced Wheatstone bridge. A well-known offset reduction technique applied to hall sensors is current spinning. This method relies on periodically altering the point where the current is injected and the terminals where the hall voltage is sensed. If this alteration occurs at a rate $f_{spin}$ then the hall signal is up-modulated to that frequency and the offset remains at DC. As shown in the figure below, a demodulator can down-convert the hall signal while the offset of the hall sensor and the readout amplifier gets up-modulated and later removed with a low-pass filter (Fig. 9.21).

Since hall sensors have sensitivity perpendicular to the plate, two hall sensors on the die of a chip can be placed on top of a current carrying trace (Fig. 9.22). With the hall sensors located at two sides of the trace a differential field sensing scheme is created that helps rejecting common-mode fields such as stray fields (Earth magnetic field, etc.)

### 9.5.3 *Fluxgate Sensors*

Fluxgate sensors are more sensitive than Hall and Magneto-resistive sensors as they can provide nT-level offset and noise. Recently, integrated fluxgate sensors (IFG) are introduced, which enable monolithic solutions for contactless current sensors with low-range accuracy [48].

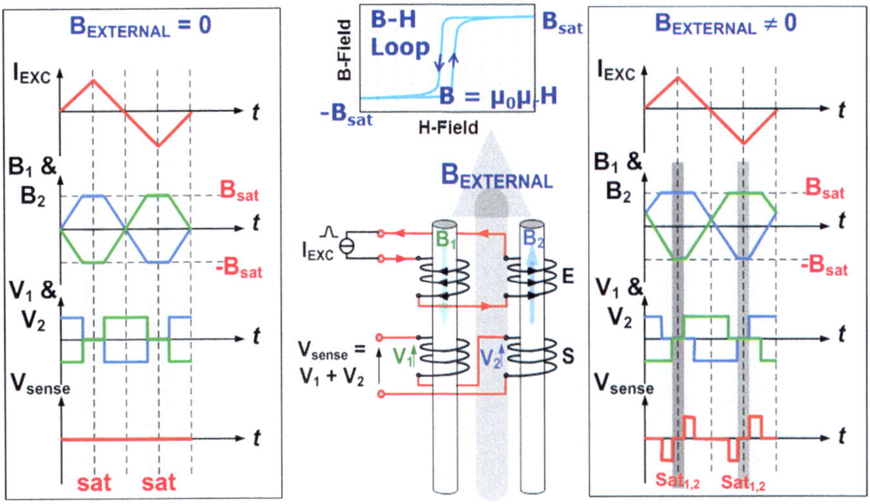


**Fig. 9.23** Time domain operation of a fluxgate sensor

Challenges for their adoption into current sensing applications, include the limited BW of their readouts that have been mainly developed for compass applications (10's Hz). Furthermore, their non-linear transfer characteristic requires the application of linearization techniques such as closed-loop operation.

An integrated fluxgate sensor is a planar structure, which can be integrated into CMOS as backend processing. This usually includes two magnetic cores and up to three coils (Fig. 9.23). An excitation coil drives the magnetic cores periodically from positive saturation to negative saturation. In transitions from saturation to saturation, the core is highly permeable with very strong coupling from excitation to sense. When there is no external field, the two opposite internal fields induced into the two magnetic cores are at opposite directions, inducing opposite polarity voltages across the two half sense coils. These cancel each other resulting in zero total sense voltage. In presence of an external field, one of the cores reaches saturation earlier than the other one (the core with external field in the same direction as its internal field). This results in a net time domain imbalance in the induced voltages, resulting in an AC signal appearing at the sense coil at two times the excitation frequency. This voltage can be demodulated and filtered, whose average will be proportional to the external field. Sometimes a third coil called a compensation coil creates a compensation field in both magnetic cores that is used to cancel the external field in closed-loop readouts.

The sense axis of a fluxgate sensor, being a planar structure in CMOS process is parallel to the die of the chip. Its transfer function has a limited linear range and dynamic range.

Fluxgate sensors can be used with open-loop readout, which involves a demodulator, and amplifier and a filter. This readout has limited dynamic range (typically

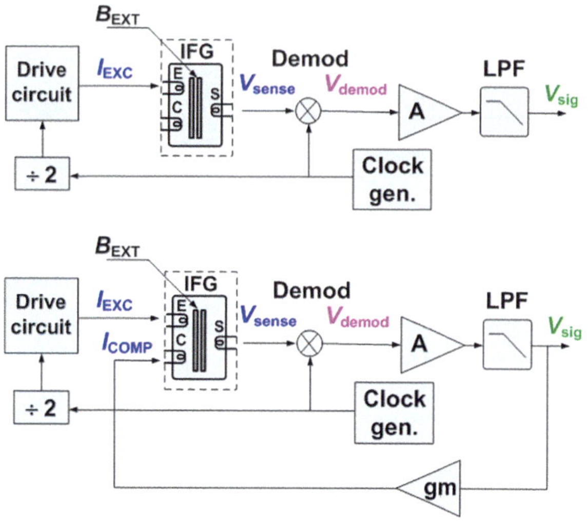


**Fig. 9.24** Open-loop and closed-loop fluxgate readout schemes

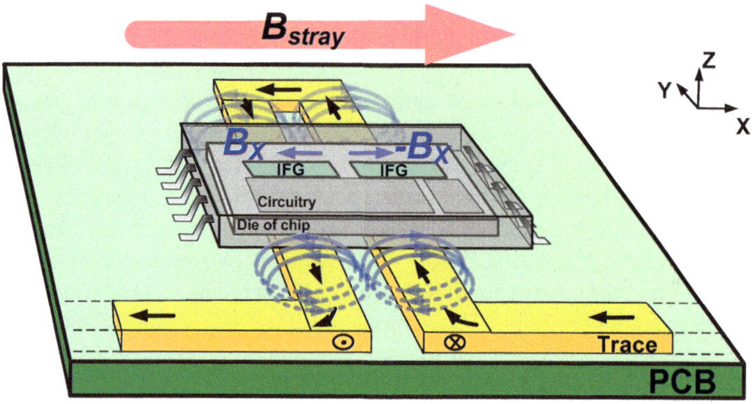


**Fig. 9.25** Differential closed-loop sensing mechanism

<1 mT) and limited linearity, however, is simple and low cost. To enhance linearity and dynamic range a feedback path can be added to drive a current into the compensation coil of the fluxgate sensor that cancels the external filed. This feedback path determines the overall readout linearity and range (Fig. 9.24).

One way of creating a contactless current sense module is to use two on-chip fluxgate sensors to create a fully integrated differential closed-loop sensing mechanism (Fig. 9.25). The current of interest can be routed under the chip including two fluxgate sensors in the form of a U-trace. This could be on a PCB or in the lead-frame of a chip. The two legs of the U carry the current bidirectionally resulting in a differential magnetic field aligned with the axis of sensitivity of the two fluxgate sensors. Closed-loop differential readout on chip compensates the differential magnetic field [48].

## 9.6 Conclusions

Current sensing has various applications in automotive, industrial, and consumer domains as well as many emerging applications [64–71]. Battery monitoring systems, motor control, and converters are some of these applications. The two mainstream configurations of current sensors include contacted (shunt-based) and contactless (induction based or magnetic field based). The former is used extensively due to lower cost and simplicity; however, the latter is required when galvanic isolation is needed accessing high voltage domains. Current sense amplifiers form the readout of the contacted (shunt) current sensors, which have various configurations such as isolation, difference, and instrumentation amplifiers. Magnetic based sensors rely on Ampere's law sensing the magnetic field created by a current. Hall effect, fluxgate, and magneto-resistive are some of the extensively used magnetometers for contactless current sensing. Inductive sensors can only measure AC currents in a transformer configuration; however, they are being integrated together with magnetic based sensors in the form of hybrid sensors that create wide-band contactless sensors.

## References

### *Introduction and Applications*


1. P. Semig, C. Wells, Texas Instruments, A current sensing tutorial, parts 1 to 4. (EE Times, 2012)
2. S. Ziegler et al., Current sensing techniques: a review. IEEE Sensors J. **9**(4), 354–376 (2009)
3. Maxim Integrated, Practical considerations for advanced current sensing in high-reliability systems, Application note 3883. (Sept 2006)
4. T. Regan et al., Current sense circuit collection, Application note 105. (Linear Technology (ADI), Dec 2005)
5. J. Douglass, Linear Technology (ADI), Battery management architectures for hybrid/electric vehicles. (Mar 2009)
6. LTC6802-1 multicell battery stack monitor, datasheet. (Linear Technology (ADI))
7. Automotive current sensing. Eur. J. Power. Electron. (3) (Apr 2008)
8. V. Yang, Peak current mode and continuous current mode DC-to-DC converter modeling and loop compensation design considerations, Technical article. (Analog Devices (ADI))
9. H.J. Zhang, Modeling and loop compensation design of switching mode power supplies, Application note 149. (Linear Technology (ADI), Jan 2015)
10. A.M. Patel et al., Advanced current sensing techniques for power electronic converters, in *IEEE Vehicle Power and Propulsion Conference* (Sept 2007)
11. H.P. Forghani-zadeh et al., Current-sensing techniques for DC-DC converters, in *IEEE MWSCAS* (Aug 2002)
12. B. Lynch, Feedback in the fast lane – modeling current-mode control in high-frequency converters, Application paper. (Texas Instruments), Available online
13. M. Cortez, U. Sengupta, Introduction to battery management, in *Battery Charging – Texas Instruments Training*, https://training.ti.com/sites/default/files/Intro_to_Battery_ Management.pptx

14. J. Lepkowski, Motor control sensor feedback circuits, Application note AN894. (Microchip, 2003)
15. W. Slowik et al., Selected current sensing circuits for motor control applications. Pomiary Automat. Robot. **21**(1), 5–12 (2017)
16. J. Bridgmon, C. Andrews, Current sensing for inline motor-control applications, Application report SBOA172. (Texas Instruments, Oct 2016)

## *Contacted (Shunt-Based) Current Sensing*

17. Texas Instruments, Application note SLYB194d, Current sense amplifiers. (2018), Available online
18. W. Kindt, Current sense amplifiers with extended common mode voltage range, in *Analog Circuit Design*, ed. by M. Steyaert, A. H. Roermund, H. Casier, (Springer, Dordrecht, 2009)
19. Texas Instruments, Reference design, Automotive shunt-based ±500A precision current sensing reference design. (Feb 2017)
20. Maxim Integrated, Tutorial 746, High-side current-sense measurement: circuits and principles. (Nov 2001)
21. N. Aupetit, ST Microelectronics, Application Note AN4835, High-side current sensing for applications using high common-mode voltage. (Nov 2016)
22. Vishay Intertechnology Inc., Shunts, current shunts, and current-sensing resistors, An overview of resistor shunt products for current sensing. (2018)
23. L. Spaziani, Using copper PCB Etch for low value resistance, Unitrode (Texas Instruments) design note, DN71. (2001)
24. F. Witte et al., A current-feedback instrumentation amplifier with 5 μV offset for bidirectional high-side current-sensing. IEEE J. Solid State Circuits **43**(12), 2769–2775
25. F. Rothan et al., A ±1.5% nonlinearity 0.1-to-100A shunt current sensor based on a 6kV isolated micro-transformer for electrical vehicles and home automation, in *IEEE ISSCC* (IEEE, 2011), pp. 112–114
26. Z. Tan et al., A fully isolated delta-sigma ADC for shunt based current sensing. IEEE J. Solid State Circuits **51**(10), 2232–2240 (2016)
27. Q. Fan et al., *Capacitively-Coupled Chopper Amplifiers* (Springer, Cham, 2017)
28. Q. Fan et al., A capacitively coupled chopper instrumentation amplifier with a ±30V common-mode range, 160dB CMRR and 5μV offset, in *IEEE ISSCC* (IEEE, 2012), pp. 374–376
29. S.H. Shalmany et al., A ±5 A integrated current-sensing system with ±0.3% gain error and 16 μA offset from −55 °C to +85 °C. IEEE J. Solid State Circuits **51**(4), 800–808 (2016)
30. S.H. Shalmany et al., A ±36-A integrated current-sensing system with a 0.3% gain error and a 400-μA offset from −55 °C to +85°C. IEEE J. Solid State Circuits **52**(4), 1034–1043 (2017)
31. L. Xu et al., A ±4A high-side current sensor with 25V input CM range and 0.9% gain error from −40°C to 85°C using an analog temperature compensation technique, in *IEEE ISSCC 2018* (IEEE, 2018), pp. 324–326

## *Contactless Current Sensing*

32. D.P. Pappas, *High Sensitivity Magnetic Field Sensor Technology Overview* (National Institute of Standards and Technology, NIST, Boulder). ftp://ftp.boulder.nist.gov/pub/pappas/CNRS-APS%20Sensors/CNRS4.ppt

## *Hall Effect Sensors*

33. Infineon, Application note, Current sensing using linear hall sensors. (Feb 2009)
34. J. van der Meer et al., A fully integrated CMOS Hall sensor with a 3.65μT 3σ offset for compass applications, in *IEEE ISSCC 2005* (IEEE, 2005), pp. 246–247
35. C. Schott et al., CMOS single-chip electronic compass with microcontroller. IEEE J. Solid State Circuits **42**(12), 2923–2933 (2007)
36. M. Motz, A miniature digital current sensor with differential Hall probes using enhanced chopping techniques and mechanical stress compensation, in *IEEE Sensors, 2012* (IEEE, 2012), pp. 1–4
37. P.D. Dimitropoulos et al., A 0.35um-CMOS, wide-band, low-noise HALL magnetometer for current sensing applications, in *IEEE Sensors 2007* (IEEE, 2007), pp. 884–887
38. V. Mosser et al., A spinning current circuit for Hall measurements down to the nanotesla range. IEEE Trans. Instrum. Meas. **66**(4), 637–650 (2017)
39. S. Gambini et al., A 10 kPixel CMOS Hall sensor array with baseline suppression and parallel readout for immunoassays. IEEE J. Solid State Circuits **48**(1), 302–317 (2013)
40. H. Heidari et al., A CMOS current-mode magnetic Hall sensor with integrated front-end. IEEE Trans. Circuits Syst. I Regul. Pap. **62**(5), 1270–1278 (2015)
41. J. Jiang et al., A continuous-time ripple reduction technique for spinning-current Hall sensors. IEEE J. Solid State Circuits **49**(7), 1525–1534 (2014)
42. J. Jiang et al., Multipath wide-bandwidth CMOS magnetic sensors. IEEE J. Solid State Circuits **52**(1), 198–209
43. J. Jiang et al., A hybrid multi-path CMOS magnetic sensor with 76 ppm/°C sensitivity drift and discrete-time ripple reduction loops. IEEE J. Solid State Circuits **52**(7), 1876–1884 (2017)

## *Fluxgate Sensors*

44. C. Maier et al., 2D magnetic micro fluxgate system with digital signal output, in *IEEE ISSCC 1999* (IEEE, 1999), pp. 130–131
45. A. Baschirotto et al., A CMOS 2D micro-fluxgate earth magnetic field sensor with digital output, in *IEEE ISSCC 2007* (IEEE, 2007), pp. 390–391
46. W. Magnes et al., A 92dB-DR 13mW ΔΣ modulator for spaceborn fluxgate sensors, in *IEEE ISSCC 2007* (IEEE, 2007), pp. 388–389
47. F. Gayral et al., A 100Hz 5nT/$\sqrt{}$Hz low-pass ΔΣ servo-controlled microfluxgate magnetometer using pulsed excitation, in *IEEE ISSCC 2007* (IEEE, 2007), pp. 384–385
48. M. Kashmiri et al., A 200kS/s 13.5b integrated-fluxgate differential-magnetic-to-digital converter with an oversampling compensation loop for contactless current sensing, in *IEEE ISSCC 2015* (IEEE, 2015), pp. 490–491
49. M.F. Snoeij et al., Integrated fluxgate magnetometer for use in isolated current sensing. IEEE J. Solid State Circuits **51**(7), 1684–1694 (2016)
50. Texas Instruments Inc., DRV401 and DRV411 data sheets. (Sept 2014), www.ti.com

## *Magnetoresistive Sensors*

51. Sensitec, Magnetoresistive current sensors, CFS1000 and CMS2000 datasheets, Available online
52. Allegro, ACS70331 datasheet, High sensitivity, 1 MHz, GMR-based current sensor IC. (May 2018), Available online

53. Y. Ouyang et al., A current sensor based on the giant magnetoresistance effect: design and potential smart grid applications. Sensors **12**(11), 15520–15541 (2012)
54. A. Nebeling, W. Schreiber-Prillwitz, Architecture of an integrated AMR current sensor (IACS) system for a wide range of automotive applications, in *Advanced Microsystems for Automotive Applications 2011. VDI-Buch*, ed. by G. Meyer, J. Valldorf, (Springer, Berlin, 2011)
55. A. Bernieri et al., An AMR-based three-phase current sensor for smart grid applications. IEEE Sensors J. **17**(23), 7704–7712 (2017)
56. R.P. Singh et al., Giant magneto resistive (GMR) effect based current sensing technique for low voltage/high current voltage regulator modules. IEEE Trans. Power Electron. **23**(2), 915–925 (2008)
57. S.J. Nibir et al., Wideband contactless current sensing using hybrid magnetoresistor-rogowski sensor in high frequency power electronic converters, in *IEEE Applied Power Electronics Conference* (IEEE, 2018), pp. 904–908
58. TMR2705, American electronic components, datasheet, https://www.aecsensors.com/components/com.../shop.../TMR2705-datasheet.pdf
59. E.G. Vidal et al., Electronic energy meter based on a tunnel magnetoresistive effect (TMR) current sensor. Materials (Basel) **10**(10), 1134 (2017)
60. TDK, TMR sensors, Tech notes, https://product.tdk.com/info/en/products/sensor/angle/tmr_angle/technote/tpo/index.html
61. T. Funk, B. Wicht, A fully integrated DC to 75MHz current sensing circuit with on-chip rogowski coil, in *IEEE CICC 2018* (IEEE, 2018), pp. 1–4
62. Si85xx, Unidirectional AC Current Sensors. (SiLabs, 2012), datasheet available online
63. D.E. Shepard, D.W. Yauch, LEM DynAmp Inc., An overview of rogowski coil current sensing technology, Available online

## *Emerging Current Sensing Applications*

## *Biomedical*

64. C.L. Hsu, D.A. Hall, A current-measurement front-end with 160dB dynamic range and 7ppm INL, in *IEEE ISSCC 2018* (IEEE, 2018), pp. 326–328
65. S. Dai et al., A 155-dB dynamic range current measurement front end for electrochemical biosensing. IEEE Trans. Biomed. Circuits Syst. **10**(5), 935–944 (2016)
66. J. Li et al., Solid-state nanopore-based DNA single molecule detection and sequencing. Microchim. Acta **183**(3), 941–953 (2016)

## *Power/GaN*

67. Y. Woo et al., Load-independent control of switching DC-DC converters with freewheeling current feedback. IEEE J. Solid State Circuits **43**(12), 2798–2808 (2008)
68. Texas Instruments, Design note: TIDA-01455, Current sense reference design for high-switching transient three-phase inverter for servo drives. (Sept 2017), Available online
69. X. Ke et al., A 10MHz 3-to-40V VIN tri-slope gate driving GaN DC-DC converter with 40.5dBμV spurious noise compression and 79.3% ringing suppression for automotive applications, in *IEEE ISSCC 2017* (IEEE, 2017), pp. 430–431

## *TFT Display/Laser Drive*

70. B. You et al., Real-time current-sensing feedback system for compensating process–voltage–temperature variations of display using double-gate oxide TFT. Electron. Lett. **53**(2), 117–119 (2017)
71. Whitepaper, iCHaus, Design and test of fast laser driver circuits, http://www.ichaus.com

# Chapter 10
# Wide Bandgap Integrated Circuits for High Power Management in Extreme Environments

**Carl-Mikael Zetterling, Saleh Kargarrazi, and Muhammad Shakir**

## 10.1 Introduction

Semiconductor materials having a bandgap energy $E_g$ larger than silicon (Si) are generally referred to as wide bandgap semiconductors (WBGS). This includes silicon carbide (SiC), gallium nitride (GaN), diamond (C), and recently gallium oxide ($Ga_2O_3$) [1, 2]. The large commercial interest in these materials is mainly connected to the large critical field for avalanche breakdown $E_c$ that is much higher than that of silicon, see Table 10.1. As a result of the higher $E_c$, the blocking region of the semiconductor switch can be made narrower (in direct proportion to the higher $E_c$) and with higher doping (proportional to $E_c$ squared). The specific on-resistance $R_{\text{on}}$ for a unipolar switch that is designed with a breakdown voltage $V_B$ can be derived to be [3, p. 280]

$$R_{\text{on}} = \left(4\ V_B{}^2\right) \Big/ \left(\mu_n\ \varepsilon_r\ \varepsilon_0\ E_c{}^3\right) \tag{10.1}$$

where $\mu_n$ is the electron mobility and $\varepsilon_r$ is the semiconductor relative dielectric constant. If the on-resistance of the switch is limited by the on-resistance of the blocking junction, the improvement can be a factor of 400 (for SiC over Si) or more. This is the case for switches with breakdown voltages of around 600 V or more. The higher the breakdown voltage is, the larger the advantage of using a WBGS.

C.-M. Zetterling (✉) · M. Shakir
Department of Electronics and Embedded Systems, KTH Royal Institute of Technology, Kista, Sweden
e-mail: bellman@kth.se

S. Kargarrazi
Department of Aeronautics and Astronautics, Stanford University, Stanford, CA, USA
e-mail: skargar@stanford.edu

© Springer Nature Switzerland AG 2020

A. Baschirotto et al. (eds.), *Next-Generation ADCs, High-Performance Power Management, and Technology Considerations for Advanced Integrated Circuits*, https://doi.org/10.1007/978-3-030-25267-0_10

**Table 10.1** Properties of semiconductors

| Semiconductor | Si | SiC | GaN | $Ga_2O_3$ | C |
|---|---|---|---|---|---|
| Bandgap energy $E_g$ (eV) | 1.12 | 3.2 | 3.4 | 4.9 | 5.5 |
| Critical field $E_C$ (MV/cm) | 0.3 | 3 | 4.9 | 10 | 13 |
| Mobility $\mu_n$ ($cm^2$/Vs) | 1350 | 950 | 1000 | 180 | 4500 |
| Rel. Dielectric constant $\varepsilon_r$ | 11.9 | 10 | 10.4 | 10 | 5.7 |

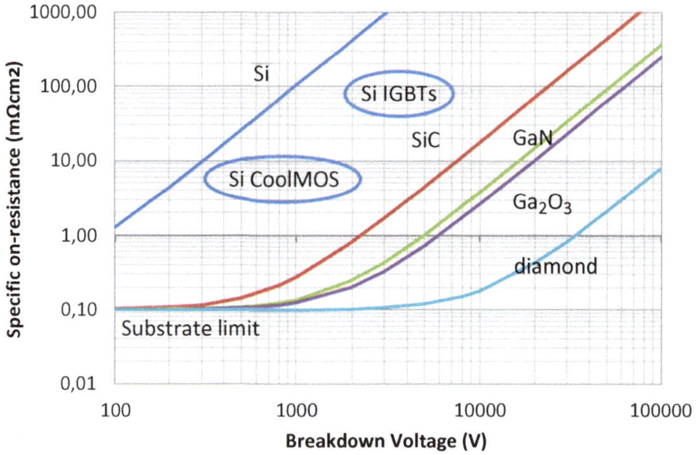


**Fig. 10.1** Unipolar specific on-resistance versus breakdown voltage for WBGS

However, for vertical high voltage devices, the specific resistance of the wafer can be in the range 0.1–0.5 m$\Omega$cm$^2$, indicated in Fig. 10.1.

Silicon CoolMOS super junction transistors are better than the unipolar limit. Silicon insulated gate bipolar transistors (IGBTs) for high voltage applications are bipolar rather than unipolar in operation, and therefore fall below the line. High injection of minority carriers will, therefore, reduce the specific on-resistance as indicated in Fig. 10.1. Due to lower minority carrier lifetime in WBGS, high injection lowering of the resistance hasn't been demonstrated except in a few cases. However, the on-resistance of unipolar WBGS switches is lower than silicon IGBTs anyway. Without the minority carrier charge injection, the WBGS switches have lower switching losses and faster switching, thereby allowing higher operating frequency. This results in further savings, since the passives, and especially magnetics, can be made smaller in volume and weight.

One advantage of WBGS that hasn't been fully exploited, although it was suggested early, is high temperature operation. When semiconductor devices are heated (from the environment or from self-heating) thermal generation of electron-hole pairs increases the intrinsic concentration $n_i$. Initially, this is seen as increased noise or leakage currents that can be handled, but at some temperature the extra thermally generated current will influence circuit operation. At high enough temperatures, some switches will inadvertently turn on, and could short-circuit the

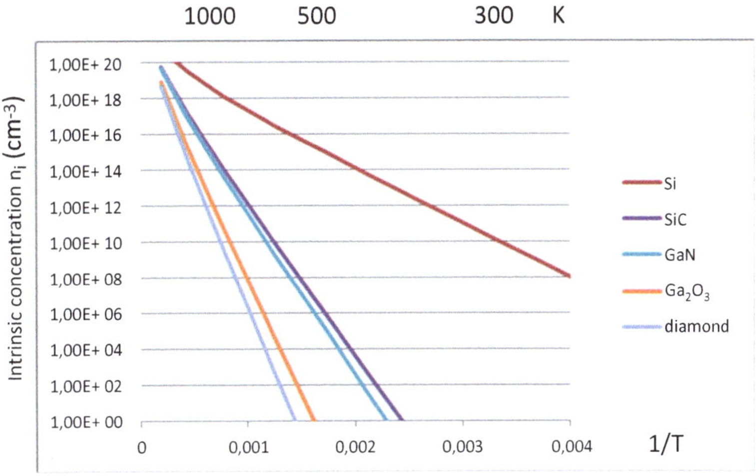


**Fig. 10.2** Intrinsic concentration versus temperature for different WBGS

supply voltage. When minority carrier generation is high enough, the p-n junctions no longer block or rectify current, and the circuit will fail destructively. Silicon circuitry can operate up to 250 °C, and SOI circuitry slightly higher thanks to the dielectric isolation of the buried oxide. However, SiC circuits have been routinely tested up to 500 or 600 °C, and there are even results of operation up to 800 °C. Since the intrinsic concentration $n_i$ depends exponentially on the energy bandgap $E_g$, see Eq. (10.2) and Fig. 10.2 [3], we can predict that SiC and other WBGS can work above 1000 °C. Instead it is the metallization that would tend to fail sooner than the semiconductor.

$$n_i = (N_C\ N_V)^{0.5}\ \exp\left(-E_g/2\,\mathrm{k\,T}\right) \tag{10.2}$$

Radiation hardness is expected to be better for all WBGS, since the ionization energy is proportional to the bandgap energy: for the same radiation energy, less carriers will be generated. The binding energy between the elements is also higher than silicon, at least for SiC and diamond, which increases the hardness to destructive displacement damage from radiation. For unbiased or low voltage ICs in SiC this has been demonstrated to some extent [4, 5], but for high voltage SiC devices under reverse bias this is not necessarily the case [6]. If the radiation hardness can be shown to be ten times higher, there are space applications in near Earth orbit that could benefit from less weight for shielding and longer operation. However, the large opportunity is probably terrestrial: oil and gas drilling, aerospace, gas turbines, and nuclear energy generation.

One extreme environment application that we propose for ICs using WBGS is space exploration, specifically Venus. With a surface temperature of 460 °C, a Venus

Lander with silicon electronics will only last a few hours before it is overheated (https://en.wikipedia.org/wiki/Venera), whereas a WBGS based lander is not limited by the extreme temperature [7–9].

## 10.2 Technological Considerations

The first technological consideration is the availability of high quality low cost wafers. Silicon technology has a head-start, presently using 300 mm wafers. SiC wafers are still 100 times more expensive per unit area, after 30 years of focused development, but now there are 200 mm wafers available. GaN was initially grown on sapphire or SiC wafers, but there is also work on improving GaN growth on silicon, which would reduce the wafer cost enormously, and make large wafer sizes available. Diamond has always suffered from expensive small diameter wafers, even with work on CVD deposition on other wafers. This is where $Ga_2O_3$ comes in. The growth process can be made from a melt, similar to silicon bulk growth, thereby potentially offering large diameter low cost wafers.

Secondly, doping control is necessary for device and circuit fabrication. Both n-type (donors) and p-type (acceptors) need to be identified for all WBGS. SiC has the best balance of dopant activation energy and carrier mobility, both n-type and p-type dopants are available with reasonable mobility. However, ion implantation requires annealing at temperatures of around 1600 °C, which has made self-aligned MOSFETs challenging [10]. In GaN, finding a p-type dopant has been a challenge, and the mobility is low. However, for GaN HEMT structures the 2D high electron mobility channel is created by using an AlGaN gate. This technique can be used for both high frequency and high voltage devices. Diamond still does not have reliable n-type doping, so device types are limited to JFETs and MESFETs, which reduces choices in terms of IC solutions. $Ga_2O_3$ has n-type dopants identified with low mobility, and unknown p-type dopants with extremely low mobility (single digits).

Since ion implantation in SiC requires high annealing temperatures to remove damage, technologies have been explored that do not need ion implantation [11]. Epitaxially grown layers with different doping are dry etched to form the devices, much like in compound semiconductor technology (GaAs, GaN), see Fig. 10.3.

The next challenge is finding a gate oxide for WBGS MOSFETs. The valence and conduction band offsets are reduced when we replace silicon with SiC in a MOSFET, which increases tunneling and gate leakage currents. As we go to wider bandgaps this drawback worsens. When the high critical electric field for breakdown is exploited in the blocking region of a switch, the electric field in the gate oxide has to be taken into account; otherwise Fowler–Nordheim tunneling will quickly degrade the oxide properties. High-K dielectrics are not necessarily a solution since they usually have lower bandgap energy and lower critical field than silicon dioxide. For GaN the HEMT structure is used instead of MOSFET. For diamond and $Ga_2O_3$ we will probably have to work with MESFET topologies.

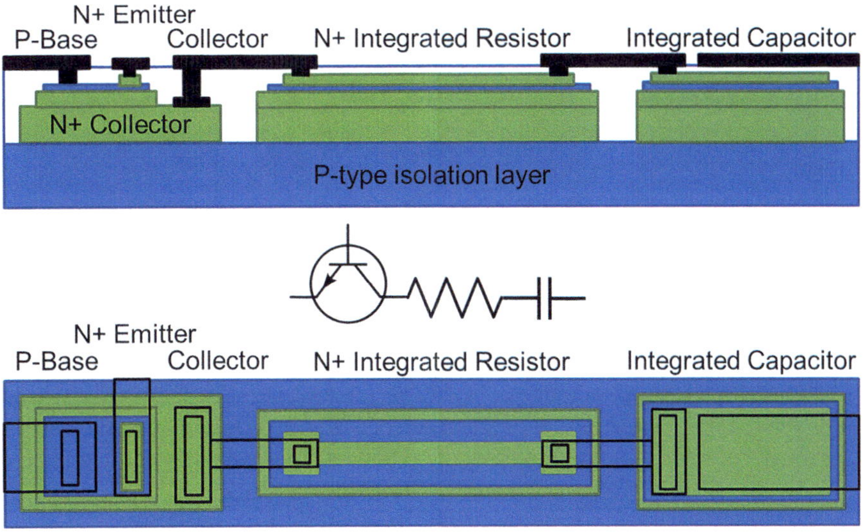


**Fig. 10.3** Bipolar transistor, resistor, and capacitor in SiC without ion implantation

Ohmic contacts are in general a challenge for WBGS since the wider bandgap in combination with midgap fermi level pinning leads to high Schottky barriers. Although SiC technology is far from the advanced processing of silicon ICs [12], some processing techniques like CMP and self-aligned contacts have been transferred [13, 14].

On the other hand, the merged p-n Schottky diode or junction barrier Schottky diode (MPS or JBS) is a very successful diode in SiC, and was the first commercial device, nowadays available in voltages from 600 V to 2500 V. The Schottky junction with a barrier of around 1 V has low leakage current and has no minority charge storage, leading to low switching losses. SiC MOSFET switches are commercially available in the voltage range 600–1700 V. The development of these devices started in the 1990s, and has taken more than 20 years. JFETs and bipolar junction transistors (BJTs) [15] have also been investigated. GaN/AlGaN was initially developed for blue and white LEDs and high frequency transistors, but has lately become interesting for high voltage devices as well. This is mainly due to the hope that less expensive silicon wafers can be used for GaN lateral power devices [16]. However, these are limited to a few kV, unless a costly partial wafer removal process is used, otherwise the breakdown will occur in the underlying silicon wafer.

A power switch for large currents (100 A) is sectioned into cells (multiple small transistors) that are connected in parallel with multiple layers of metal for minimum series resistance. The edges of the devices have complex edge termination to support the reverse voltages. By adding device isolation, the step to making ICs is not that large. Passive devices can be added, for instance resistors utilize doped regions of the transistors, and capacitors can be made between metal layers, see Fig. 10.3.

ICs in WBGS are mostly considered for high temperature operation, since in comparison to silicon technology they are inferior at room temperature. The size of transistors is much larger, current levels higher, and gain and efficiency lower. It is only GaN circuits for high frequency (a few transistors with matching networks) that can compete at room temperature. $Ga_2O_3$ and diamond will not be considered in the following sections, since circuit topologies are limited by only having one channel polarity.

Finally, packaging is a challenge, both for high voltage devices (new types are appearing) and for high temperature operation. Ceramic packages can probably operate above 250 °C, but they have not been certified for this use since there has not been a need from industry.

Operating ICs at 500 °C is a reliability nightmare, and testing them is even worse. Most reliability testing is done as highly accelerated stress test (HAST), but this implies designing circuits that could operate at higher temperatures than 500 °C. Although most WBGS would allow this, most metallization does not. Refractory metal silicides have to replace Al and TiW, and this has led to development of self-aligned metallization similar to silicon technology [13, 14].

## 10.3 Design Considerations

Several IC technologies have been explored in SiC: Bipolar [11], CMOS [17], NMOS [18], JFET [8–10], and MESFET [19], and they each have their advantages and drawbacks. In GaN, IC technologies are limited to HFET [20] and MOSHEMT [20, 21]. Circuit topologies are limited if only one channel conductivity type is available, and devices are normally on. The many technologies being investigated is because self-aligned SiC CMOS is not available, otherwise it would probably dominate high temperature ICs just like silicon CMOS ICs dominate at room temperature.

Another challenge is that for operation in a wide temperature range from room temperature (RT) up to 500 °C, many device parameters change much more than for electronics designed in commercial or military temperature ranges. For bipolar junction transistors (BJTs) the built in voltage of the base-emitter junction is reduced by 2 mV/°C, or 1 V for 500 °C. For BJTs this voltage is around 2.5 V at RT, compared to 0.7 V for silicon at RT, leading to the need for larger supply voltages. Threshold voltages for MOSFETs can be reduced by as much as 1 V per 100 °C, due to interface trap activation. For normally off operation at the highest operation temperature, this implies much higher threshold voltages at RT, and thereby higher supply voltages and more power are needed.

Current gain for BJTs is reduced by about 50% from RT to 300 °C and then stabilizes due to competing mechanisms for dopant activation and minority carrier lifetime increase in the base region; see Fig. 10.4. The sheet resistance of the doped layers used for integrated resistors varies with temperature due to dopant activation and mobility reduction at higher temperatures; see Fig. 10.4. The largest variation

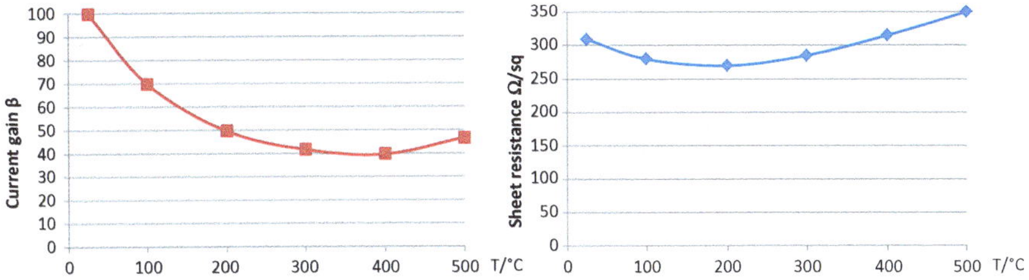


**Fig. 10.4** SiC BJT current gain and sheet resistance temperature dependence

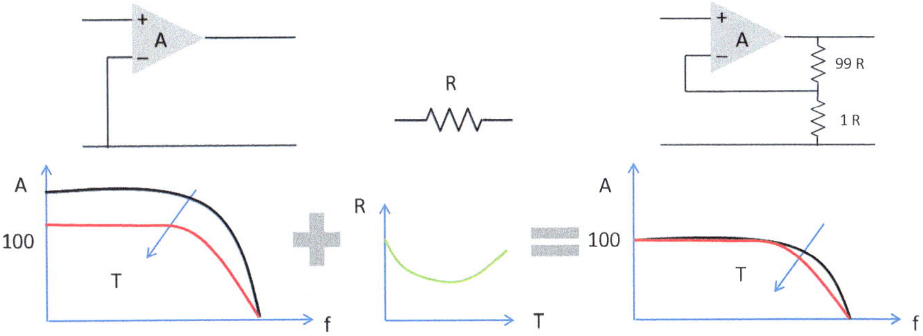


**Fig. 10.5** Principle of analog feedback used in SiC analog ICs

of sheet resistance occurs for p-type resistors. Capacitors can be made more or less temperature independent if they are made using interlayer dielectric and metal, but if voltage controlled capacitors are needed (either MOS- or p-n junction-based), there will be a temperature dependence to take into account in the modeling.

For simulation, compact device models are needed that include the temperature dependence over the entire range [22]. An alternative route is to use binned models for discrete temperatures (RT, 100, 200, 300, 400, and 500 °C) and to run several simulations to ensure proper operation at all temperatures.

For analog designs, standard negative feedback techniques are used to stabilize gain in amplifiers and other circuits. On-chip resistor divider networks can be used, but the resistors must be made with a similar layout and in the same device layer, so that temperature coefficients are identical, see Fig. 10.5. As long as the minimum open-loop gain for any intended temperature is much higher than the requested closed-loop gain for the circuit, this is achievable [23, 24].

For digital designs, the good news is that noise margins are not an issue, see Fig. 10.6. Since larger supply voltages will be used, several volts of noise margin can be achieved over the entire temperature range [25]. The drawback is that this is at the expense of power consumption. In the SiC community, ICs are considered low-voltage designs (at 15–50 V) but this is certainly not so in comparison to silicon (1 V).

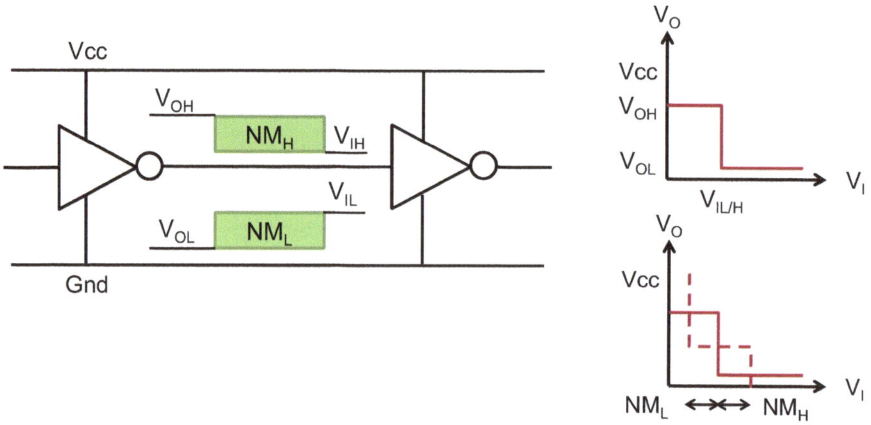


**Fig. 10.6** Digital noise margins

## 10.4 WBGS IC Examples

In any application where a WBGS switch would be used in an extreme environment (high temperature or radiation), it is of interest to develop gate drivers and other control circuitry using WBGS ICs that can be mounted close to or co-packaged with the switch. In this way, the inductive loss in long cables can be avoided. Applications include photoelectric energy conversion, electric or combustion motor control, nuclear energy generation, and power conversion in general. There are also low power applications with sensors instead of switches, in extreme environments. In this case amplifier and analog-to-digital conversion (ADC) ICs close to the sensor can improve signal to noise ratios.

The first example is the driver needed for a high voltage switch to supply the gate voltage for a FET, see Fig. 10.7, or base current for a BJT. The driver can be made in CMOS technology [27] or bipolar technology [28]. The challenge is to be able to supply enough current at voltages in the range of 48 V [29]. Lateral MOSFETs [27] or lateral BJTs [30] have been used.

Another example is the linear voltage regulator with external or integrated switch [18, 26], see Fig. 10.8. For better conversion efficiency, but at the cost of external capacitors and inductors, DC/DC conversion can be used. The possible savings have been demonstrated at room temperature with SiC switches and silicon control circuitry [31], but SiC and GaN control circuitry have also been demonstrated at lower total power [32, 33].

If the power converter needs advanced protection from overvoltage, short circuit currents or high temperature, or other digital control, digital building blocks are readily available. CMOS, JFET, and bipolar technologies (see Fig. 10.9) have been demonstrated in the range 400–800 °C [8, 9, 17, 25].

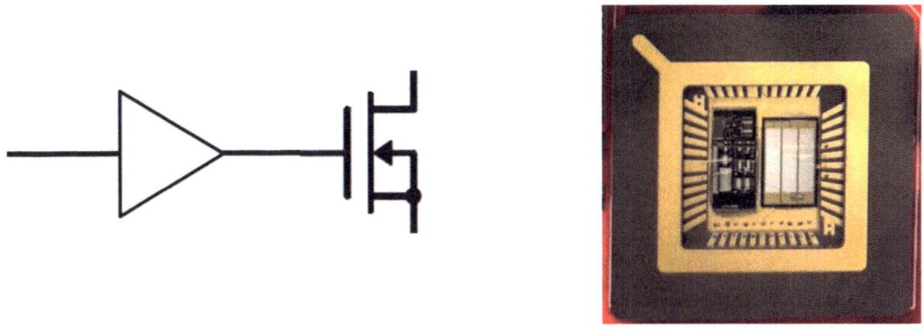

Fig. 10.7 A driver circuit and packaged switch

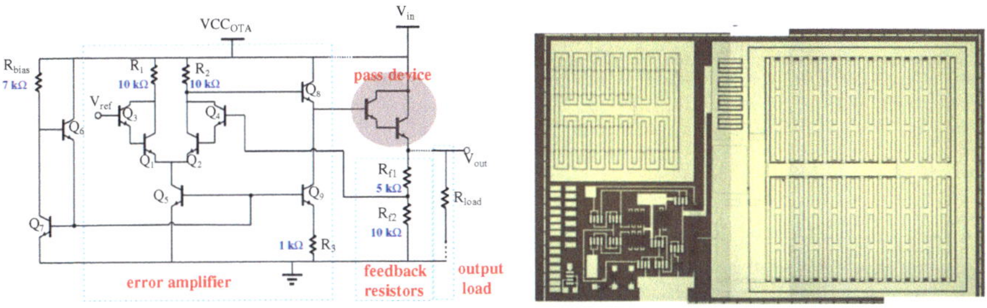


Fig. 10.8 Linear voltage regulator [26]

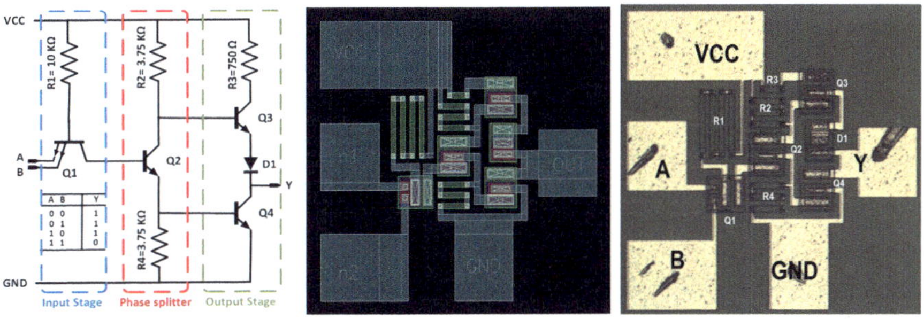


Fig. 10.9 TTL inverter

For analog-to-digital conversion (ADC) the relevant building blocks have been demonstrated [34], including digital-to-analog conversion (DAC) [34, 35], bandgap voltage references [36], comparators. Even UV detectors [37] and building blocks for radios have been demonstrated at 500 °C [38]. Higher levels of integration have also been achieved. To demonstrate the possibility of VLSI circuits, a 2-bit arithmetic-logic unit (ALU) with 720 transistors and resistors was made and characterized up to 500 °C [39], see Fig. 10.10. However, for complex mixed signal building blocks, it is clear that SiC ICs can never compete with silicon at room temperature [40]. It is only in applications requiring temperatures above RT that SiC or GaN ICs can be the choice.

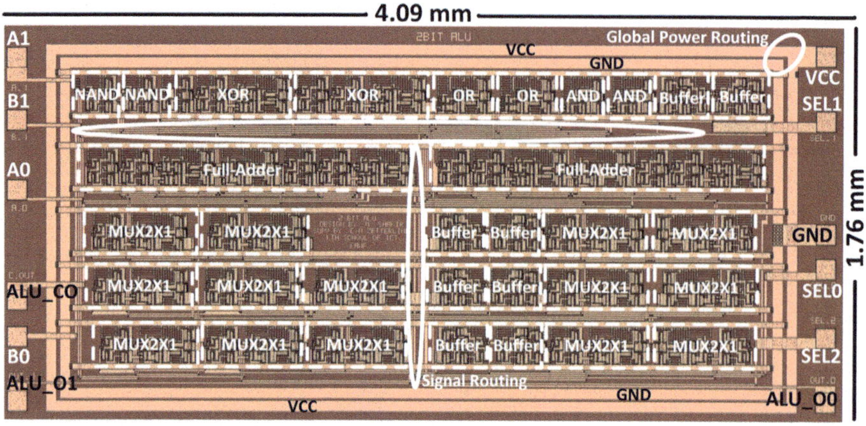


**Fig. 10.10** TTL ALU with 720 devices [39]

## 10.5 Conclusion

Although IC technologies in WBGS, including SiC and GaN, cannot compete with silicon at room temperature, in extreme environments where silicon no longer operates it can be an enabler for power management and other applications in aerospace, oil and gas drilling, and nuclear energy generation.

## References

1. S.J. Pearton, F. Ren, M. Tadjer, J. Kim, Perspective: Ga2O3 for ultra-high power rectifiers and MOSFETS. J. Appl. Phys. **124**, 220901 (2018)
2. J.Y. Tsao et al., Ultrawide-bandgap semiconductors: research opportunities and challenges. Adv Electron. Mater. **4**(1), 1600501 (2018)
3. T. Kimoto, J.A. Cooper, *Fundamentals of Silicon Carbide Technology* (Wiley, Singapore, 2014)
4. S.S. Suvanam, L. Lanni, B.G. Malm, C.-M. Zetterling, A. Hallén, Effects of 3 MeV protons on 4H-SiC bipolar devices and integrated OR-NOR gate. IEEE Trans. Nucl. Sci. **61**, 1772–1776 (2014)
5. S.S. Suvanam, S.I. Kuroki, L. Lanni, R. Hedayati, T. Ohshima, T. Makino, A. Hallén, C.-M. Zetterling, High gamma ray tolerance for 4H-SiC bipolar circuits. IEEE Trans. Nucl. Sci. **64**, 852–858 (2017)
6. A. Akturk, J. McGarrity, N. Goldsman, D.J. Lichtenwalner, B. Hull, D. Grider, R. Wilkins, The effects of radiation on the terrestrial operation of SiC MOSFETs, in *2018 IEEE International Reliability Physics Symposium (IRPS)*, (IEEE, Piscataway, 2018)
7. C.F. Wilson, C.-M. Zetterling, W.T. Pike, *Venus Long-life Surface Package*, arXiv:1611.03365, (2016)
8. P.G. Neudeck, D.J. Spry, L. Chen, N.F. Prokop, M.J. Krasowski, Demonstration of 4H-SiC digital integrated circuits above 800 °C. IEEE Electron Device Lett. **38**(8), 1082–1085 (2017)

9. P.G. Neudeck, et al., Operational testing of 4H-SiC JFET ICs for 60 days directly exposed to venus surface atmospheric conditions in *IEEE Journal of the Electron Devices Society,* (IEEE, New York, 2018)
10. C.-M. Zetterling, Integrated circuits in silicon carbide for high-temperature applications. MRS Bull. **40**, 431–438 (2015)
11. L. Lanni, B. Malm, M. Östling, C.-M. Zetterling, 500°C bipolar integrated OR/NOR gate in 4H-SiC. Electron Device Lett. IEEE **34**(9), 1091–1093 (2013)
12. A.L.S. Loke et al., Analog/mixed-signal design in FinFET technologies, in *Hybrid ADCs, Smart Sensors for the IoT, and Sub-1V & Advanced Node Analog Circuit Design*, ed. by P. Harpe, K. Makinwa, A. Baschirotto, (Springer, Cham, 2018), pp. 259–280
13. M. Ekström et al., Low temperature Ni-Al ohmic contacts to p-type 4H-SiC using semi-salicide processing, in *ICSCRM 2017, Materials Science Forum*, vol. 924, (Trans Tech Publication, Uetikon, 2018), pp. 389–392
14. H. Elahipanah et al., A wafer-scale Ni-salicide contact technology on n-type 4H-SiC. J. Solid State Sci. Technol. **6**, 197–200 (2017)
15. A. Salemi et al., 15 kV-class implantation-free 4H-SiC BJTs with record high current gain. IEEE Electron Device Lett. **39**, 63 (2018)
16. K.J. Chen, O. Häberlen, A. Lidow, C.l. Tsai, T. Ueda, Y. Uemoto, Y. Wu, Gan-on-si power technology: Devices and applications. IEEE Trans. Electron Devices **64**(3), 779–795 (2017)
17. D. Clark, E.P. Ramsay, A. Murphy, D.A. Smith, R. Thompson, R. Young, J.D. Cormack, S. Finney, et al., High temperature digital and analogue integrated circuits in silicon carbide, in *Materials Science Forum*, vol. 740, (Trans Tech Publication, Uetikon, 2013), pp. 1065–1068
18. J. Valle-Mayorga, A. Rahman, H. Mantooth, A SiC NMOS linear voltage regulator for high-temperature applications. IEEE Trans. Power Electron. **29**(5), 2321–2328 (2014)
19. J. Mogniotte, D. Tournier, C. Raynaud, M. Lazar, D. Planson, B. Allard, Silicon carbide technology of MESFET-based power integrated circuits. IEEE J. Emerg. Sel. Top Power Electron **6**(2), 539–548 (2018)
20. A.J. Suria, A.S. Yalamarthy, H. So, D.G. Senesky, DC characteristics of ALD-grown Al2O3/AlGaN/GaN MIS-HEMTs and HEMTs at 600 °C in air. Semicond. Sci. Technol. **31**(11), 115017 (2016)
21. A.M.H. Kwan, X. Liu, K.J. Chen, Integrated gate-protected HEMTs and mixed-signal functional blocks for GaN smart power ICs, in *2012 International Electron Devices Meeting*, (IEEE, Piscataway, 2012), pp. 7.3.1–7.3.4
22. Y. Tian et al., SiC BJT compact DC model with continuous- temperature scalability from 300 to 773 K. IEEE Trans. Electron Dev. **64**, 3588–3594 (2017)
23. R. Hedayati, L. Lanni, S. Rodriguez, B.G. Malm, A. Rusu, C.-M. Zetterling, A monolithic, 500°C operational amplifier in 4H-SiC bipolar technology. IEEE Electron Device Lett. **35**, 693–695 (2014)
24. S. Kargarrazi, L. Lanni, And C. M. Zetterling, "A study on positive -feedback configuration of a bipolar SiC high temperature operational amplifier," Solid State Electron., Vol. 116, 2016, pp. 33–37
25. M. Shakir, S. Hou, B.G. Malm, M. Östling, C.-M. Zetterling, A 600°C TTL-based 11-stage ring oscillator in bipolar silicon carbide technology. IEEE Electron Device Letters **39**(10), 1540–1543 (2018)
26. S. Kargarrazi, H. Elahipanah, S. Rodriguez, C.-M. Zetterling, 500°C, high current linear voltage regulator in 4H-SiC BJT technology. IEEE Electron Device Letters **39**(4), 548–551 (2018)
27. M. Barlow, S. Ahmed, H.A. Mantooth, A.M. Francis, An integrated SiC CMOS gate driver, in *2016 IEEE Applied Power Electronics Conference and Exposition (APEC)*, (IEEE, Piscataway, 2016), pp. 1646–1649
28. S. Kargarrazi, L. Lanni, A. Rusu, C.M. Zetterling, A monolithic SiC drive circuit for SiC power BJTs, in *2015 IEEE 27th International Symposium on Power Semiconductor Devices IC's (ISPSD)*, (IEEE, Piscataway, 2015), pp. 285–288

29. B. Wicht, J. Wittmann, A. Seidel, A. Schindler, High-voltage fast-switching gate drivers, in *Wideband Continuous-Time $\Sigma\Delta$ ADCs, Automotive Electronics, and Power Management*, ed. by A. Baschirotto, P. Harpe, K. Makinwa, (Springer, Cham, 2017), pp. 155–176
30. H. Elahipanah, S. Kargarrazi, A. Salemi, M. Östling, C.M. Zetterling, 500°C high current 4H-SiC lateral BJTs for high-temperature integrated circuits. IEEE Electron Device Lett. **38**, 1429 (2017)
31. J. Rabkowski, D. Peftitsis, H.-P. Nee, Design steps towards a 40-kVA SiC inverter with an efficiency exceeding 99.5%, in *IEEE Applied Power Electronics Conf. and Exposition (APEC)*, (IEEE, Piscataway, 2012), pp. 1536–1543
32. S. Kargarrazi, H. Elahipanah, S. Saggini, D. Senesky, C.-M. Zetterling, 500°C SiC PWM integrated circuit. IEEE Trans. Power Electron. **34**(3), 1997–2001 (2019)
33. H. Wang, A.M.H. Kwan, Q. Jiang, K.J. Chen, A GaN pulse width modulation integrated circuit for gan power converters. IEEE Trans. Electron Devices **62**(4), 1143–1149 (2015)
34. A. Rahman, L. Caley, S. Roy, S. Kuhns, A. Mantooth, J. Di, A.M. Francis, J. Holmes, High temperature data converters in silicon carbide CMOS. IEEE Trans. Electron Devices **64**, 1426–1432 (2017)
35. R. Hedayati, L. Lanni, B.G. Malm, A. Rusu, C.-M. Zetterling, A 500°C 8-bit digital-to-analog converter in silicon carbide bipolar technology. IEEE Trans Electron Dev. **63**, 3445–3450 (2016)
36. R. Hedayati, L. Lanni, A. Rusu, C.-M. Zetterling, Wide temperature range integrated bandgap voltage references in 4H–SiC. IEEE Electron Device Lett. **37**, 146–149 (2016)
37. S. Hou, P.-E. Hellström, C.-M. Zetterling, M. Östling, A 4H-SiC BJT as a switch for on-chip integrated uv photodiode. IEEE Electron Device Lett. **40**(1), 51–54 (2019)
38. M.W. Hussain, H. Elahipanah, J.E. Zumbro, S. Schröder, S. Rodriguez, B.G. Malm, H.A. Mantooth, A. Rusu, A 500°C active down-conversion mixer in silicon carbide bipolar technology. IEEE Electron Device Lett. **39**(6), 855–858 (2018)
39. M. Shakir, S. Hou, B.G. Malm, M. Östling, C.-M. Zetterling, Towards silicon carbide based VLSI circuits for extreme environment applications. MDPI J. Electron. **8**(5), 496 (2019)
40. C.-M. Zetterling et al., Bipolar integrated circuits in SiC for extreme environment operation. Semicond. Sci. Technol. **32**, 034002 (2017)

# Chapter 11
# On the Limits of Driving Wide-Bandgap Transistors

**Jef Thoné and Mike Wens**

## 11.1 Introduction

The push for decreasing carbon emissions and the increased electrification of our society drive the need for more efficient power conversion technologies. Over the last decade, the semiconductors silicon carbide (SiC) and gallium nitride (GaN) have found their application in wide-bandgap power transistors. They potentially offer large efficiency improvements for switching power converters over traditional silicon semiconductors, based on their faster switching properties, lower gate drive losses combined with higher breakdown voltages, lower specific $R_{\mathrm{dson}}$, and ability to function at elevated temperatures.

The faster switching slopes that are characteristic for wide-bandgap transistors (100 V/ns and higher) yield specific challenges for the design of the gate driver, levelshifting, isolation, and overall (isolated) supply system. The lower specific $R_{\mathrm{dson}}$ yields physically smaller output stages, which pushes the need for a good thermal interface and higher efficiency.

Design of the right gate driver for wide-bandgap transistors is a complex art, putting well-mastered gate drive techniques out of their comfort zone. This pushes the driver designer to look for new state-of-the-art topologies that require a combination of high-end analog design, time-critical and high-current high-power design. This article treats some of the primary challenges related to wide-bandgap gate driver design.

J. Thoné (✉) · M. Wens
MinDCet NV, Leuven, Belgium
e-mail: jef@mindcet.com

© Springer Nature Switzerland AG 2020

A. Baschirotto et al. (eds.), *Next-Generation ADCs, High-Performance Power Management, and Technology Considerations for Advanced Integrated Circuits*,
https://doi.org/10.1007/978-3-030-25267-0_11

## 11.2 Switching Losses Revisited

The switching losses incurred when hard switching a power half-bridge can be generalized as follows:

- *Gate loss*: the power needed to drive the transistor gate;
- $C_{oss}$ *loss*: output capacitance loss;
- $R_{dson}$ *loss*: the Joule loss in the transistor on-state resistance;
- *Switching loss*: the power lost in the switching edges at non-zero output current;
- *Reverse recovery loss*: power lost through reverse recovery charge.

### 11.2.1 Gate Loss $P_g$

When designing a gate driver for a power transistor, the gate charge $Q_g$ is an important parameter. It is the total charge needed to charge the gate between its off- and on-state with a specified gate-source and drain-source voltage swing. As the $C_{gs}$ and $C_{gd}$ are non-linear capacitors, the gate charge is the best parameter to assess the complete charge to be delivered to the gate. The power needed to drive a gate with a gate charge $Q_g$, gate voltage $V_g$, at a switching frequency $f_{sw}$ is presented in Eq. (11.1). Note that the loss is independent of the rise- and fall-time of the gate voltage.

$$P_g = Q_g.V_g.f_{sw} \tag{11.1}$$

As shown later, a popular FOM for benchmarking power transistors is $R_{dson}{\cdot}Q_g$. As $Q_g$ depends on the physical area of the device (assuming similar $t_{ox}$), this FOM will tell how area efficient a transistor is for a given $R_{dson}$.

The power delivered to the predriver circuit to drive the gate is dissipated—50/50 by the pull-up and the pull-down predriver. Throwing silicon area at the predriver is typically cost limited, which is why for larger gate losses ($P_g > 0.5$ W), it is economically wise to use external gate drive resistors, which will burn the majority of the gate loss power, depending on their ratio with the predriver impedance.

### 11.2.2 Output Capacitance Loss $P_{Coss}$

In a switching transistor, $C_{oss}$ (equivalent $C_{gs} + C_{gd}$) is charged and discharged with the switching voltage $V_{sw}$ at a switching rate $f_{sw}$, forming the output capacitance loss as presented in Eq. (11.2). The $C_{oss}$ parameter is also found in a typical power FET datasheet, and allows the designer to estimate the power loss contribution by its inherent capacitance. In a buck converter half-bridge configuration, depending on the switch timing, half of the $P_{C_{oss}}$ is charged by the load, leading to half the $P_{C_{oss}}$ losses as compared to an unloaded half-bridge.

$$P_{\text{Coss}} = C_{\text{oss}}.V_g^2.f_{\text{sw}} \tag{11.2}$$

### 11.2.3 $R_{dson}$ Loss $P_{\text{Rdson}}$

$R_{\text{dson}}$ loss is due to the rms drain-source current flowing through the power FET in on-state:

$$P_{\text{Rdson}} = R_{\text{dson}}.I_{\text{ds\_rms}}^2 \tag{11.3}$$

The specific $R_{\text{dson}}$ [$\Omega \cdot \text{mm}^2$] of the device is another important FOM to compare transistors—the smaller the more area efficient. Unfortunately, for commercial FETs this is a parameter that can only be estimated by the package size XRAY or decapping.

### 11.2.4 Output Switching Losses $P_{sw}$

For hard switching applications, where the outputs exhibit a limited rise- and fall-time ($t_r$ and $t_f$), $I_{\text{out}}$ being the momentary current flowing through the switch, $V_{\text{sw}}$ being the voltage swing at the output, and $f_{\text{sw}}$ the switching frequency dissipates an average power as presented in simplified Eq. (11.4):

$$P_{\text{sw}} = V_{\text{sw}}.I_{\text{out}}.\left(t_r + t_f\right).f_{\text{sw}}/2 \tag{11.4}$$

Wide-bandgap technologies like GaN and SiC provide potential benefits with respect to output switching losses: the shorter the switching slopes at the output, the lower the losses for a given frequency. In a real world setup with track inductance, capacitor ESL and bridge parasitic inductances, faster—more harmonic rich—slopes will induce more and higher magnitude ringing. This is unwanted from an EMC point of view, as well as transistor voltage rating point of view. A trade-off needs to be made between the switching slopes, the parasitic inductance present in the circuit, and the circuit voltage rating.

### 11.2.5 Reverse Recovery Losses

Silicon diodes exhibit a reverse recovery current when going from forward conduction to reverse conduction [1]. This is visible as a reverse recovery charge $Q_{\text{rr}}$ that is flowing while the diode is turning off, which is flowing on top of the average current. This will cause additional switching losses in the switch turning on.

The reverse recovery current $I_{rr}$, reverse recovery time $t_{rr}$ and reverse recovery charge $Q_{rr}$, are intrinsic diode parameters, and depending on

- The forward diode current before turning off;
- The slope of the diode current while turning off;
- The applied reverse diode voltage;
- Ambient temperature.

$$P_{rr} = Q_{rr}.V_{SW}.f_{sw} \tag{11.5}$$

The (idealized) charges leading to switching and reverse recovery losses are depicted in Fig. 11.1, with $V_q$ the transistor drain-source voltage, $I_q$ the transistor current, and $I_d$ the diode current.

The five main charge contributors to the switching losses are depicted in the colored areas of Fig. 11.1.

- Area 1 and 4 represent the normal switching charge.
- Area 2, 3, and 5 represent additional charges that are due to reverse recovery conduction of the diode.
- Area 3 and 5 represent the reverse recovery.

The rise- and fall-time of the current switching from the diode to the transistor is limited by the parasitic inductance in the construction of the power module, which has a negative impact on the overall switching losses [2].

Reverse recovery current is much less in SiC as compared to Si (about 30% of comparable Si diodes) [3], and inherently not present in GaN transistors.

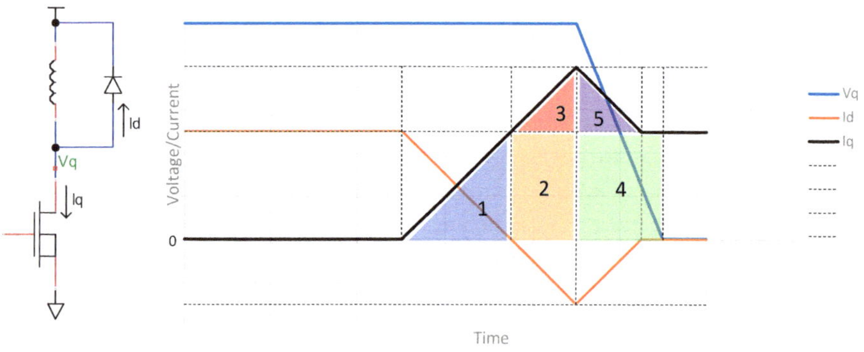


**Fig. 11.1** Switching and reverse recovery losses

## 11.3 Benchmarking

SiC and GaN transistors provide technical advantages compared to traditional Si transistors, but benchmarking is needed to make an impartial comparison. The most widely used FOMs are gate charge vs $R_{\rm dson}$ [4], and breakdown voltage vs $R_{\rm dson}$ [5], typical graphs are depicted in Figs. 11.2 and 11.3. These parameters on itself are not so relevant if not compared in a tangible and normalized case.

### *11.3.1 Benchmarking Based on $Q_g$*

A selection of six high voltage transistors (two Si, two SiC, and two GaN devices) was made for benchmarking the losses for a given load case, with a given $R_{\rm dson}$ vs $Q_{\rm g}$ as presented in Fig. 11.4.

In general, the following FOM = $R_{\rm dson}{\cdot}Q_g$ holds (valid for 1.2 kV transistors):

- SiC devices = 2–10 mΩ·uC
- Si devices > 100 mΩ·uC
- GaN devices < 0.5 mΩ·C

This suggests that GaN devices are far more efficient in combining low gate driver losses with low output impedance.

As use case, one SiC, Si, and GaN FET were selected from Fig. 11.4, and normalized to a fixed $R_{\rm dson}$ of 5 mΩ, 600 V, and 100 A load, to provide a fair loss comparison. $P_g$ vs. $f_{\rm sw}$ is depicted in Fig. 11.5. In general, the power devices are liquid cooled, in contrast to the gate driver board, which may be enclosed and

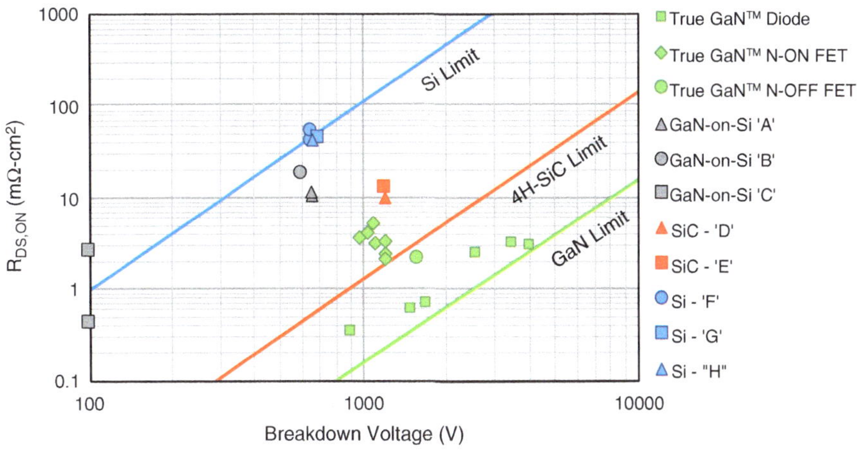


**Fig. 11.2** $R_{\rm dson}$ vs Breakdown voltage for Si, SiC, and GaN [5]

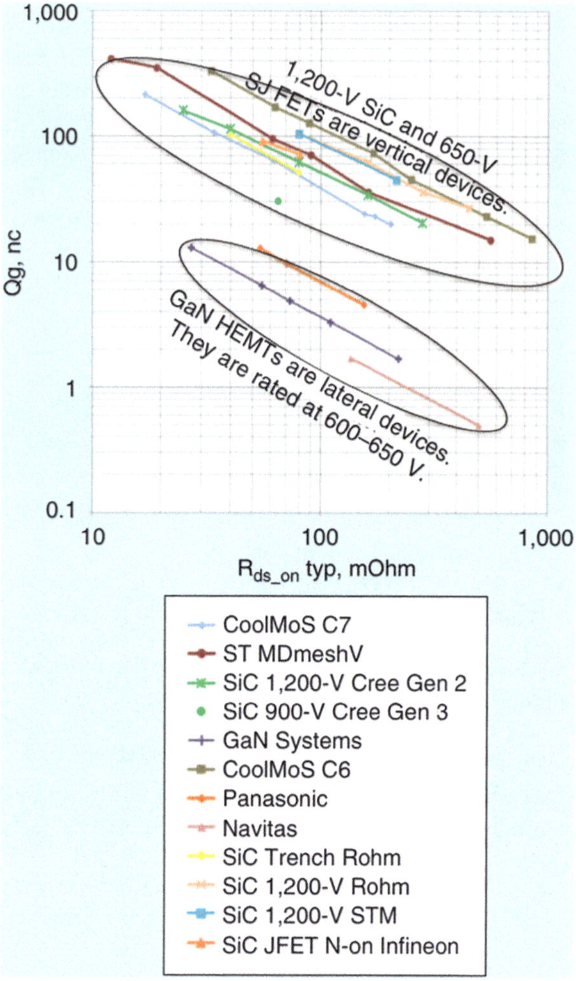


**Fig. 11.3** Gate charge vs $R_{dson}$ for Si, SiC, and GaN [4]

does not have active cooling. Practical gate driver board thermal constraints limit the gate driver dissipation to 1–3 W.

## 11.3.2 Benchmarking Based on $P_{loss}$

The same devices used in Sect. 11.3.1 were used to provide a fair loss comparison. $P_{\text{loss}}$ consists of $P_g$, $P_{\text{sw}}$, $P_{\text{Coss}}$, and $P_{\text{Rdson}}$. $P_{\text{rr}}$ loss was not included in this comparison, as it is only representative for Si power transistors.

**Fig. 11.4** Gate charge vs $R_{dson}$ for Si, SiC, and GaN for six off-the-shelve (OTS) devices

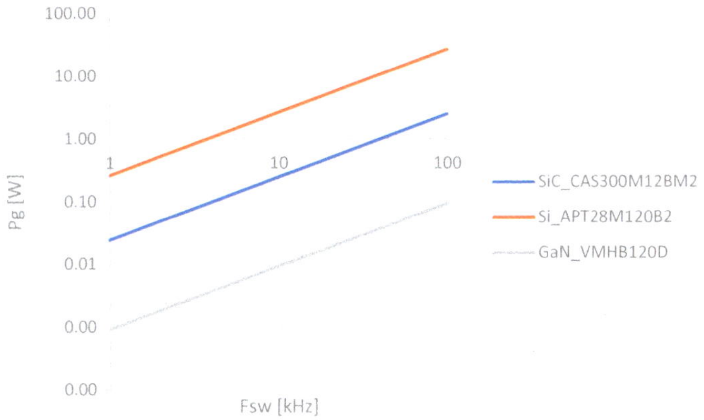


**Fig. 11.5** Gate losses for Si, SiC, and GaN for three OTS devices

Figure 11.6 shows that at low switching frequency (1–2 kHz), there is little value in GaN and SiC vs Si. Above 10 kHz, the loss behavior for SiC and GaN is very similar, with the main difference found in the switching losses.

### *11.3.3 Benchmarking Based on Specific Area*

As noted in Sect. 11.2.3, the specific $R_{dson}$ is seldom available for commercial power devices, and an estimate based on the package size needs to be done. When comparing the three reference devices of Sects. 11.3.1 and 11.3.2, the following specific $R_{dson}$ values are estimated:

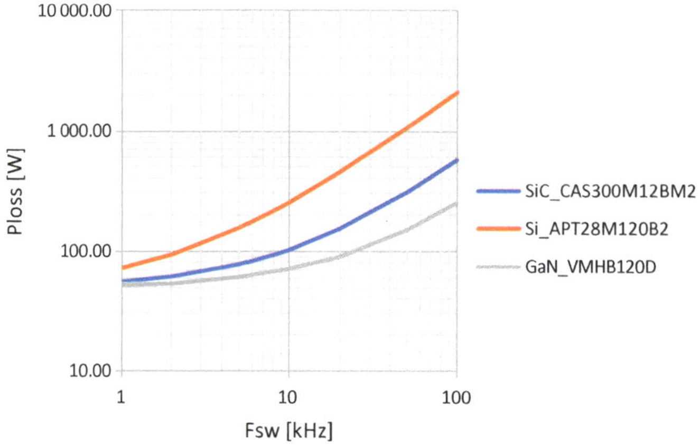


**Fig. 11.6** $P_{\mathrm{loss}}$ vs $f_{\mathrm{sw}}$ for 5 mΩ normalized SiC, GaN, and Si FETs, switching 600 V/100 A

- SiC (SKM350MB120SCH15): 19.5 Ω·mm$^2$
- GaN (VM40HB120D): 125 Ω·mm$^2$
- Si (APT28M120B2): 244 Ω·mm$^2$

This shows that for 1.2 kV applications, the largest power density can be obtained using SiC based devices (5–10 times smaller than GaN or Si). For 600 V devices, GaN devices have an unbeatable specific resistance, but the thermal system design becomes increasingly challenging.

## 11.4 Design Trade-Offs

### *11.4.1 Challenges for Wide-Bandgap Drivers*

Gate drivers for wide-bandgap transistors have their specific design challenges, which were less prominent in the era of "old-school" power semiconductors pre-drivers. The following table lists a (non-exhaustive) list of challenges and how they can be tackled for GaN and SiC based power converters.

The focus of the next paragraphs will be on slew-rate control of the output stages, which forms one off the key techniques to trade-off efficiency versus EMI.

### *11.4.2 Gate Driving Principle*

A gate driver controlling the $V_{gs}$ of a power transistor will pass different three different phases in its off-to-on (or on-to-off) transition, as depicted in Fig. 11.7.

| Challenge | Limits | GaN | SiC |
|---|---|---|---|
| Gate loop inductance | Achievable $dI_g/dt$ and excessive ringing | Low inductive layout, keeping the gate-source loop inductance small | Low inductive layout, laminated gate-source connections to limit the self-inductance |
| | | Current programmable slope-control of the gate trading efficiency vs EMI performance | Resistor programmable slope-control of the gate voltage trading efficiency vs EMI performance |
| | | Integrated predriver + power stage design | Miller clamp and negative gate bias to avoid parasitic turn-on, at the cost of increased gate loss |
| Supply loop inductance | Achievable slew-rates (>100 V/ns) and required derating of the power devices | Integrated predriver + power stage design | Laminated bus-bar designs, limiting the self-inductance |
| | | Co-design of thermal and power interface | Low inductance and parallelized supply decoupling |
| | | Active slew-rate control | Multiphase control |
| Accurate gate voltage control | Achievable power density and robustness versus gate oxide breakdown | Low gate swing (typically 5 V) requires regulated and preferably programmable gate voltage control | SiC behaves as a voltage controlled resistor, so keeping the $V_{gs}$ as close as possible to breakdown is preferred |
| Short-circuit capability | Higher power density and inherent smaller thermal mass limit the short-circuit capability | Fast short-circuit detection circuitry | Fast DESAT detection circuitry ($\ll 1\,\mu$s), or combination with Rogowski sensing |
| High slew-rates | Levelshifter and isolator common-mode transient immunity (CMTI) performance | Custom levelshifter designs | High CMTI capable isolators |

1. In the *precharge phase A*, the gate is charged until it reaches the transistor threshold $V_{th}$.
2. Beyond the threshold $V_{th}$, the load current will start to flow in the power transistor, and when high enough $V_{ds}$ will start to slew down. The slewing $V_{ds}$ injects a charge into gate through $C_{dg}$, counteracting a further increase of $V_{gs}$. A balance exists between the charge delivered from the gatedriver and the charge evacuated through $C_{dg}$, keeping the $V_{gs}$ quasi constant at the *Miller plateau phase B*.

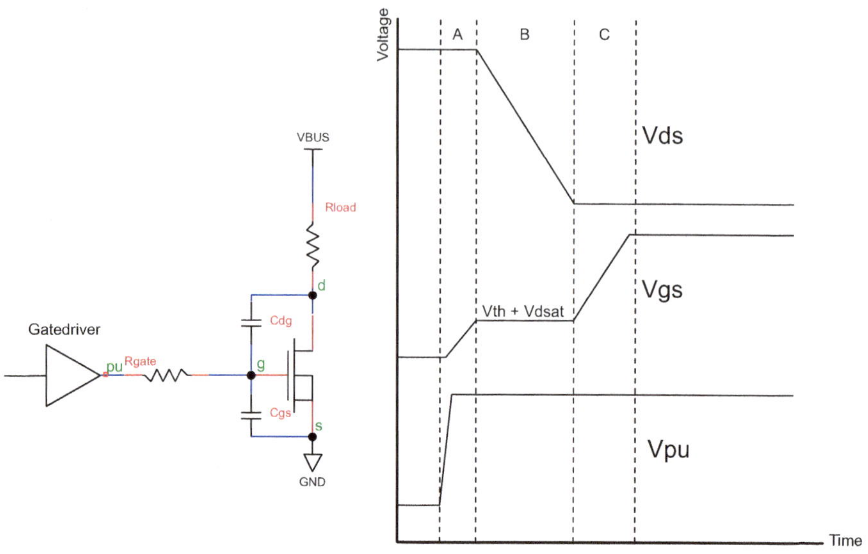


**Fig. 11.7** Principle gate driver voltage waveforms

3. Once the $V_{ds}$ has completely slewed down, the $C_{dg}$ charge injection stops, and $V_{gs}$ can further charge through its *post-charge phase C*, where the power transistors reaches its nominal $R_{dson}$.

The precharge A and post-charge C phases are ideally minimized at the benefit of:

- Maximal dutycycle range;
- Minimizing free-wheeling losses;
- Minimal joule losses in the power transistor, by minimizing the time to reach its nominal $R_{dson}$.

The Miller plateau phase B is a trade-off between predriver strength, board layout quality, and supply system quality [5], and can be estimated from the specified gate-drain charge (at a specified $V_{ds}$ swing) and predriver impedance, as shown in Eq. 11.5.

$$t_B = R_g . \frac{Q_{GD}}{V_{DD} - (V_{th} + V_{dsat})} \tag{11.6}$$

Note that a Miller plateau is not necessarily present: in case of inductive loads, depending on the current direction, the output is pulled high/low during the non-overlap of low and high-side drivers, and both $C_{dg}$ as $C_{ds}$ are (dis)charged through the load—before the $V_{gs}$ transient has occurred.

A gate resistor—either in the form of an effective resistor or the inherent predriver impedance is a trade-off:

- Dissipating the gate loss in the gatedriver is costly (die area) or impossible (lack of cooling), and a gate resistor keeps the power outside of the die package.
- Inherent gate-source parasitic inductance may cause excessive ringing of the gate. Since for OTS power FETs the inductance is not controllable, adding a gate resistor will limit $dI_g/dt$ [6].
- Lack of Kelvin gate connections may also cause excessive ringing beyond the safe gate-source voltage levels. Similarly, adding a gate resistor will slow down the Miller plateau, and slow down the $I_{ds}$ changeover.

### *11.4.3 Multiphase Turn-on/off and Slew-Rate Control*

Keeping the precharge and postcharge time minimal makes sense to keep the non-overlap times minimal and dutycycle range maximal. This can be implemented in several ways, depending on how much gate loss is expected. For $P_g < 0.5$ W, and a decent thermal interface for the gate driver, it is possible to drive without external gate resistors, using programmable and dynamically switched current sources (e.g., Texas Instruments DRV8301). For $P_g > 0.5$ W external gate resistors are mandatory to keep the predriver area contained.

Multiphase turn-on and turn-off schemes can be devised that modulate the gate resistance $R_g$ during phase A and C. This can be done in closed- and open-loop. An example of closed-loop control use gate voltage comparators is referenced in [7]. In open-loop programmable time delays can be used to dynamically modulate $R_g$, but will require upfront knowledge about the power transistor dynamics. Clearly, using external gate resistors puts a practical limit to the granularity of gate strength control that can be applied: each resistor requires a separate (internal) driver + IO, with its own board space and inherent parasitics.

The inherent fast switching capability of GaN and SiC transistors goes hand in hand with a low-inductive gate driver and supply system design. On one hand, fast switching edges are required to minimize switching losses, on the other hand the inherent package and supply inductances and involved ringing will require a method to keep the slew-rate under control, trading-off efficiency versus EMI performance.

### *11.4.4 Solutions for GaN Gate Drivers*

A custom gate driver for GaN transistors was developed by MinDCet, the MDCD074, keeping in mind the right trade-offs to meet the stringent requirements of high $dV/dt$ (>100 V/ns) operation characteristic for GaN transistors. The design was done in 0.18 μm SOI, featuring both high-side and low-side pre-drivers. The internal gate driver supply generation is implemented with a chargepump assisted bootstrap circuit, allowing 100% dutycycle and safe start-up. The resistive levelshifters were specifically developed to meet the fast common mode transients,

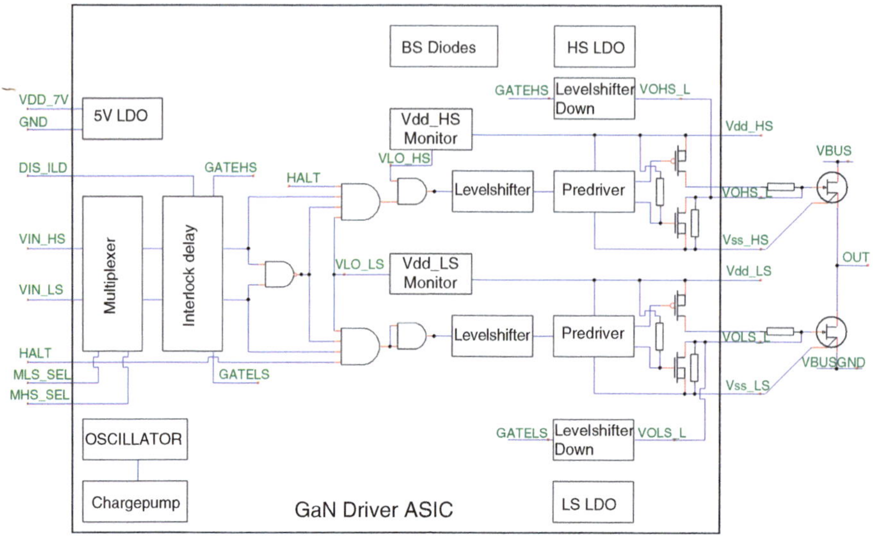


**Fig. 11.8** Block diagram of the MDCD074 GaN driver ASIC

combined with short propagation delay (<10 ns) and low and high-side delay matching to within one 1 ns.

Tight control of the gate voltage is an inherent reliability requirement for GaN transistors, and is achieved by using on-chip programmable floating LDOs, which avoid unwanted overcharging through the bootstrap circuits. Undervoltage lockout circuits make sure the gate driver goes into a safe state in case of a power failure.

The block diagram of the MDCD074 is presented in Fig. 11.8.

As one of the first adopters of GaN in switching power stages, we combined the MDCD074 with GaN power transistors in a power module called GaNyMAD. It features Alu-core PCB with CoB (Chip on Board) assembly, which allows for an optimal thermal design—respecting low inductance driver design in a very compact floorplan. The GaNymad module is presented in Fig. 11.9.

## *11.4.5 Solutions for SiC Gate Drivers*

A custom SiC gate driver was developed in 0.18 μm SOI, keeping in mind the challenges as listed in the table of Sect. 11.4.1. As the majority of SiC gate drivers are simple buffers without intelligence—and SiC is challenging to drive—intelligent gate drivers are becoming a necessity. For that reason many programmable configuration and safety related features were implemented. One of the key features is the internal diagnostics, which allows for realtime cross-checking of the driver state versus its control inputs, fast desaturation detection (<1 μs) with programmable thresholds, and fast high dynamic range gate comparators. The

**Fig. 11.9** MCD074 GaN driver combined with 90A GaN FETs into the GaNyMAD module

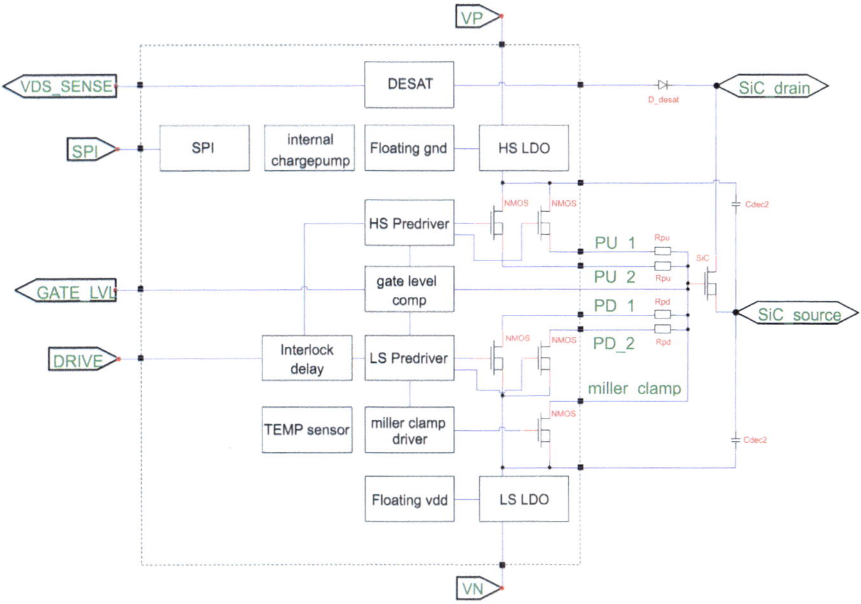


**Fig. 11.10** SiC gate driver for 3uC SiC FETs, featuring multiphase turn-on and off, and automatic Miller clamp driving

floorplan of the driver was developed to limit parasitic inductance to an absolute minimum, allowing to achieve gate drive currents in excess of 10 A with propagation delays below 50 ns. A self-latching separate Miller clamp driver was implemented, tying the gate voltage of the SiC FET negative to its source as a remedy for parasitic turn-on.

A block diagram of the developed driver is depicted in Fig. 11.10.

## 11.5 Conclusions

There is a clear efficiency benefit for GaN and SiC when switching at higher frequencies, which will lead to longer battery life and more compact and lighter drives with less heat to be evacuated.

Getting the maximal performance out of wide-bandgap transistors requires a custom gate driver, due to inherent parasitics in the build-up or the combination of gate driver and power transistor [2]. Neglecting one of the aspects (gate-loop, parasitic turn-on . . . ) will result eventually in decreased lifetime of the system.

The optimal gate driver solution combines speed with high-drive strength. Integration of the complete gate driver system is a necessity to cope with the high switching node $dV/dt$. Full integration of the driver with the module is the key to a successful power module, combining low parasitics with a good thermal interface.

Future improvements that will stretch the efficiency boundary for GaN and SiC even further will require integration of the driver inside the power module. This can be done fully monolithic, as implemented in the SloGaN project, (https://www.imec-int.com/en/what-we-offer/research-portfolio/slogan), Navitas and Panasonic [8], or requires a custom SiC power module development. Especially for SiC, a lot of performance gain is expected by optimizing the power delivery to the driver stages, through lamination, as well as lamination of the predriver connections [9]. Further developments are ongoing that will focus on isolators with high CMTI, and isolated supplies with ultra-low parasitic coupling.

## References

1. ST, *Calculation of turn-off power losses generated by an ultrafast diode*, Application note AN5028
2. P. Rutter, Optimizing power MOSFET behavior for high frequency switching, in *PwrSOC 2014*
3. X. Zhang, Unleash SiC MOSFETs – extract the best performance, in *APEC 2018*
4. IEEE, IEEE ITRW working group position paper—System integration and application: Gallium nitride. IEEE Power Electron. **5**(2), 34–39 (2018)
5. Vishay Siliconix, *Power MOSFET basics*, Application note AN608A
6. P. Brohlin, Electrical and thermal packaging challenges for GaN devices, in *PwrSOC 2016*
7. A. Paredes, V. Sala, H. Ghorbani, L. Romeral, A novel active gate driver for silicon carbide MOSFET, in *IECON 2016 - 42nd Annual Conference of the IEEE Industrial Electronics Society, Florence*, (IEEE, 2016), pp. 3172–3177
8. T. Morita, Monolithic integration of GaN power transistors integrated with gate drivers, in *PwrSOC 2016*
9. G. Königsmann, R. Herzer, High power, high frequency gate driver for SiC MOSFET modules, in *PCIM 2016 Nürnberg*

# Chapter 12
# Challenges in Driving New Generations of Power Switches for Motor Drive: A dV/dt Self-Adjusting Architecture for Superjunction Power Devices

**Sergio Morini, Davide Respigo, and Martina Arosio**

## 12.1 Introduction

In recent years, Superjunction (SJ) technology has revolutionized the industry of high voltage power devices. Power IGBTs and MOSFETs based on SJ technology are able to overcome the trade-off relationship between breakdown voltage and on-resistance, better known as the "Silicon-Limit" [1].

The Superjunction structure is characterized by a voltage sustaining layer composed by the alternation of vertical $p$ and $n$ doped regions, called pillars, instead of the $n$- drift region of conventional devices. In Fig. 12.1, the vertical SJ sustaining layer is used in a trench field effect transistor (FET) structure compared with the structure of a standard MOSFET [2, 3].

The basic rule for SJ layer design is to satisfy the charge balance condition:

$$Y_N N_D = Y_P N_A$$

where $Y_N$ and $Y_P$ are the widths of the two pillars while $N_D$ and $N_A$ are their dopants concentration.

The same total amount of $n$ and $p$ dopant (i.e., the charge balance rule) makes the sustaining layer behave exactly like the intrinsic $n$- layer of conventional power devices in reverse bias. The condition for this is that the pillar region must be completely depleted before reaching the breakdown voltage for a simple planar $N_D$-$N_A$ diode.

S. Morini (✉) · D. Respigo
Infineon Technologies Italia S.r.l., Pavia, Italy
e-mail: sergio.morini@infineon.com

M. Arosio
Department of Physics, University of Milano-Bicocca, Milan, Italy

© Springer Nature Switzerland AG 2020

A. Baschirotto et al. (eds.), *Next-Generation ADCs, High-Performance Power Management, and Technology Considerations for Advanced Integrated Circuits*, https://doi.org/10.1007/978-3-030-25267-0_12

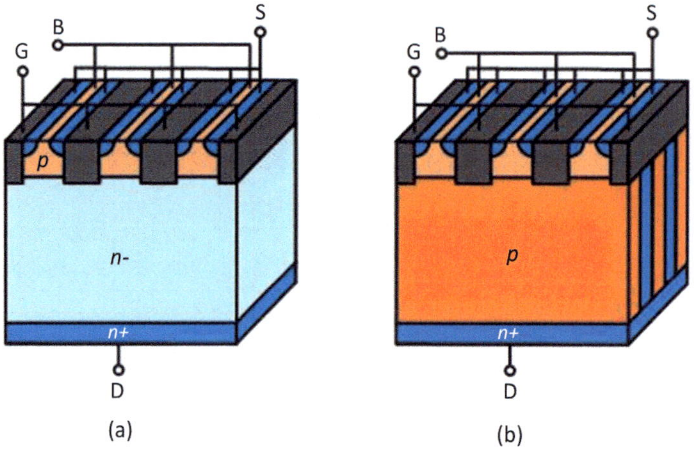


**Fig. 12.1** (**a**) Standard MOSFET structure, (**b**) SJ FET structure

So if the technology is able to build very thin pillars, the doping on both sides can be proportionally increased. When the pillars region is completely depleted, the electric field is almost linear, slowly increasing with respect to the applied voltage and behaving substantially as the *n*- layer from breakdown voltage point of view.

Highly doped pillars provide a low resistance path for the current flow in direct biasing condition of the junction, making the on-resistance drastically lower than the usual *n*- layer. The total p-n junction area is also greatly increased by the folded shape generated by alternate pillars doping. SJ technology dramatically improves the on-resistance versus breakdown voltage trade-off with respect to conventional planar technology.

The SJ structure constitutes the drain-bulk or the collector-base junctions where it is needed to sustain high voltage. However, SJ power switches have also an adverse effect. SJ devices, in fact, show a steep jump or discontinuity in the output capacitance. They are characterized by a gate-drain ($C_{GD}$) or gate-collector ($C_{GC}$) capacitance which is extremely non-linear along the drain-source or collector-emitter voltage range. This happens because the depletion of the sustaining layer is non-linear too. When the pillars region is not fully depleted, as in Fig. 12.2a, even a small voltage increase requires a large movement of charge, which results in a large measured output differential capacitance. The junction is a "huge highly doped diode." When the edges of the depletion region along the pillars are instead connected, as in Fig. 12.2c, and this happens quite abruptly, the further increase of the applied voltage can occur with a very small movement of charge which results in a very low measured output capacitance. The junction is equivalent to a thick depleted n- region, where capacity depends on the thickness of the pillars, which is an independent parameter of the technology.

This capacitance is fundamental for switching speed transients. Many applications (e.g., motor drives) require to turn on the switch when $V_{DS}$ is high, in other

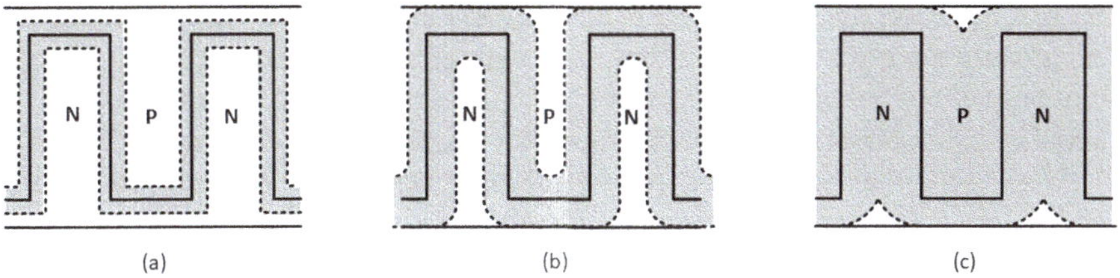


**Fig. 12.2** The non-linear depletion of the SJ structure at (**a**) small applied reverse voltage, (**b**) medium applied reverse voltage, and (**c**) high applied reverse voltage

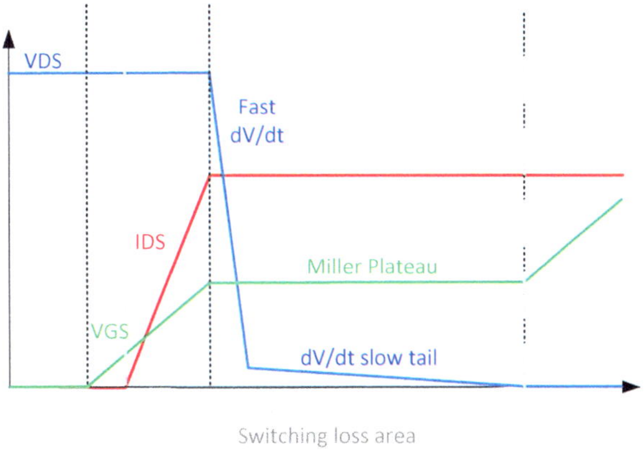


**Fig. 12.3** dV/dt characteristics for SJ MOSFETs

words there is an overlap between high voltage and high current during the turn-on switching event. In those applications, called "hard-switching," a very steep $V_{DS}$ voltage transient, called dV/dt is present.

The power switch turn on transient is usually performed by applying the full $V_{GS}$ to a P-channel driving MOSFET (see Fig. 12.4) whose source is connected to local supply and drain to the gate of the power switch itself. So initially, the pmos driver is in saturation and gives his maximum current called Io+. After its drain rises, the pmos enters into triode region and the drain current starts decreasing. However, Io+ is usually almost stable until the external power transistor threshold. Consequently, the Io+ value is fundamental for the switching speed.

The fast transient is mitigated by the $C_{GD}$ Miller multiplication factor in standard planar power transistor technology, this fact has been implicitly used in most motor drive systems for a coarse, but effective dV/dt setting. In fact, the Io+ current almost fully flows into $C_{GD}$ during the dV/dt and the gate voltage appears flat forming the "Miller plateau." If dV/dt doubles his speed, the $C_{GD}$ doubles its apparent value and the system is somehow self-regulating.

The output switching starts with very high dV/dt in SJ devices (when $V_{DS}$ voltage is high) and it finishes with a long "slow tail" in the last few volts (when $C_{GD}$ becomes huge) as it can be seen in Fig. 12.3.

This dynamic behavior is typical of every SJ device and both very fast and slow tail transients are not wanted. Furthermore, $C_{GD}$ being a parasitic capacitor, its value is not usually controlled in production causing a wide lot to lot variation, up to 30%–300% of the nominal value. So it is important to adjust the voltage slope dV/dt.

Driving a motor with power switches always carries two constraints to be taken into account. On one hand we would like an as fast as possible dV/dt to limit the power dissipation, on the other hand we have to not exceed a certain speed limit to meet electromagnetic interference (EMI) and electromagnetic compatibility (EMC) specifications and to protect the motor.

Usually the Io+ driving capability of the gate driver is precisely selected to find the best compromise between these two opposite constraints. By using SJ devices, instead, is impossible to find a correct single Io+ current value to turn on the device: the switching speed appears to be too fast and too slow at different moments of the same switching event.

In this paper, a system is discussed that is able to work in the optimum self-adjusting operating point for the fast dV/dt portion and to avoid the final slow tail. According to the motor insulation requirements and to EMI/EMC limitations it is typically recommended not exceeding a dV/dt limit of approximately 5 V/ns.

## 12.2 dV/dt Classification

Most electrical motors used in the main appliance market need to be driven with a sinusoidal phase current with variable frequency and amplitude. The usual way to provide it is to generate a sinusoidal voltage waveform in the lower frequency spectrum using pulse width modulation technique (PWM).

The AC signal with variable voltage and frequency is provided to the load by using power-switching elements (bipolar transistor, MOSFET, IGBT), used to modulate the voltage applied to the motor, connected in the inverter-leg configuration illustrated in Fig. 12.4. The "Gate Driver" shown in the figure provides the way to charge the gate of the power-switching elements towards Io+/Io− current [4].

As an example, for the sake of simplicity, two IGBT transistors are considered. The IGBT is a unidirectional device, unlike MOSFET's that have bidirectional current switching capabilities. In general, the current in the inverter-leg is bidirectional, and can flow from the leg into the motor or vice versa thanks to the use of freewheeling diodes in parallel to the IGBTs.

In normal operation, the two power-switching devices of the inverter-leg switch on and off alternatively, but since power-switching devices are not ideal, the turn on time and turn off time are not strictly identical. Having both devices conducting at the same time will result in an abnormal rise of current. In order to avoid this dangerous bridge shoot-through, a so-called dead time (DT) is always inserted into the control scheme. With this additional time, which is always much shorter than the semi-period of the PWM, one switch will be always turned off first and the other will be turned on after DT is expired. Providing DT can on one side avoid bridge

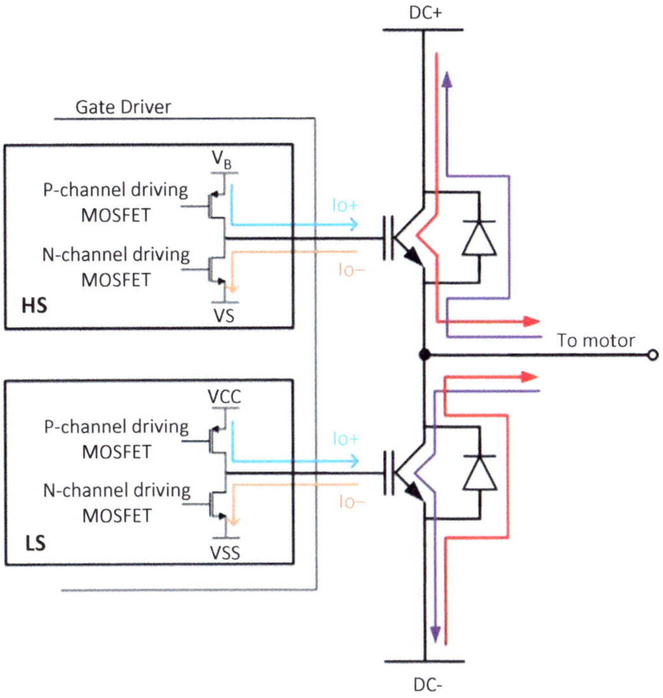


**Fig. 12.4** Inverter-leg configuration with gate driver output stage

shoot-through but on the other side has also an adverse effect. The load current in an inverter-leg can be positive (first row in Fig. 12.5) or negative (second row in Fig. 12.5). In both cases, the inverter-leg can be in three situations (as shown): High switch off, Low switch off; High switch on, Low switch off; High switch off, Low switch on. The current path is also shown in red.

Depending on the $I_{LOAD}$ direction, the current can flow into the IGBT or in the diode. If before DT the current was passing through the diode, when the IGBT is turned off the phase remains stable and the dV/dt, which happens after the switching on of the complementary IGBT, is defined *active*. If before DT the current was passing through the IGBT, when it is turned off, the motor phase has a slope dependent on $I_{LOAD}$, i.e., on how much current is flowing into the motor. If the current is high enough, the phase moves from DC+ to DC− or vice versa inside DT and the dV/dt is defined *totally passive*. If the current is lower than a certain level the phase starts to move but does not reach the opposite DC bus. At the switching on of the complementary IGBT the phase changes slope and becomes an active dV/dt. In this situation, the dV/dt is defined *partially passive*.

The three different types of dV/dt are represented in Figs. 12.6 and 12.7: *totally passive* (red and yellow lines), *partially passive* (green and blue lines), and *active* (violet line).

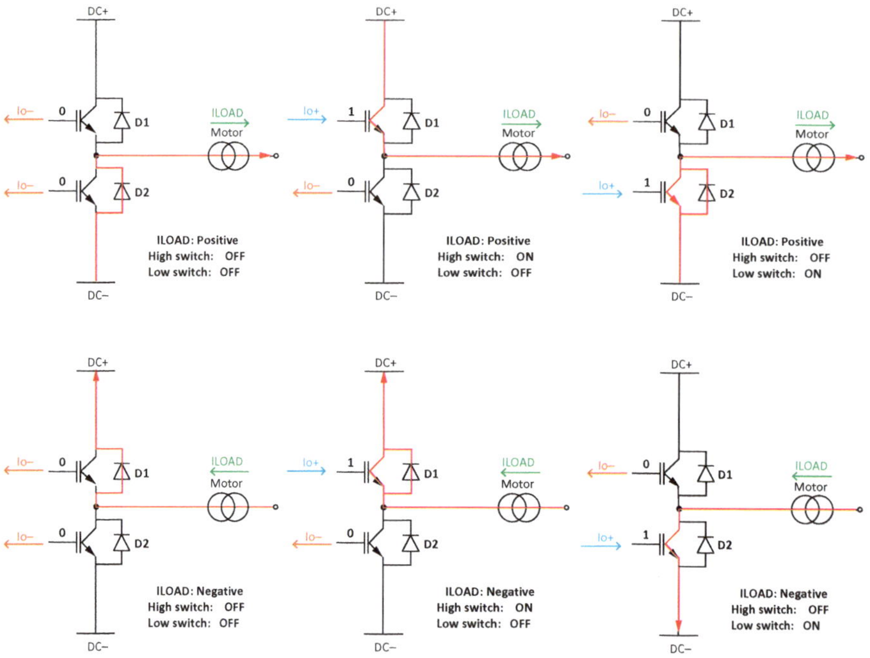


**Fig. 12.5** Possible configurations of the inverter-leg with positive and negative $I_{\text{LOAD}}$

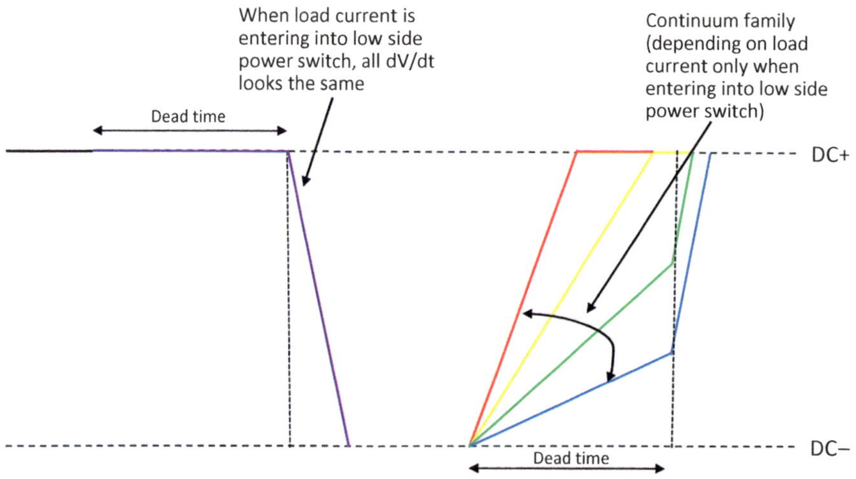


**Fig. 12.6** Negative $I_{\text{LOAD}}$, i.e., flowing from the motor to the inverter-leg

In particular, negative active dV/dt is driven by the switching on of the low side switch, while the positive active one is driven by the switching on of the high side switch. In totally passive dV/dt, the slope is entirely inside DT and before the

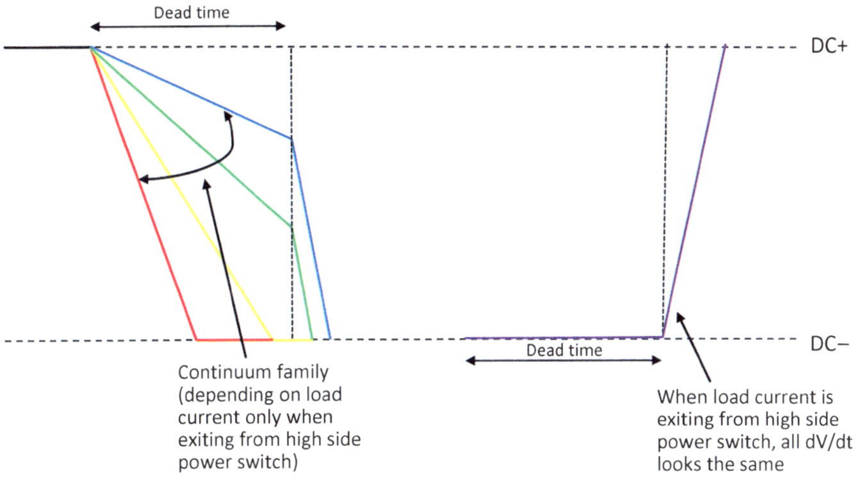


**Fig. 12.7** Positive $I_{LOAD}$, i.e., flowing from the inverter-leg to the motor

complementary power device turns on. This kind of dV/dt is out of the control of the gate driver, as previously said it mainly depends on the instantaneous current in the load and parasitic capacitances at motor phase node. It is usually sufficiently slow and no power is dissipated in the active devices. It is always slower than active ones because current sank from a turned-on power device is greater than the maximum motor current. Active dV/dt is the only one that has to be regulated.

In Fig. 12.8, a single phase of a motor drive stage and its driver are illustrated in order to show the nomenclature used in the following paragraphs.

In particular, $V_B$ is the high side floating supply voltage; VS is the high side floating ground voltage; VCC is the low side supply voltage; HO is the high side floating output voltage; LO is the low side output voltage; HIN and LIN are the logic inputs voltage. $D_{boot}$ and $C_{boot}$ provide the high side floating bootstrap supply.

## 12.3 Architecture Design

The core of the proposed dV/dt self-adjusting architecture is shown in Fig. 12.9. It comprises an HV capacitance, two resistances, two diodes, and six comparators. The concept behind it is very simple: using a linear HV capacitance, integrated on chip, it is possible to measure the voltage slope (dV/dt) in hard-switching applications and consequently regulate the next active switching event adapting the Io+ current. After the "measure" of the fast portion of dV/dt has expired, the current is increased to a much higher "uncontrolled" level $Io^{+}_{MAX}$ to "cut" the long secondary slow tail effect.

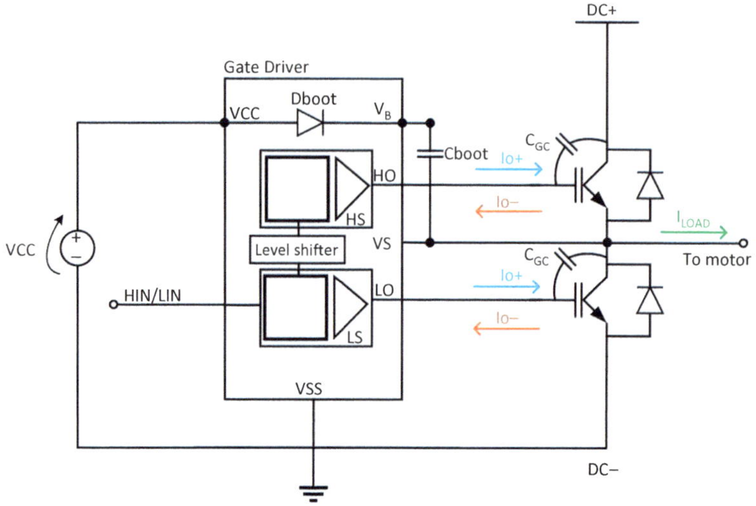


**Fig. 12.8** A single phase of a motor drive stage and its driver

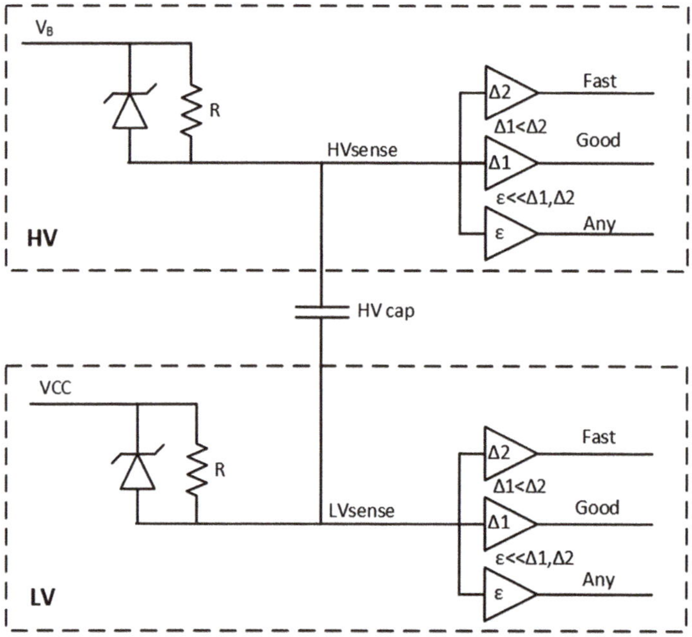


**Fig. 12.9** Architecture design of the proposed dV/dt self-adjusting architecture

### 12.3.1 High Voltage Capacitance

We need an HV (200–1200 V) capacitance connected between HV and LV sides. During the dV/dt, the charging and discharging of this capacitor requires a current that can be measured. In our solution, we target a 0.1 pF capacitance, which at 5 V/ns gives 500 μA, a current easily detectable and measurable. The high voltage capacitance can be external, even if one integrated on chip is preferred. In our solution we use an integrated one for simplicity, reliability, and cost reasons.

How an HV capacitance can be monolithically built in an IC strongly depends on specific integration technology, only generic ideas are described here.

Usually a linear high voltage integrated capacitance uses silicon dioxide as dielectric. To ensure lifetime reliability of this isolation layer it is required roughly a 3–4.5 μm thickness range for 600 V application, 6–9 μm for 1200 V.

Most HVIC technologies, where HV lies in the 200–1200 V range, are built with two or more voltage "domains," where each of them has a local supply in the 3.3–20 V range. At least one of those domains is free to move up to the HV range admitted by the technology referenced to another domain. As an example, a first domain, defined as low voltage, can be tied to 0 V (supplied by 15 V), and a second domain, the high voltage one, can be freely moved by the application in the range 0–600 V and supplied by a floating 15 V.

The voltage domains are isolated by a "termination" area between them (20–300 μm wide), which usually cannot be crossed by metal or other interconnection layers.

Many solutions to build such a small HV capacitance on chip can be used. All of them can be distinguish in two possible alternative approaches:

- Vertical approach: thick oxide (e.g., 4 μm) must be found. Electrodes can be top metal and poly or top metal and n+ silicon, as an example. Since it is often not possible to cross the termination to connect one cap side to LV and the other to HV, an internal bonding (i.e., both ends of bonding wire on the same chip) can be used. Eventually two or even more metal pad-poly caps can be connected in series to withstand the voltage with a floating bonding as interconnection (see Fig. 12.10). In case a 4 μm oxide is available between top metal and some conductive layer below, a vertical cap for 600 V range is feasible in a reasonable area. In fact a 120 μm × 100 μm capacitance gives a 100 fF capacitance, considering the silicon dioxide dielectric constant $\varepsilon_r = 3.8$. While in case of 8 μm oxide, suitable for 1200 V applications, 240 μm × 100 μm gives the same 100 fF capacitance.
- Lateral approach: use of long top metal wires at the minimum distance needed by the voltage rating all along the termination (see Fig. 12.11). Be aware of vertical voltage limits between wires and termination or substrate below. In this case, using the parallel plate capacitance formula can be misleading since fringe capacitance is by far dominant (e.g., $\varepsilon_r = 3.8$, 2 wires 1 μm thick, 2400 μm long (termination rings are often in the mm range), 4 μm distance gives 20 fF

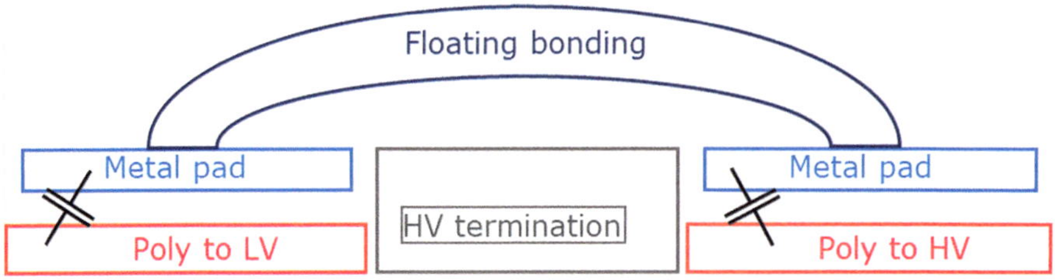


**Fig. 12.10** Vertical approach to build an HV capacitance integrated on chip

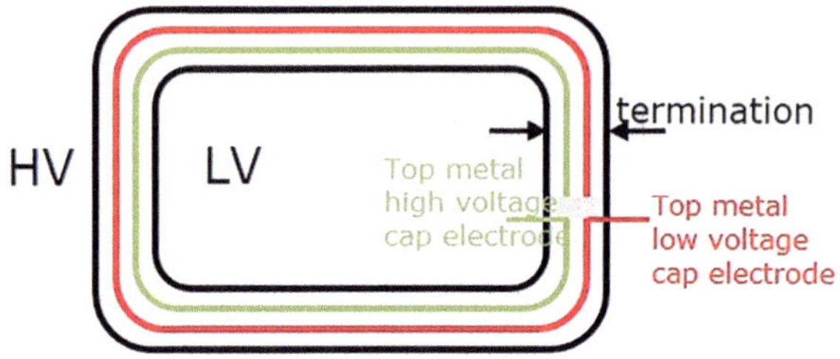


**Fig. 12.11** Lateral approach to build an HV capacitance integrated on chip

using the parallel plate capacitance formula. Instead, 101 fF is calculated when including dominant fringe effect contributions.)

## *12.3.2 dV/dt Self-adjusting Sensing Circuit*

The dV/dt validation and measurement are performed connecting the HV capacitance to local supply (VCC for the low side and $V_B$ for the high side) through a resistance and a protection Zener diode both in high side and low side. The positive active dV/dt slope is determined by the turning on of the high side power transistor, so its measurement is performed locally in the high side. During the positive slope, the HVsense node is pulled down and the voltage difference compared to the static value $V_B$ is proportional to the slope. The situation is specular for the low side. The RC time constant, where $C$ is the HV cap and $R$ the resistor to the local supply, is always much less than the active dV/dt duration: roughly 0.1 pF $\times$ 10 K$\Omega$ = 1 ns, while dV/dt lasts 50–200 ns.

The three comparators, placed on each side to sense the level reached by HVsense and LVsense nodes, have three different levels of thresholds, being $\Delta 1 < \Delta 2$ and $\varepsilon \ll \Delta 1, \Delta 2$. In this way, the logic signal *Good* ($\Delta 1$ threshold) is active at slower dV/dt with respect to *Fast* ($\Delta 2$ threshold), while the logic signal *Any* ($\varepsilon$ threshold) is active on any active and any passive dV/dt, even the slower ones, but it is not active during the slow tail. 5 V/ns is the target for active dV/dt, passive strongly depends on the application, but a possible range is 0.2–3 V/ns. Slow tail is often something radically different: 40 V/μs, as an example, so it is easy to distinguish it. Actually

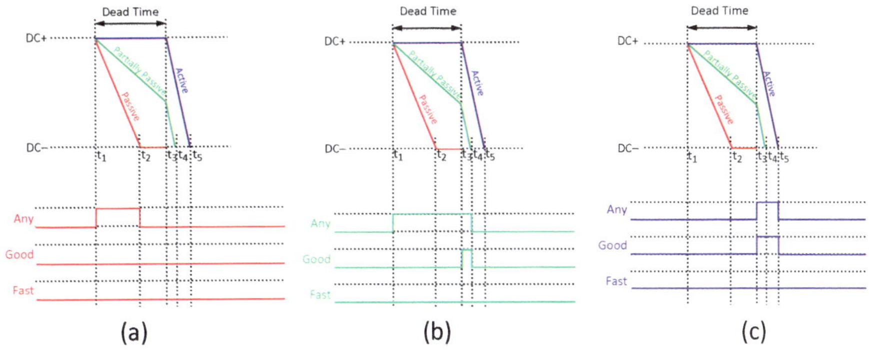


**Fig. 12.12** Monitoring of Any to categorize dV/dt types: (**a**) Totally passive, no comparator is active after the turn on, (**b**) Partially passive, Any is active before the turn on, and (**c**) Active, Any is active only after the turn on

*Any* signal does not detect strictly speaking the slow tail, it distinguishes between active and passive on one side and slow tail or steady state on the other.

As previously specified, we want to measure and regulate only active dV/dt, so during the measurement we need to distinguish between active, partially passive, and passive dV/dt. This is performed by monitoring the *Any* signal in comparison with the status of the locally driven power switch (Fig. 12.12). Be aware that the high side ignores the status of the low side switch and the exact DT position, it receives only the turn on and off timing information to drive the high side switch.

If a partially passive happens, *Any* is active during the complete dead time before turning on the power switch, an easy detectable situation. If a totally passive happens, *Any* is active during a first part of dead time, then is inactive for the remaining part of the dead time and remains inactive after the switching on of the power switch. Care must be put in detecting these situations. In both previous cases, the measurement is ignored.

Furthermore, when a partially passive dV/dt happens, comparators output can be unreliable in the valid window, because the active part of the pulse on HVsense or LVsense nodes has a smaller duration. For example, the duration can be so short that *Good* and *Fast* comparators do not activate at all and using only two comparators, we cannot distinguish this case from the one where dV/dt is slower than the target.

Since the width of the impulse sensed at the LVsense (HVsense) node is equal to the duration of the negative (positive) active dV/dt, which is in the order of hundred nanoseconds ($\tau = \frac{600\text{V}}{(5\text{V/ns})} = 120$ ns), the comparators input signal lasts sufficiently long to allow the use of a low-speed comparator. Furthermore, the height of the impulse sensed on LVsense (HVsense) node is set by the values of the HV capacitance and the resistors, which can be freely chosen; input signals can be in the Volt range by choosing appropriate resistors. The comparators can thus be made very simple, we can even use simple CMOS inverters with different switching thresholds as comparators. CMOS inverters have the great advantage of not dissipating static power because the static input levels are equal to the supply. This is a relevant benefit in bootstrap supplied systems, where static power consumption in HV needs to be very low.

Therefore, in an active dV/dt event, we can have only four valid combinations of the three logic signals:

- *Good*, *Fast*, and *Any* signals are active: dV/dt is faster than target.
- *Good* and *Any* are active and *Fast* is not: dV/dt is inside the target range.
- *Good* and *Fast* signals are not active, but *Any* is: dV/dt is slower than target.
- *Good*, *Fast*, and *Any* signals are not active: dV/dt is in the slow tail range.

### *12.3.3 dV/dt Self-adjusting Regulating Circuit*

The Io+ memorization and regulation is performed using a digital register. The register has an n-bits "active part" whose bits select binary weighted Io+ current component (i.e., binary weighted width P-channel driving MOSFETs). Each bit activates or deactivates one of the n P-channel driving MOSFETs connected in parallel forming the total driving capability required as it can be seen in Fig. 12.13. There is at least one driving P-channel that cannot be deactivated by the register because Io+ must have a low level of current that cannot be turned off. Otherwise Io+ could vanish, the switching event cannot take place, no dV/dt can be measured and the system could be blocked.

The register can also have an m-bits "buffer part" used to "stabilize" Io+ level in order to avoid ping-pong effect. These m bits are the least significant ones and have no direct influence on the Io+ value. An overflow bit can be used for fault condition detection, in case the max Io+ current available is not sufficient.

To converge towards the correct Io+ needed by the application, the measurement result can be, as an example, digitally integrated into a register using this simple algorithm in the three measurement cases:

- Case faster: register value is decreased by 1 LSB.
- Case target: register value is kept unchanged.
- Case slower: register value is increased by 1 LSB.

The signal *Any*, when it finished to be active, is used to force a much higher Io+ to reduce the slow tail effect. Forcing an $Io^{+}_{MAX}$ also helps to keep the power switch fully on by rejecting coupling noise induced by other legs switching.

Conditions which determine the correct Io+ value are slowly changing in time (due to IC and switches temperature, aging, supply) or not changing at all (switch used, parasitic capacitance at the phase node), so the system is able to reach the target and track the changes if needed.

To avoid to strobe wrong measurement results into the register, the dV/dt measurements must be validated. Each measurement is valid only in a short time window (500–1000 ns) after the corresponding power device gate has started to rise

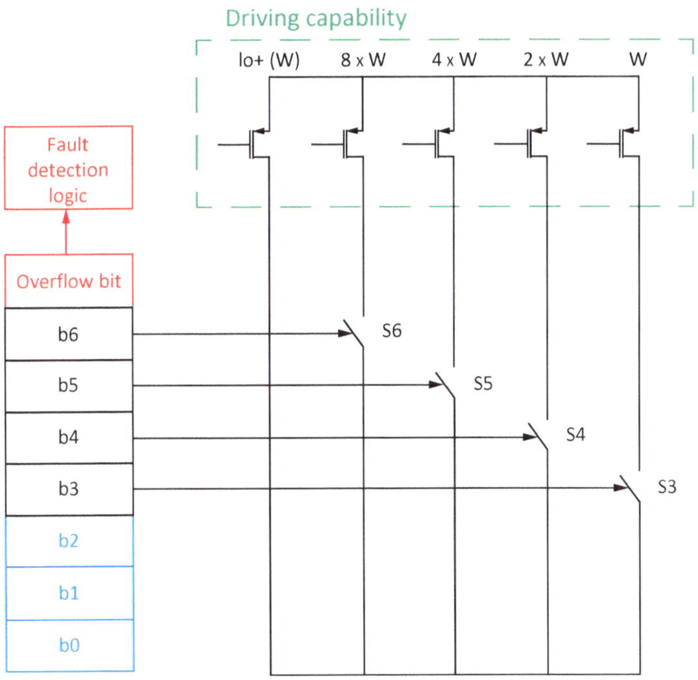


**Fig. 12.13** Digital register selecting binary weighted *W* of P-channel driving MOSFETs

or after Lin/Hin turn on commutation. This will avoid most false measurements due to, as an example, a huge DC bus noise sometimes induced by other legs switching.

If the *Any* signal is active before the validation windows (during DT), it means that a total or partial passive dV/dt event happens. In this case the data must be ignored and no data is strobed into the register.

The philosophy behind this is that in case of doubt, strobing the result in the register must be avoided. Strobing a small percentage of wrong measurements can induce the dV/dt to drift away from the optimum operating point. Ignoring the same percentage of valid measurements, instead, does not affect the system.

## 12.4 Simulation Results

The simulations of the proposed dV/dt self-adjusting architecture were conducted using the 600 V CoolMOS™ CFD7 SJ MOSFET, which is an Infineon's high voltage superjunction MOSFET technology with integrated fast body diode. The simulated system includes also a gate driver and the dV/dt self-adjusting sensing circuit. The gate driver has independent high and low side referenced output

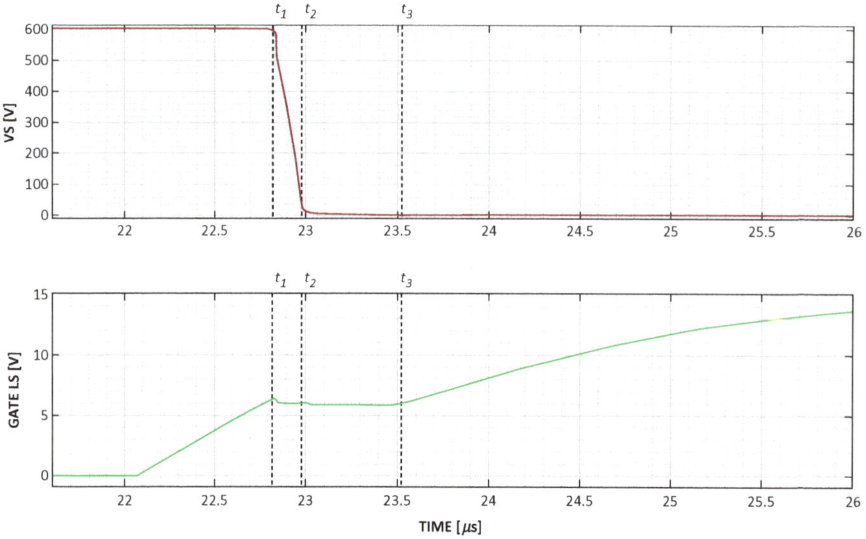


**Fig. 12.14** Low side active turn-on switching event

channels and it is configured to generate an Io+ current during turn-on switching events to drive both high side and low side transistors of the inverter-leg.

Each turn-on switching event starts with very high dV/dt (between $t_1$ and $t_2$) and it finishes with a long slow tail (between $t_2$ and $t_3$) as it can be seen in Fig. 12.14.

During the almost flat Miller plateau, the drive current is flowing into $C_{GD}$, which varies as a function of VS. The duration of the $t_2$–$t_3$ interval is 535 ns, a much larger value than that of the $t_1$–$t_2$ one, which is 152 ns. As it can be calculated, the slope of the slow tail in the $t_2$–$t_3$ interval is 0.039 V/ns. As it can be seen in Fig. 12.15, the power dissipation contribution due to the slow tail is not negligible.

The voltage across the transistor decreases from 600 V to 379 mV (the low side switch $V_{DS}$ voltage at the given current) during the turn-on switching event. The initial slope is detected by *Fast*, *Good*, and *Any* comparators. In Fig. 12.16, the output signals of the three comparators are shown together with the LVsense analog signal. In this example, we can see that the signals are not well stable along the whole width of the Miller plateau. This happens because we are very close to the dV/dt target of 5 V/ns.

In Fig. 12.17 the three $\Delta 1$, $\Delta 2$, $\varepsilon$ thresholds of the respective comparators are shown on the LVsense signal. When the LVsense signal is lower than $\Delta 2$, dV/dt is faster than target; when it is between $\Delta 1$ and $\Delta 2$, dV/dt is in the target range; when it is between $\Delta 1$ and $\varepsilon$, dV/dt is slower than target; finally when it is above $\varepsilon$, dV/dt is in the slow tail range. As it can be calculated, dV/dt is equal to 8.59 V/ns in the $t_1$–$t_2$ interval, 3.58 V/ns in the $t_2$–$t_3$ interval, and 4 V/ns in the $t_3$–$t_4$ interval.

Figure 12.18 represents the negative active dV/dt level with a too fast dV/dt. As it can be seen from the figure when the low side switch is turned on by the low voltage

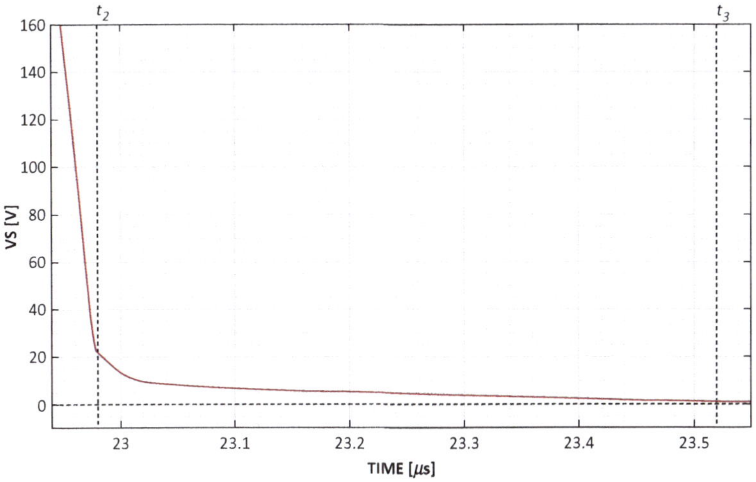


**Fig. 12.15** Zoomed portion of the slow tail

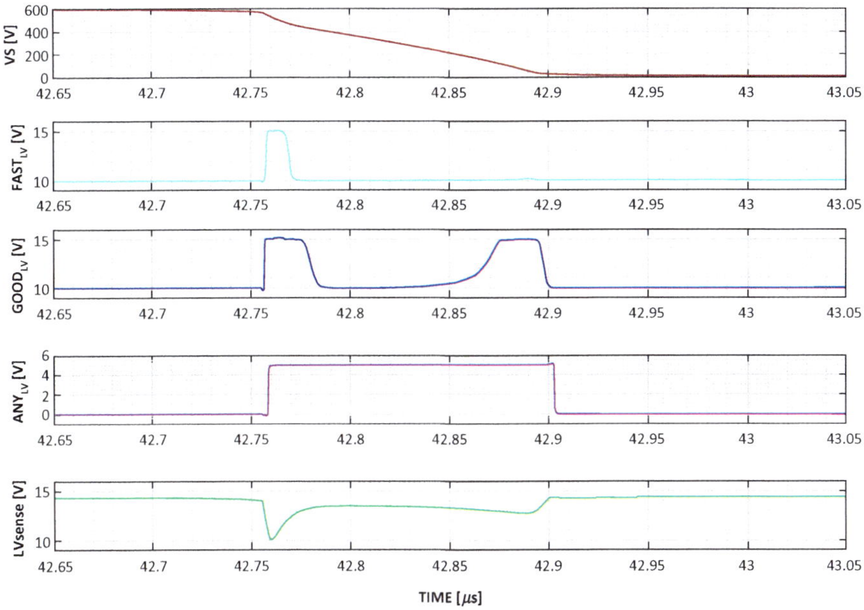


**Fig. 12.16** Comparators output signals with a close to target dV/dt

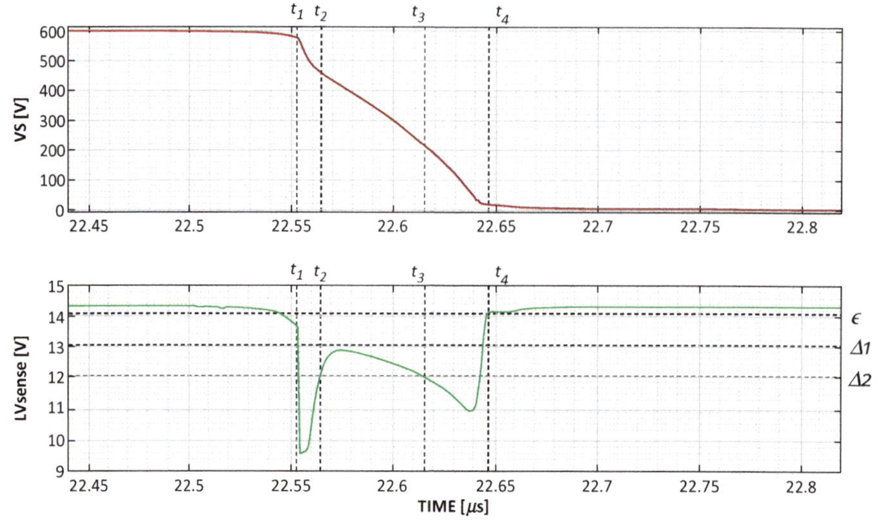


**Fig. 12.17** Comparator thresholds on analogue LVsense signal

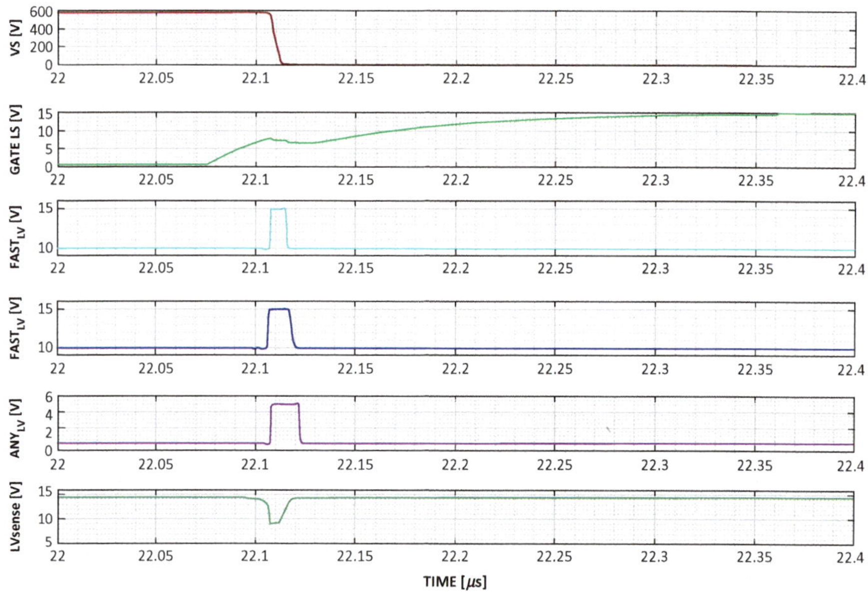


**Fig. 12.18** Comparators output signals with a too fast dV/dt

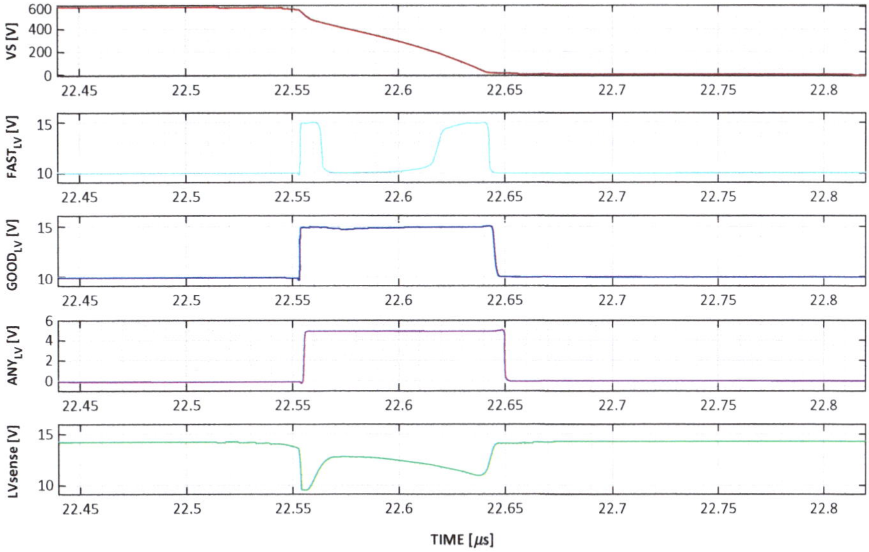


**Fig. 12.19** Comparators output signals with target dV/dt

side of the gate driver, the gate of the low side switch starts to rise. During the Miller plateau, LVsense node is pulled down compared to the static VCC voltage.

All the three comparators detect the active slope, this means that dV/dt is faster than target. As it can be seen, the expiration of the *Any* signal will be used to detect the end of the Fast-dV/dt portion of the turn-on switching event.

Figure 12.19 represents the negative active dV/dt level with a dV/dt in the target range. *Good* and *Any* comparators detect the total slope: this means that dV/dt is in the target range, but *Fast* comparator is also active in the first and last portions of dV/dt, this means we are still close to the target range. In these situations, comparator outputs are not well stable along the total slope and this has to be taken into account. As in previous cases, the *Any* signal will be used to detect the start of the slow tail portion of the turn-on switching event.

Figure 12.20 represents the positive totally passive dV/dt. As it can be seen it is slower than active ones, it has a slope of 2.37 V/ns, making *Any* the only signal active during the switching event. The positive dV/dt in the figure is a passive one because the VS node starts to rise immediately after the turn off of the low side switch and before the turn on of the high side switch.

Figure 12.21 represents the positive partially passive dV/dt with a too fast active portion. It has been simulated with the same condition as the totally passive one but with a reduced $I_{LOAD}$ current. As it can be seen, this leads to a dual-slope VS with a too fast active portion easily distinguishable both in *Good* and *Fast* comparator output signals and HVsense node.

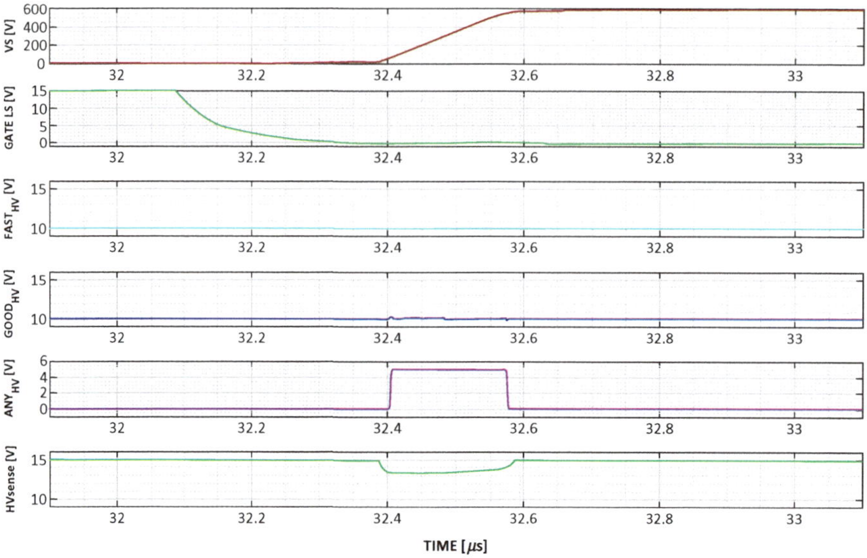


Fig. 12.20 Comparators output signals with a totally passive dV/dt

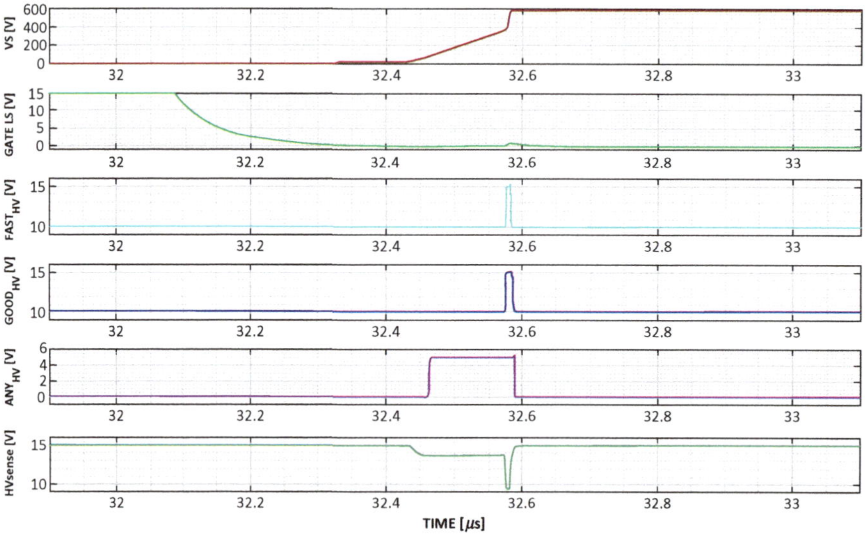


Fig. 12.21 Comparators output signals with a partially passive dV/dt

## 12.5 Conclusions

A dV/dt self-adjusting architecture composed by a sensing circuit and a regulating circuit is proposed and detailed in this paper. The sensing circuit was also validated through software simulations. The obtained results show that the idea is effective in mitigating SJ adverse effects. Once a complete system-level simulation comprising the gate driver, the sensing circuit, and the regulating circuit will be accomplished, the experimental results on a testchip will be conducted to confirm the effectiveness of the new driving strategy for hard-switching inverters stages proposed.

## References


1. E. Napoli, Superjunction, in *Wiley Encyclopedia of Electrical and Electronics Engineering*, ed. by J. Webster, (Wiley, Chichester, 2014)
2. T. Fujihira, Theory of semiconductor superjunction devices. Jpn. J. Appl. Phys. **36**, 6254–6262 (1997)
3. T. Fujihira, Y. Miyasaka, Simulated superior performances of semiconductor superjunction devices, in *International Symposium on PowerSemiconductor Devices & ICs, Kyoto*, (IEEE, Tokyo, 1998), pp. 423–426
4. Y. Lobsiger, J.W. Kolar, Closed-loop di/dt and dv/dt IGBT gate driver. IEEE Trans. Power Electron. **30**(6), 3402–3417 (2015)

# Part III
# Technology Considerations for Advanced IC

The third part of this book is dedicated to recent developments in the field of technology innovation in particular where efficient synergy is achieved between the external interfaces and the internal circuitry for specific application. The different papers address the aspects relative to different applications.

Chapter 13 by Baljit Chandhoke (GlobalFoundries) talks about 5G mmWave applications, how 5G new radio standards are driving increasing radio complexity and cost, and what are the technology requirements of mmWave applications. This paper concludes with technologies to meet the requirements of 5G mmWave applications.

Chapter 14 by Maurizio Pagani et al. (Huawei) presents different aspects in the development of millimeter-wave (mmWave) transceivers. His discussion introduced critical points due to economic aspects (including market penetration) and technological challenges. This paper concludes opening new scenarios for future millimeter-wave applications beyond the present transceiver use.

Chapter 15 by Andrea Onetti (STMicroelectronics) discusses recent micro-electromechanical systems (MEMS) development. This paper presents the main opportunities for MEMS-based devices in the fields of smart industry and smart driving, and discusses the key parameters that qualify high-accuracy sensors: noise, stability, and tolerance. Some design cases validate the proposed discussion.

Chapter 16 by Arnaud Laville et al. (Melexis Technologies) proposes the development aspects of position sensors, in volume, the most prevalent sensor in cars. The case of fully integrated Hall CMOS sensors is deeply studied since they have to face challenging operating conditions including wide temperature range ($-40^{\circ}$C to $+160^{\circ}$C) and electromagnetic compatibility (EMC). These developments are possible, thanks to design techniques with great synergy between Hall elements (HEs) and low and high-voltage transistors.

The final two papers deal with the performance reduction in integrated circuit exposed to radiation, which could be reduced by means of two different approaches.

In Chap. 17, Cristiano Calligaro and Umberto Gatti (RedCat Devices) shows that in space and avionics environments, the electronic components have to perform a certain level of radiation hardness, in terms of total ionizing dose (TID) and single event effects (SEE). To achieve this, various requirements, issues, and solutions related to the design of digital and analog ICs for such applications are applied. This paper shows that radiation-hardened-by-design (RHBD) approach allows to achieve the target radiation hardness, even with standard CMOS technologies (from 180 nm to 65 nm) where radiation is extremely affecting the single device performance.

In the last paper (Chap. 18), Marcello De Matteis et al. (University of Milano-Bicocca) shows how devices realized in scaled technology exhibit good radiation hardness. This reduces the design effort for achieving the target performance radiation hardness. This is demonstrated by means of measurement on single devices irradiated up to 1Grad-TID and with a design case of a pixel read-out chain, all of them realized in bulk-28 nm technology.

# Chapter 13
# Silicon Technologies for the Next Age of Wireless Connectivity

**Baljit Chandhoke**

## 13.1 Introduction

5G is not just a new generation, but a new kind of network. 5G needs scalability and adaptability across extreme variations in use cases. For IoT Applications which need to connect to a massive number of things, ultralow cost and ultralow energy consumption is a requirement. For Ultra Mobile Broadband Applications, extreme broadband due to applications such as 4K video, gaming, and ultra-high capacity to support the data rates needed for these applications is a requirement. For Ultra-reliable mission critical applications such as services to remotely control vehicles, medical procedures with ultra-reliable links, use cases like autonomous vehicles, remote medicine, ultra-low latency, and ultra-high reliability is a requirement. Deep Awareness is needed for proximal and aware services to empower and optimize new user experiences.

B. Chandhoke (✉)
GlobalFoundries, Santa Clara, CA, USA

© Springer Nature Switzerland AG 2020

A. Baschirotto et al. (eds.), *Next-Generation ADCs, High-Performance Power Management, and Technology Considerations for Advanced Integrated Circuits*,
https://doi.org/10.1007/978-3-030-25267-0_13

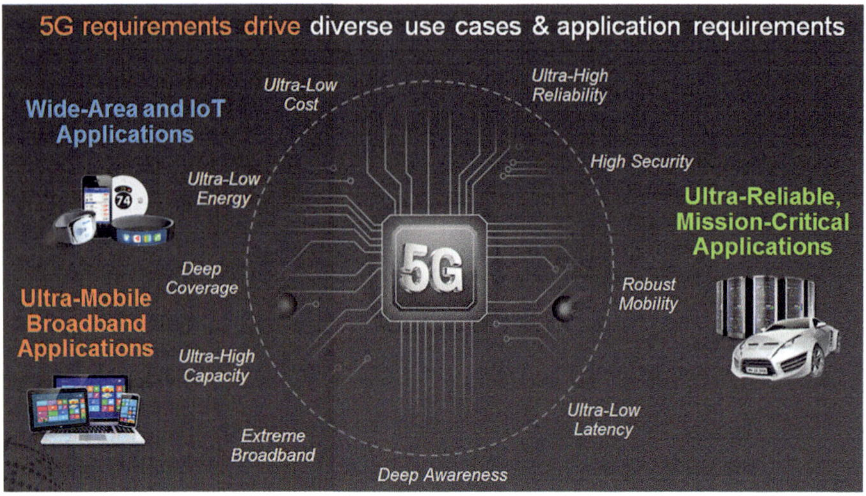


## 13.2 Phased Arrays and mmWave Applications

Most of the applications for mmWave use Phased Array Antenna Systems. Instead of a single Antenna, Phased Array Antenna systems use multiple elements or array of antennas and do constructive interference for receiving and transmitting the signal by focusing the energy from the beam in a particular direction. mmWave applications including Advanced Driver Assistance Systems ADAS Auto RADAR for automotive collision detection, mmWave back haul, fixed wireless, 5G handset and infrastructure devices like small cells, femto cells, pico cells, micro cells all use phased array systems for beam forming. WiGig—Wireless Gigabit Alliance (WiGig) which provides multi-gigabit per second speed wireless communications technology operating over the unlicensed 60 GHz frequency band also uses mmWave beam forming. Satellite communications uses beam forming for Geo Synchronous satellite communication for mobile airborne applications. Phased arrays are also being used in Low Earth Orbit LEO Satellite communication for broadband communications to focus the beam to the satellite and vice versa.

### *13.2.1 Benefits of Phased Arrays*

The benefits of using phased arrays for these applications are to focus the beam which allows spatial multiplexing where the beam that is being broadcasted is very narrow beam and is targeted to a single user. On the transmit side each element of the array generates power, in addition the array has array gain. Equivalent Isotropic Radiated Power (EIRP) is the product of transmitter power and the antenna gain in a given direction relative to an isotropic antenna of a radio transmitter. The EIRP increases by a factor of N square or 20 Log N. Each element is adding power and the array adds the array gain to get the N square factor. As a result for the same EIRP the power per phased array element is lower. At mmWave frequencies, the antenna dimension shrinks and power levels transmitted by each antenna elements reduce drastically. This happens because the energy is being concentrated in one direction. As the transmitter power requirement is reduced, the requirement for each power amplifier connecting to each phased array element is reduced. This allows the Silicon based technologies to be used for phased array antennas instead of the three five semiconductors like GaN, GaAs which are used for high power amplifiers [1].

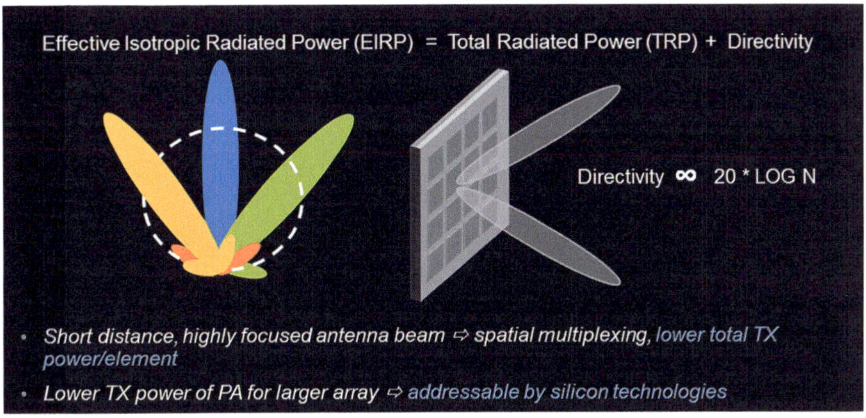


## 13.3 Beamforming Architectures

Now let's look at different types of Beamforming architectures for Phased Array Antenna Systems. There are three different types of Beamforming architectures

1. Analog Beamforming
2. Digital Beamforming
3. Hybrid Beamforming

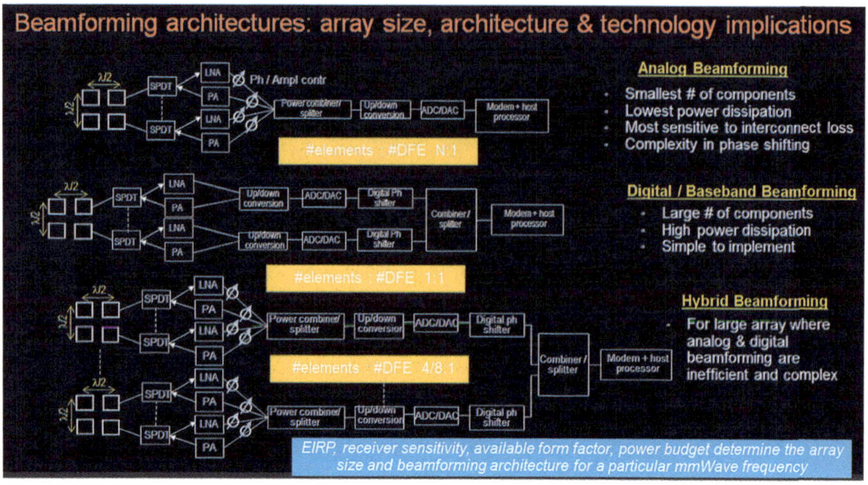


### 13.3.1 Analog Beamforming

For any Phased Array systems, the ideal separation between elements is λ/2. In the example shown above, there are four phased array elements separated by wavelength λ/2 distance. So for a 30 GHz, signal, it will be 5 mm separation between phased array elements. The front end module (FEM) is comprised of the Switch, LNA, and Phase and amplitude controller at the receiver and Phase shifter, PA, and Switch at the transmitter. In Analog Beamforming, the phase shifter does the beam forming by changing phase to do constructive interference for receiving and transmitting the signal by focusing the energy from the beam in a particular direction. This is done at RF Frequency, hence is most sensitive to interconnect losses. Then the signal from the FEM goes to the power combiner/splitter, followed by up down converter and ADC/DAC to the base band. In this case for N Phased Array elements there is only one Digital Front end. The benefits of this architecture are the smallest number of components, lowest power dissipation. However as the phase shifting is done in RF bands—this type of beamforming architecture is most sensitive to interconnect losses and complexity in phase shifting.

### *13.3.2 Digital Beamforming*

In Digital Beam forming, the phase and amplitude conversion in done in the digital domain—beamforming is accomplished by digital phase shifting after up down conversion to from baseband. Full RX chains are replicated hence digital beamforming has large number of components and high power consumption compared to Analog Beamforming. The benefit of digital beamforming is that it is easy to create multiple beams and generate true time delays—so much more granularity in beamforming compared to Analog Beamforming.

### *13.3.3 Hybrid Beamforming*

Hybrid beamforming is used for large arrays like in SatCom applications where array size can be 256 or even larger. In this case using only Analog Beamforming or only Digital beamforming is inefficient and complex. So a hybrid approach is adopted where there is analog Beamforming used in the RF front end part of the array and Digital Beamforming is used in the remaining array to get the most efficient solution for beamforming for large array sizes.

## 13.4 5G New Radio (NR) Standards Driving Increasing Radio Complexity

5G is being rolled out in stages leveraging the existing 4G/LTE backbone to help set 5G communication services. Several enhancements to the 4G/LTE system, such as LAA (Licensed Assisted Access) and LTE over WiFi will be employed to make the transition. An initial rollout of sub-6 GHz bands will be deployed and will use techniques such as massive MIMO to provide individual users and groups of users high data rate services. A second rollout expanding network capacity and driving even higher data rates will be deployed by leveraging mmWave bands where there is little or no interference or congestion. Both of these rollouts will mean enabling a more complex radio which will need to not only work with the new network protocols but also support the legacy protocols and bands as well.

5G NR standards are driving higher data rates and improved mobile user experience resulting in greater complexity in the radio and higher cost.

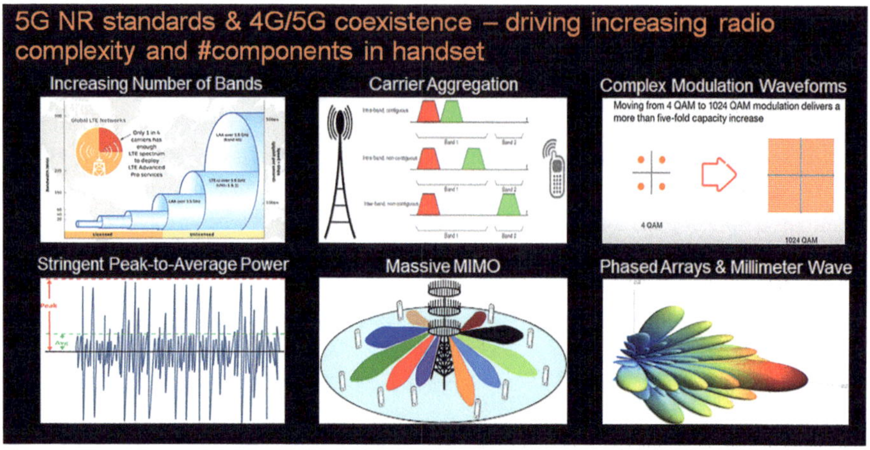


- Increasing number of bands—pursuit of clear spectrum to add capacity is adding more filters and switches and more complex PA structures.
- Massive MIMO and carrier aggregation—channel bonding and driving individualized high BW services to the edge of the network require new radio architectures.
- Complex modulation (moving from 4 AM to 1024 QAM delivers a more than fivefold capacity increase)—requires much better VCO phase noise, NF, EVM, and linearity from various blocks in the radio to maintain the fidelity of the signal.
- Phased arrays—drive a much more complex architecture for performance and control.

## 13.5 Impact of Semiconductor Technology for 5G

Let's look at impact of Semiconductor Technology on meeting the requirements of 5G NR standard. Antenna gain-to-noise-temperature (G/T) is a figure of merit in the characterization of antenna performance, where G is the antenna gain in decibels at the receive frequency, and T is the equivalent noise temperature of the receiving system in kelvins.

- For an active phased array of N elements
  - Every element added to an array:

    G/T of a Rx array increases by $10 * \log(N)$
    EIRP of a Tx array increases by $20 * \log(N)$

  - The performance of semiconductor technology with reference to the PA output power and the LNA noise figure determined the size of array needed to meet system requirements.

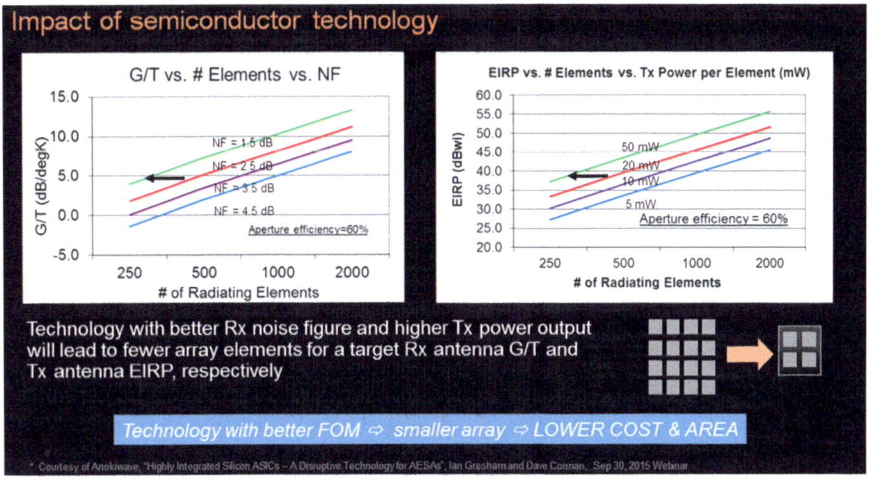


On the Receive side, G/T—Antenna Gain over the thermal noise is the figure of merit. For a given system target—G/T or NF

- NF of 2.5 dB, system target is 5 for G/T it will take 500 elements
- More chips, larger phased array system
- If the NF of the technology is 1.5 dB, system target is 5 for G/t it will only take 250 elements
- Less chips, smaller phases array system

On the transmit side, for an EIRP of 40 dBwi requirement, technology with PA capable of outputting 20 mW will need 500 radiating elements, whereas technology with PA capable of outputting 50 mW will only need 250 radiating elements of the phased array antenna system.

Technology with lower Noise figure and Higher Power output per phased array antenna element will result in lower number of elements, hence smaller phased array and hence lower cost and area [1].

## 13.6 Semiconductor Technologies for 5G mmWave

Essential Elements for a mmWave Technology are

1. High-performance technology
   - Higher performance enables design techniques for a more robust design
   - $f_T/f_{MAX}$ should be at a mininum 3× and preferably >5× application frequency

2. Low loss BEOL (metal and dielectric stack)
   - Thick top metal(s)
   - Distance to substrate
   - Substrate resistivity
3. Well-modeled technology
   - RF model-to-hardware correlation is key to minimizing design iterations
4. Good parasitic extraction deck and Electro-Magnetic (EM) simulation tools
   - Scalable Transmission Line and mmWave passives Process Design Kit (PDK) library minimize time spent in custom EM simulation

The different technologies that GlobalFoundries has for 5G mmW radio interface are shown in the table with the key features for each technology.

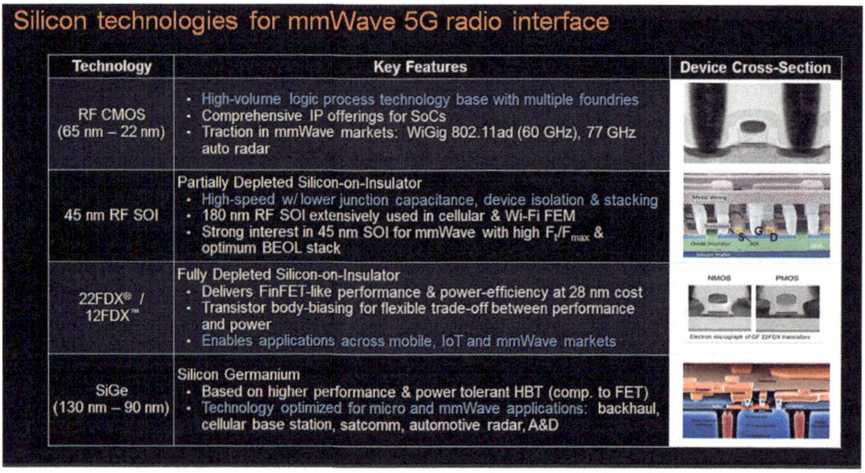
Silicon technologies for mmWave 5G radio interface

| Technology | Key Features | Device Cross-Section |
|---|---|---|
| RF CMOS (65 nm – 22 nm) | • High-volume logic process technology base with multiple foundries<br>• Comprehensive IP offerings for SoCs<br>• Traction in mmWave markets: WiGig 802.11ad (60 GHz), 77 GHz auto radar | |
| 45 nm RF SOI | Partially Depleted Silicon-on-Insulator<br>• High-speed w/ lower junction capacitance, device isolation & stacking<br>• 180 nm RF SOI extensively used in cellular & Wi-Fi FEM<br>• Strong interest in 45 nm SOI for mmWave with high $F_t/F_{max}$ & optimum BEOL stack | |
| 22FDX® / 12FDX™ | Fully Depleted Silicon-on-Insulator<br>• Delivers FinFET-like performance & power-efficiency at 28 nm cost<br>• Transistor body-biasing for flexible trade-off between performance and power<br>• Enables applications across mobile, IoT and mmWave markets | |
| SiGe (130 nm – 90 nm) | Silicon Germanium<br>• Based on higher performance & power tolerant HBT (comp. to FET)<br>• Technology optimized for micro and mmWave applications: backhaul, cellular base station, satcomm, automotive radar, A&D | |

## 13.7 Architecture for mmWave 5G Radio: Partitioning Options

mmWave phased array system helps reduced Tx power/element, enabling silicon technologies to play key roles in different applications including 5G. Unique features of SOI technologies like transistor stacking, reduced junction parasitics, low loss interconnect, and high $F_t/F_{\max}$ help build an efficient mmWave phased array system by reducing the array size. Different architecture and chip partitioning options exist depending on application specs and constraints, which can be addressed using GlobalFoundries technologies [2].

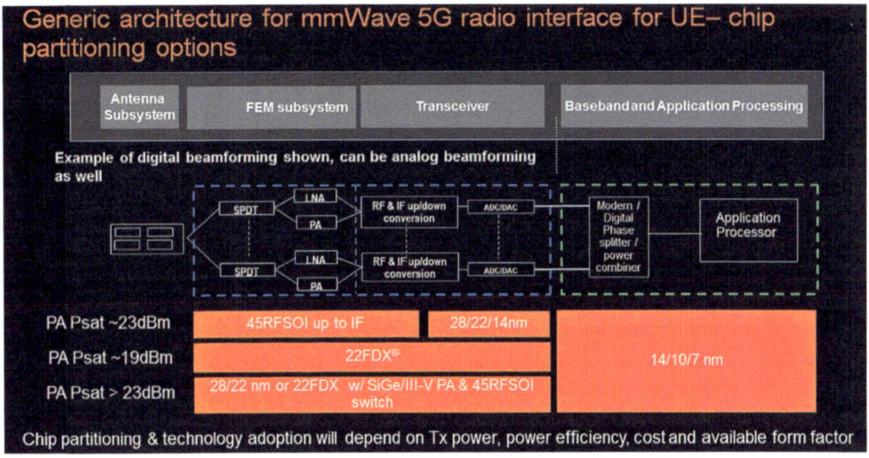


## 13.8 Conclusions

This paper showed the 5G mmWave applications, how 5G new radio standards are driving increasing radio complexity and cost, and what are the technology requirements of mmWave applications. Importance of technology in meeting the requirements of 5G mmWave applications was shown. Technology with lower Noise figure and Higher Power output per phased array antenna element will result in lower number of elements, hence smaller phased array and hence lower cost and area. mmWave phased array system helps reduced Tx power/element, enabling silicon technologies to play key roles in different 5G applications. Overview of different types of beam forming architectures for phased array antenna systems was shown. The benefits of Analog Beamforming architecture are the smallest number of components, lowest power dissipation. However as the phase shifting is done in RF bands, this type of beamforming architecture is most sensitive to interconnect losses and complexity in phase shifting. In Digital beamforming, full RX chains are replicated hence digital beamforming has large number of components and high power consumption compared to Analog Beamforming. The benefit of digital beamforming is that it is easy to create multiple beams and generate true time delays—so much more granularity in beamforming compared to Analog Beamforming. Hybrid beamforming is used for large arrays like in SatCom applications where array size can be 256 or even larger. In this case using only Analog Beamforming or only Digital beamforming is inefficient and complex. The different technologies that GlobalFoundries has for 5G mmW radio interface were shown with the key features for each technology. Unique features of SOI technologies like transistor stacking, reduced junction parasitics, low loss interconnect, and high $F_t/F_{\max}$ help build an efficient mmWave phased array system by reducing the array size. Different architecture and chip partitioning options exist depending on application specs and constraints, which can be addressed using GlobalFoundries technologies.

## References

1. B. Chandhoke, Design innovations in 5G mmWave FEMs & phased arrays (Microwave Journal Webinar, 2018)
2. B. Chandhoke, A powerful technology for 5G applications comes to market (GlobalFoundries Blog, 2018)

# Chapter 14
# IC Technologies for MM-Wave Applications

**M. Pagani, P. Rossi, and R. Lombardi**

## 14.1 Introduction

The mm-wave technology is mostly used in bandwidth-intensive high-speed wireless communications. In the past, due to the propagation characteristics and the limitation with respect to the achievable distance caused by high path loss and interference issues, most of the communications industry has been focused on the use of mm-wave transmissions only for point-to-point links in line-of-sight connections. This fact has hindered the market growth of mm-wave ICs, but the continuous research and development in telecommunications and microelectronics industry has led to an evolution of mm-wave technology such as to create new promising opportunities for mm-wave semiconductor business.

Today, the mm-waves technology market is well consolidated [1], in terms of applications and covers different segments, as shown in Fig. 14.1.

So far, mm-wave technology has been mostly applied in the area of telecommunication infrastructures (Fig. 14.2), which is still the largest commercial application of mm-wave systems [2].

However, the microelectronics industry has involved an incessant search for new technological applications capable to intensify the economic exploitation of existing market segments and to establish new business areas. The next frontier is to bring mm-wave technology in smartphones and make it available for the masses.

Today a number of manufacturers are producing components that can handle mm-waves, and the most advanced semiconductor technologies are capable of operating at frequencies up to 200 GHz, with applications in V-band (57–66 GHz), E-band (71–86 GHz), and D-band (130–175 GHz). The 60 GHz unlicensed

M. Pagani (✉) · P. Rossi · R. Lombardi
Huawei Italy Research Center, Segrate, Italy
e-mail: maurizio.pagani@huawei.com

© Springer Nature Switzerland AG 2020

A. Baschirotto et al. (eds.), *Next-Generation ADCs, High-Performance Power Management, and Technology Considerations for Advanced Integrated Circuits*,
https://doi.org/10.1007/978-3-030-25267-0_14

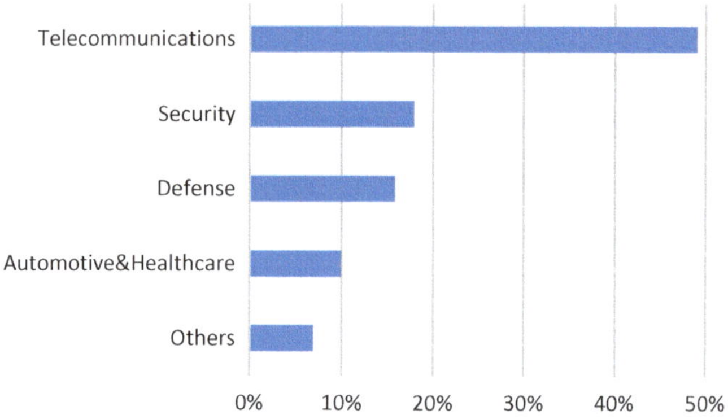


**Fig. 14.1** mm-wave technology market by application [1]

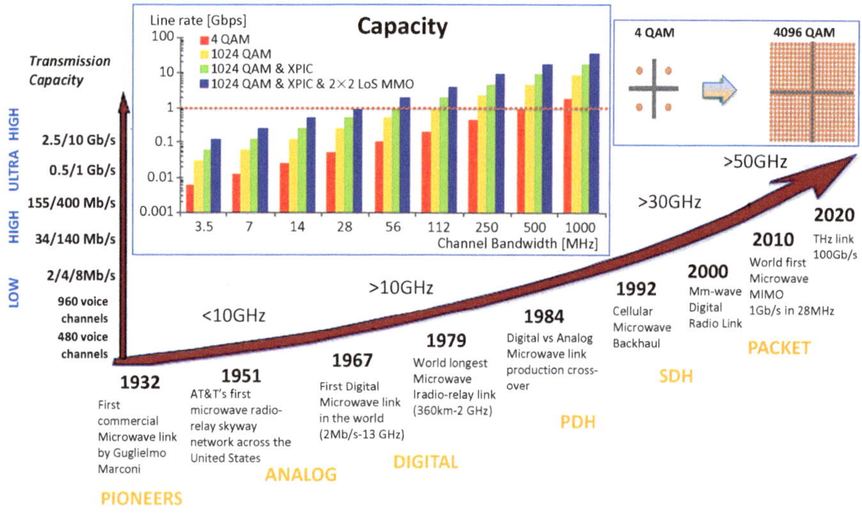


**Fig. 14.2** Evolution of wireless infrastructures

industrial-scientific-medical (ISM) band has recently got lots of attention, especially for short-range wireless technologies.

The nature of electromagnetic waves at mm-wave frequency band proved to be well suited also for the development of many applications [1], and the advancement in technologies over the past decade has enabled a widespread use of mm-waves to address the challenges of Wireless Local or Personal Area Networks (WLAN/WPANs), short-distance high data rate wireless links, automotive radars, high-resolution imaging systems, and micro-satellites for remote sensing or communications.

The growing demand for millimeter components has pushed manufacturers to develop technologies and affordable components with new semiconductor materials,

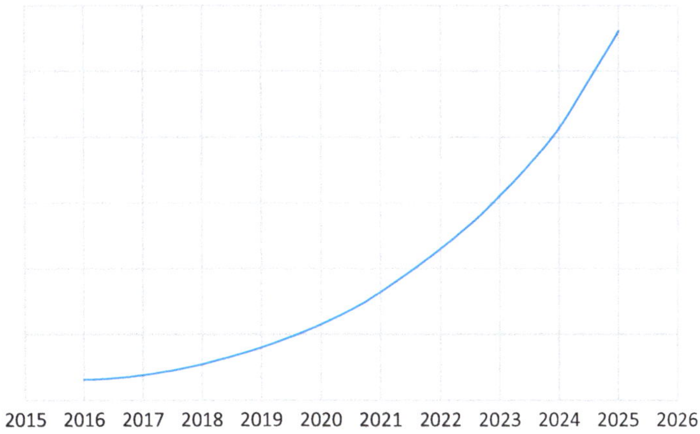


**Fig. 14.3** mm-wave technology market growth (forecast, [1])

such as gallium nitride (GaN), gallium arsenide (GaAs), silicon germanium (SiGe), and indium phosphide (InP).

Increased funding and R&D activities have driven the semiconductor industry to provide a path for commercial applications and to design mm-wave circuits less expensive, more reliable, and smaller, making them potential consumer items.

This makes existing products cheaper and opens the door to exciting new future applications. The outlook is that mm-wave technology will continue to gain ground as effective solution in many wireless applications acting as a major driver for the growth of the mm-wave IC market in many segments, mainly defense, security, satellites, telecommunications, automotive radars, medicine, and health care [1] (Fig. 14.3).

## 14.2 Industry Snapshot

Fiber optic communications have long been established as the preferred transmission technology for data transmission in telecommunication fixed networks. However, this technology is quite expensive to deploy, the network installation is rather complicated and not very flexible.

For this reason, mm-wave links have been considered as alternatives to fiber-optics medium. A significant emerging application is the fixed broadband access where mm-waves transmission is representing a very convenient solution in many use cases [3]. The last mile, in fact, is particularly attractive because at mm-waves you can provide gigabit data rates without running cable or fiber.

Using mm-wave frequencies allows carriers to reach customers with broadband connections directly from towers to Customer Premises Equipment (CPE) placed on the side of homes.

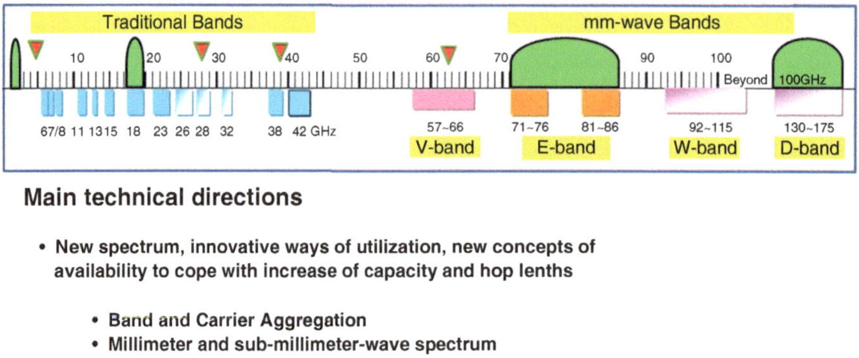


**Fig. 14.4** Technology trends for x-haul applications

In a mobile network infrastructure, point-to-point microwave links have been traditionally the preferred solution for backhauling. Many cell sites of distributed wireless base stations have been and still are connected to the main network switching centers (core network) through this technology [4].

Since the traditional frequency spectrum currently used for microwave systems is rather limited and crowded, mm-wave communications are increasingly becoming a popular and complementary technology which is expected to be extensively used also in wireless access besides backhaul; hence, we can expect mm-wave links will influence the main technology trends of mm-wave IC technology within the next 10 years and beyond [3, 5].

Over the last few years, high interest in mm-wave bands has risen also due to the enormous amount of under-utilized bandwidth that lies in this part of the electromagnetic spectrum.

The significant advantages offered by the propagation characteristics in terms of frequency reusability and large channel bandwidths make mm-waves very suitable for transmitting high capacities of data for the connection of macro base stations in rural areas and, at the same time, the preferred solution for the backhaul of small cells in dense urban environment, because of the very compact antenna size that makes products more environmentally friendly [3].

Future requirements for x-haul (front- and backhaul) of 5G will require capacities achievable only by using very large amounts of contiguous spectrum, possible only in the mm-wave frequency bands (Fig. 14.4) [3, 5].

Today more than 85% of the point-to-point communication links are for mobile backhaul, which represents the main driver for the technical and technological evolution of high-frequency systems and related ICs [4].

The basis for the deployment of a point-to-point link is very simple: two radio transceivers at each end point of the hop which employ high-gain antennas to increase the range of the link and transport the information in a reliable and cost-efficient way.

### *14.2.1 Point-to-Point Communication Links*

For over 20 years, microwave technology has been the primary solution for the rapidly rolling out of cost-effective mobile backhaul infrastructures worldwide. More recently, new requirements coming from mobile broadband have challenged microwave technologies to be able to scale up in terms of capacity and reduce all related costs.

Besides, with the evolution of telecommunication networks, the availability of spectrum has become an extremely important aspect to consider. For mobile operators, spectrum is like prime real estate: it's expensive and difficult to get the license, but with mm-wave frequency bands this problem can be partially solved by providing more room for expansion at the cost of a much higher path loss than traditional microwaves.

The adoption of smaller cells in the access network seems ideal for scenarios where extra high bandwidth is needed. This network densification is happening in the most advanced cellular networks worldwide and results in shorter hops; this is making the use of mm-wave bands even more suitable for the transmission of high capacity [3].

Emerging wireless applications at mm-waves are increasing significantly, specifically for high data rate point-to-point digital radio links at 71–76 GHz and 81–86 GHz (E-band) for both commercial wireless backhaul and military uses.

The E-band point-to-point high-bandwidth communication links in mobile networks, by operating in a licensed band which is free from oxygen absorption, represent today the mm-wave application with the highest growth rate.

In the coming years higher frequency bands like 92–114.5 GHz (W-band) [6] and 130–174.8 GHz (D-band) will be eventually adopted for applications in both backhauling and fronthauling, to support services requiring very high-speed transmission.

Today, data rate up to 10 Gbps per carrier can be achieved in V-band and E-band. In the near future, even 100 Gbps per carrier will be possible over distances in the range of 1 or 2 km with point-to-point links in D-band by adopting beam steering functionality.

This technological trend in the telecommunication industry is posing many complex challenges to the evolution of the mm-wave IC technology [5].

### *14.2.2 5G mm-Wave*

Strong technological innovations are required for 5G to deliver the promised very high transmission rate required by enhanced Mobile Broadband (eMBB) services in ultra-dense urban scenarios (Fig. 14.5).

In view of the fact that the currently used frequency spectrum for mobile applications has limited capacity, planning denser deployments with small cells

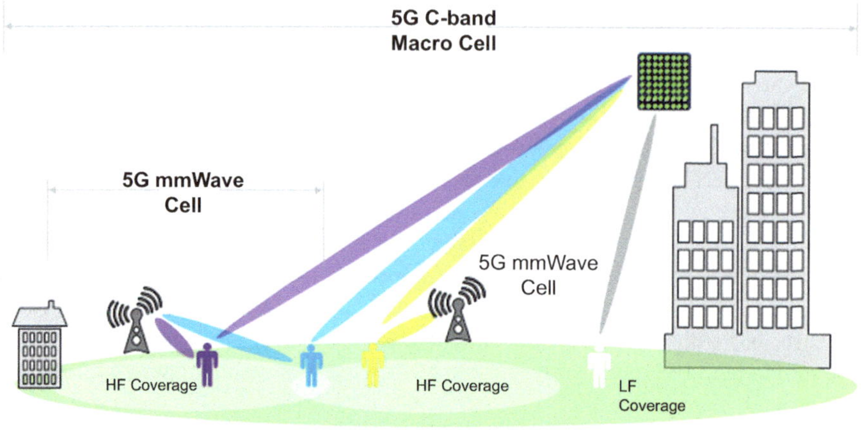


**Fig. 14.5** 5G deployment scenario to support eMBB use case

for high-capacity coverage seems to be unavoidable, but the economic feasibility and widespread availability will depend on the technical solutions and associated cost. The number of base stations per area determines the cost for equipment, deployment, and operation. For this reason, mm-wave access technology can play a vital role in the development of 5G broadband mobile services, owing to the need for wider channel bandwidths.

In reality, the reason for the adoption of mm-wave technology in future cellular networks is twofold. On one hand, lower frequency bands are congested by the continually increasing number of subscribers for mobile or other wireless communications networks. On the other hand, data rates must increase in order to support broadband internet services, and mm-waves propagation characteristics may lead to higher spectrum reuse which can facilitate sharing with existing services [3].

The use of mm-wave technology can then offer many advantages: large spectrum availability which permits wider channel bandwidth; high antenna gain which allows better signal-to-noise ratio (SNR), directional link and spectrum reuse; highly integrated architectures with beamforming/nulling capability.

In order to meet the requirements of massive connections and ultra-high data rates, 5G mobile access networks have been designed to be deployed in mm-wave bands, such as 26, 28 and 39 GHz, in addition to more traditional sub-6 GHz bands.

Evolution of 5G is driving cellular market growth, and the demand for mm-wave technology is expected to boost across the telecom industry. Thanks to that, mm-wave ICs for 5G have the possibility to emerge in the coming years significantly [1].

### 14.2.3 Other Applications

Other applications of particular interest for their great potential of high volume production are automotive radars, Wireless High-Definition Multimedia Interface (HDMI), security scanning, and imaging.

These applications can exploit the precision that can be achieved with short wavelengths, or use the ability of various mm-wave frequencies to penetrate different types of materials at different distances.

Today, many systems for mm-wave imaging for security screening applications in airport are already working in the 94-GHz band, where many materials, such as clothing, are transparent.

The short wavelength is also very beneficial because it reduces the size and the form factor of the device and gives an opportunity for very high resolution.

This makes possible to create holographic images which provide much more information for analysis in medical imaging applications.

Automotive radar is another emerging application of mm-waves, owing to its better performance when compared to many alternative sensors for advanced driver-assistance systems (ADAS). The reduced size of the device allows to mount these sensors in tricky positions to enable essential safety and comfort features.

For automotive collision avoidance radar applications, the range resolution depends on the signal bandwidth. The larger the signal bandwidth, the finer the range resolution and the more scatter points of an object of the same size are visible compared to measurements with coarse range resolution.

In this way the object gets more defined and can be classified more easily. These applications use long-range radar (LRR) systems operating in the 77 GHz frequency band (76–81 GHz).

This frequency band has several advantages for automotive use: the wide bandwidth available improves accuracy and object resolution; with a wavelength of 3.9 mm, the antenna can be small; and atmospheric absorption limits interference with other systems.

Driving aid radar systems specifically in the mm-wave frequency band are among the key enabling technologies in the development of autonomous vehicles.

The millimeter-wave radar mainly consists of the MMIC (monolithic microwave integrated circuit) which transmits and receives the high frequency signal, the signal processing circuit, and the control circuit.

MMIC, especially, has a big impact on the performance and the cost of the millimeter-wave radar. Generally, the silicon germanium (SiGe) and the gallium arsenide (GaAs) semiconductor technologies have been used for a MMIC, because of its high frequency performance; however, they are high-cost solutions.

The forecast of the production volume for this application is in the range of millions per year, as multiple units are built into every new automobile. The cost/volume benefits of CMOS technology can enable this to happen if adequate performance is realized and semiconductor industry will benefit from application of integrated high-performance RF and digital circuitry on the same silicon substrate. It will open up valuable new consumer markets in the communication and transportation sectors.

In recent years, micro-satellites have been utilized by the scientific community in a number of areas of remote sensing and Earth observation. Due to the low cost and fast development schedules [7], the small satellite architecture applied to areas of mm-wave remote sensing is rapidly expanding.

The new demand for bandwidth driven by applications such as connected cars, artificial intelligence (AI), and augmented reality (AR) is further transforming satellite communications. The low-earth-orbit (LEO) satellite technology providing reliable satellite services at a fraction of the cost of traditional space-based systems [8] can enable the delivery of high throughput satellite (HTS) services and a broadband super-highway in space.

Because Inter Satellite Links do not face the problem of atmospheric attenuation, 94 GHz, 60 GHz, and 36 GHz frequency bands can be used for medium- to long-range satellite communications, keeping in view of the smaller size of the system components and less power requirement.

With a wide range of applications, mm-wave IC technology has surely a great potential in many future commercial business opportunities.

## 14.3 Semiconductor Technologies for mm-Waves

The key solid-state processes currently available for mm-wave applications are summarized in Table 14.1.

For commercial applications, the potential of each semiconductor technology in terms of performance and integration levels must be balanced against the maturity of the process and the potential return on investment (ROI) for chipset development in the context of the market size for each use case.

**Table 14.1** Semiconductor technologies for mm-wave IC

| | GaAs HEMT | InP HEMT | GaN HEMT | RF CMOS | SiGe BiCMOS | SiGe HBT | InP HBT |
|---|---|---|---|---|---|---|---|
| Gate size | 100 nm | 70 nm | 60 nm | 45 nm | 55 nm | 0.18 μm | 0.5 μm |
| Wafer size | 6″ | 3″ | 3″ (6″) | 12″ | 12″ | 8″ | 4″ |
| Complexity | Basic | Basic | Basic | VLSI | VLSI | VLSI | LSI-MSI |
| fT/fmax [GHz] | 135/200 | 300/350 | 170/250 | 300/300 | 320/370 | 200/250 | 330/350 |
| VBr [v] | 10 | 4 | 25 | 1 | 3.2 | 1.7 | 5 |
| Cost/mm$^2$ | Medium | Medium | High | Low | Low | Low | High |
| NRE cost | Low | Low | Medium | High | High | High | High |
| Time to market | Short | Medium | Medium | Long | Long | Long | Medium |

For this reason the choice of technology is a careful balance of performance, system complexity, and cost metrics.

Each foundry process exhibits its particular strengths and weaknesses with respect to specific application areas that need to meet different system requirements with different demands on semiconductor components and different cost targets.

Since many of the semiconductor technologies suitable for mm-wave applications were originally developed using defense research funding, export control imposed in the manufacturing countries may limit the availability of certain technologies in some emerging markets [5].

### *14.3.1 Economic Considerations*

Microelectronic industries are continuously investing in research and development of new technologies and processes to meet the demand for mm-wave integrated circuits with better performances, lower development complexity and costs.

In this sense, the progress of semiconductor technologies for mm-waves IC featuring an integration with digital processing is promising the development of entirely new product concepts unfeasible just a few years back.

From the economic and strategic points of view, the integration of mm-wave circuits on silicon permits to reduce the cost and complexity of system hardware as the RF circuitry can then be combined with digital circuits on the same silicon chip.

Also thanks to that, the semiconductor business is likely to witness a strong increase of the use of mm-wave integrated circuits (ICs) in the coming years.

With its extensive application in telecommunication networks, E-band products are today generating the highest revenue for the IC technology market in the mm-wave segment [1, 2, 4, 5].

Nevertheless, the present volume for commercial applications in E-band is in the region of hundreds of thousands of units per year and represents a small opportunity for silicon-based technologies.

In fact, this production volume is not high enough to justify the large investment for production masks (Fig. 14.6).

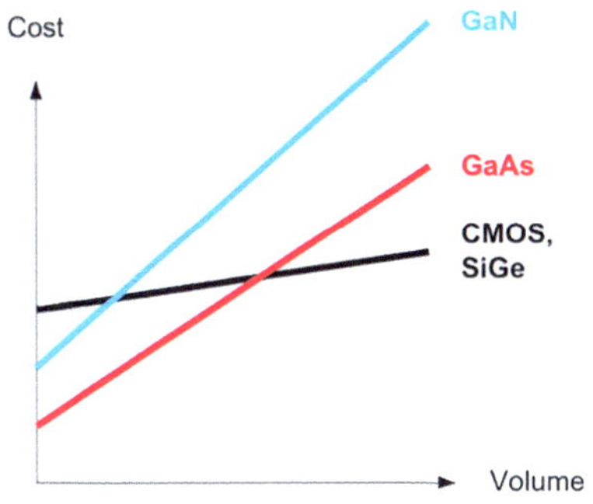


**Fig. 14.6** Scaling of production cost with volume for different IC technologies

This may present a significant commercial barrier for semiconductor manufacturers to develop chipsets for this specific application [5].

For this reason, GaAs MMICs at mm-waves are commonly and widely adopted today because the relatively low initial investments required makes them well suited to backhaul volume ranges, but a key enabling factor for the long-term volume uptake of mm-wave IC is the availability of more cost-effective solutions.

Research and development have been mostly focused on compound semiconductors in order to improve the performance of high-frequency integrated circuits in terms of phase noise, linearity, power efficiency and to move to the highest possible operating frequencies delivering the required performance at E-, W-, and D-band) [5].

The adoption of small cells which is expected in 5G could however provide the necessary high volume for commercial justification of mm-wave IC silicon-based solutions. Here, low-cost silicon chips providing frequency conversion functions and moderate levels of performance as suitable for small-cell applications could be supplemented with GaAs devices in order to provide the power and noise figure performances which are needed for the more demanding use cases [5].

### *14.3.2 Present Technology Status*

Historically the mm-wave IC application domain has been dominated for many years by III–V compound semiconductors, mainly gallium arsenide (GaAs), due to the superior electronic properties of these materials, ideally suited for high frequency operation [4].

The enhanced performance of GaAs-based solutions, particularly low noise characteristics, linearity, and power-handling capabilities, together with lower development costs made this the technology of choice on which to develop key functions for mm-wave front-end applications, thus relegating these applications into some niche market, because of the high manufacturing cost and reduced integration scale of these solutions [4, 5].

Recently, GaAs foundries have focused on employing many of the processing techniques and equipment used in the high volume applications of mobile terminals to provide low-cost solutions targeted at these higher frequencies [6]. This has resulted in the release of high performance GaAs-based technologies capable of delivering significant gain and power up to 160 GHz [3, 5]. The enhanced performance of GaAs-based solutions, particularly in terms of power and linearity, together with lower development costs makes this the technology of choice on which to develop key functions for present E- and W-band point-to-point applications.

Silicon-based IC, even being traditionally the preferred semiconductor technology for its lower cost, comes up short in many comparisons to III-V semiconductors for mm-wave applications. Silicon carrier mobility and critical electrical field are relatively low and so key device-level FOMs of raw performance appear to be inferior. Furthermore, silicon substrates present poorer isolation and higher losses in interconnects and passive devices. Each of these introduces serious challenges to implementing mm-wave functions of good quality.

The situation has recently changed, and very advanced silicon-based IC technology has gained increasing interest for emerging mm-wave markets, with ft. and $f_{MAX}$ of devices exceeding 300 GHz.

Nevertheless the market for mm-wave technology is still dominated by point-to-point applications and is not yet developed in terms of high volume deployments. The majority of mm-wave ICs in production are still fabricated in GaAs or indium phosphide (InP) materials, which are 3–5 times more expensive than silicon.

### *14.3.3 Technological Trends*

Other III–V foundry processes are under development for mm-wave IC applications, like gallium nitride (GaN) and metamorphic high electron mobility transistor (mHEMT). Compound semiconductors are essential for high performance systems at mm-waves but none of these advanced technologies is currently ready and available for commercial volume production.

Due to its physical properties of high-breakdown voltage and high-thermal conductivity, GaN is ideally suited for the development of highly efficient and high power amplifiers. Until now, the benefits of GaN technology have not been pursued in any commercial application at mm-waves, but GaN mm-wave market is developing fast, driven by mm-wave, 5G, and defense applications.

Integration and low cost have always been the key selling points for silicon technologies and continue to be the primary argument used by company to displace GaAs from mm-wave applications. The performance of very-scaled silicon-based transistor (HBT or MOSFET) is no longer the limiting factor for mm-wave transceiver front-end integration, at least for applications with limited output power and a reasonable trade-off between performance and cost.

Silicon-based technologies offer a perfect solution for low-cost short-range mm-wave systems but demand high volumes to justify the initial huge investment. Today, there are new emerging applications for silicon-based technology at mm-waves, which requires a much higher integration level and a simplification of the manufacturing processes for commercial and high volume applications such as for example automotive radars.

CMOS implementation can be realized with the higher speed of scaled technology permitting the highest levels of integration at the lowest possible cost, if volumes reach several million parts per year. To reach similar frequencies, bulk CMOS designs have to use much smaller process nodes. This makes it necessary to compromise on the design and most of the time leads to lower performance and higher cost. In the last few years, RF CMOS technology has entered radar chipsets market for long-range radar (LRR) applications. Next-generation wideband and high-resolution radar sensor chipset for ADAS featuring two transmitters and four receivers, designed for low power consumption and unit cost, has been realized using embedded wafer level ball grid array (eWLB) packaging to reduce parasitic signal, as a compact and powerful solution.

Compared to bulk CMOS, BiCMOS with silicon germanium (SiGe) heterojunction bipolar transistors (HBT) allows a much higher cut-off frequency at a given technology node together with a higher voltage capability.

One of the advantages of SiGe-BiCMOS technologies is that combining high-performance bipolar transistors with CMOS technology in a single chip. Bipolar transistors provide the high speed and gain that are critical for high-frequency analog sections, while CMOS technology is excellent for building simple, low-power logic gates.

## 14.4 Challenges at mm-Waves

Although mm-wave systems promise to offer larger bandwidth and high peak data rates, their practical implementation at low-cost faces several hardware challenges compared to more traditional communication systems.

In backhaul links, for example, the transmitter must provide a high output power with a good linearity and the design of mm-wave transceivers exhibits key challenges, such as intermodulation distortion in the transmitter and noise in the receiver.

Many challenges come from the fact that the implementation of mm-wave systems with integrated circuits raises new technical issues that do not exist at lower frequencies, due to both the physical dimensions defined by the wavelength, and the device properties which open up a new level of complexity in IC development.

The trend in mm-wave backhaul radios is towards multi-level modulation schemes, such as 512 and 1024 QAM, in order to increase the data rate and spectrum efficiency. This sets a strict constraint on the phase noise of the transceiver local oscillators. The higher carrier frequency induces larger multiplication factors, which in turn, degrades the phase noise with detrimental effects on the bit error ratio (BER) of the link.

At these high frequencies the IC design becomes more difficult because active components have higher noise and lower output power with lower efficiency and low gain; the effects of layout parasitic elements become stronger and all components suffer from increased losses with frequency.

Accurate nonlinear modeling of III–V FETs for microwave circuit design should also account for low frequency dispersive phenomena of the electrical characteristics due to deep level traps and surface state densities. These phenomena cause considerable deviations between “static” and “dynamic” operations of the device. Since microwave large-signal performance prediction involves accurate modeling of both “static” and “dynamic” components of the drain current, efforts have been made to take into account low-frequency dispersion both in mathematical and equivalent circuit models.

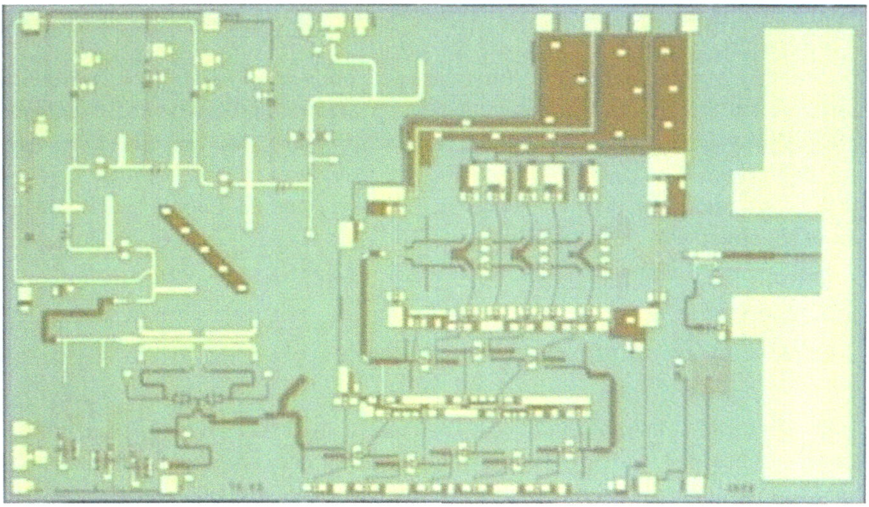

**Fig. 14.7** D-band integrated Transmitter chip

All these impairments result in limited transmit power, high receiver noise, high power consumption, and expensive components and therefore require specialist skills and experience in very high frequency transceiver design.

Another problem for mm-wave systems is represented by the increased path loss induced by the shorter wavelength that needs to be compensated by adopting a higher power amplifier in the transmitter and a lower noise amplifier in the receiver. In order to maintain the hardware cost and power consumption at reasonable levels, a suitable solution for mm-wave systems is to exploit the short wavelength to deploy a larger antenna with a much higher gain.

A very popular alternative is becoming the use of an increased number of smaller antenna elements connected to a bank of phase shifters such as to form phase-arrayed systems with hybrid and digital beamforming, where the additional power can be shared across multiple beams and users.

The high performance and high integration become essential for implementing competitive mm-wave communication systems which can benefit from the large reduction in size, weight, and power consumption that comes as a result of packaging the transmitter and receiver functions in highly integrated circuits [9]. An example is the D-band transmitter chip presented in Fig. 14.7.

Other key factor for the success of mm-wave communication system is the use of advanced digital signal processing (DSP) techniques. The benefit of relaxed requirements coming from the advanced DSP algorithm applied in modem functions can be exploited by integrating all RF front-end circuits in highly integrated semiconductor solutions with less demanding requirements. Highly integrated transceiver solutions with flexible architectures are preferred for simplifying the supply chain: same product for many market sectors will benefit of a larger production scale which allows minimizing unit cost.

Whether for high data rate communications links or automotive radar systems, millimeter-wave signals can be conveniently steered and controlled: mm-wave phased array systems will have a key role supporting the massive data rates planned for 5G networks and self-driving cars. However, unfavorable propagation and technical limitations could challenge the feasibility and adoption of such systems in large volume.

### *14.4.1 Power Amplifiers for mm-Waves*

Most advanced communication systems require power amplifiers (PAs) with high linearity and high efficiency [10]. Such high performance amplifiers are responsible for boosting the transmit power to overcome the increased path loss at mm-waves between the transmitter and the receiver. Increasing the power amplifier (PA) efficiency is crucial in high-capacity mm-wave radios, where the thermal issues are critical due to the compact radio module configuration.

Ideally, a power amplifier should exhibit linear transfer characteristics and provide at its output a perfect replica of the input signal multiplied by a scalar, but practical amplifiers suffer from various nonlinearities which degrades the quality of the carried information.

Two main technical issues related to power amplifier designs for mm-wave transmission are limited output power and poor efficiency. These technical issues together with the cost issue are creating much debate about he required output power per PA, the optimum number of transmit channels in a multi-channel system architectures together with the best choice for the semiconductor technology.

With the typical modulation and coding schemes adopted in high data rate mm-wave links, modulated signals to be transmitted present large peak-to-average power ratios (PAPR).

As well known, the highest efficiency point of a power amplifier is normally located close to the saturated power region. However, the linearity requirement forces the PA to operate at large output back-off (OBO) from the saturated output power, to maintain an acceptable linearity.

This operating mode degrades considerably the amplifier efficiency, and power amplifiers become very critical components for mm-wave systems with high spectral efficiency digital modulation schemes. Improving trade-offs between linearity and power-added efficiency (PAE) becomes extremely important for this kind of applications.

This trade-off requirement must be inherent to the design process from the outset, and an extensive load-pull assessment is usually undertaken using an accurate large-signal model or an experimental set-up to determine the optimum quiescent bias point and RF impedance terminations for each transistor.

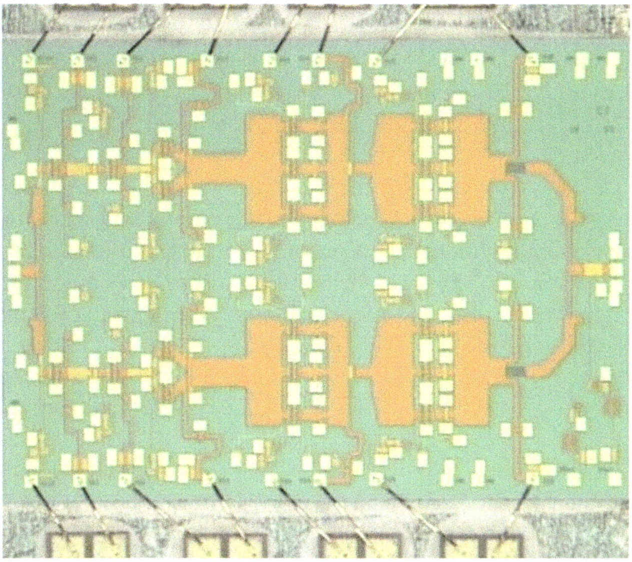

**Fig. 14.8** E-Band (71–76 GHz) 2 W Power Amplifier chip layout

In modern E-band point-to-point systems, typically very high saturated output power levels of at least 27 dBm (i.e. 500 mW) can be required and in the future the need is expected to exceed over 2 W, to extend the hop length in high-capacity communication links [6, 11].

Conventional IC technologies such as GaAs pHEMT are capable of providing lower power densities compared to GaN HEMT, prevent meeting the high power needs [6]. Only power densities as high as 0.45 W/mm have been achieved by GaAs MMIC technology so far [3], which signifies the inadequacy to compete with GaN.

Recent studies on GaN-based MMICs have reached output power densities of more than 1 W/mm at E-band [12] as well as at W-band [13–15] frequency bands.

Since GaN-based power amplifier MMICs bear the potential to provide required high power densities, they have become desirable for W- and E-band applications. As an example, in Fig. 14.8 is reported the layout of a 2 W power amplifier chip designed in 100 nm GaN on silicon technology for E-band point-to-point application (71–76 GHz).

Satellite communication is another of the millimeter-wave applications where the utilization of the E-band frequency window has enabled achieving multi-Gigabit transfer rates [16].

It can be seen that, as output power capability increases, so does cost per unit area. However, the power density per unit gate width from the active devices also increases, so that a PA on a GaN process would be at least four times smaller than a similar die on a GaAs PHEMT process.

For these reasons the choice of technology is a careful balance of performance, system complexity, and cost metrics.

## 14.5 Conclusions

Great progress has been made by IC technology over the last years which has allowed to introduce mm-wave IC components in many system applications, with major improvements in radio functionality, reliability, and cost.

Some emerging applications with potential for high volume which suffer cost pressures are still requiring cost reduction in ICs themselves, and further improvements in performance to reduce the number and complexity of supporting circuits.

The implementation of mm-wave circuits introduces new engineering challenges due to both the physical dimensions defined by the wavelength and device properties that are at the boundary of electronic behaviors.

Nevertheless, in the near future, mm-waves are expected to be among the key technologies that could enable many advanced applications in different market segments.

## References

1. Grand View Research, Millimeter Wave (MMW) Technology Market Analysis By Product (Telecom Equipment, Imaging & Scanning And Radar & Satellite Communication Systems), By Component, Application, Frequency Band, & Segment Forecasts, 2018–2025. https://www.grandviewresearch.com/industry-analysis/millimeter-wave-mmw-technology-market
2. M. Pagani, R. Lombardi, Monolithic Solutions for Digital Radio Link Transceivers, in *IEEE MTT-S (WSJ), 22-24 May, 2016, San Francisco, California, USA*, (IEEE, Piscataway, 2016)
3. R. Lombardi, Millimeter-wave technology trends for 5G and wireless transmission applications and technologies, in *IEEE MTT-S (MWS-AMP), 20-22 Sept. 2017, Pavia, Italy*, (IEEE, Piscataway, 2017)
4. V. Camarchia, R. Quaglia, M. Pirola, *Electronics for Microwave Backhaul* (Artech House, Norwood, 2016). ISBN: 1630810150, 978-1630810153
5. U. Rüddenklau et al., mm-Wave semiconductor industry technologies: status and evolution. ETSI white paper No. 15 – Second edition (2018) – ISBN No. 979-10-92620-24-5
6. J. Mayock, Cost effective high power E- and W-band amplifier MMICs. Microw. J. **55**(5), 170–182 (2012)
7. D.T. Jangam, Mm waves role in satellite communication. IOSR-JECE **4**(4), 01–04 (2013). ISSN: 2278-2834, ISBN: 2278-8735
8. J. Browne, Miniature satellite speeds communications from space. Microwave & RF - Feb 13, 2019. https://www.mwrf.com/analog-devices/miniature-satellite-speeds-communications-space
9. X. Gu et al., W-band scalable phased arrays for imaging and communications. IEEE Commun. Mag. **53**(4), 196–204 (2015)
10. M. Bignamini et al., Efficient development of highly linear MMIC power amplifiers, in *Proc. GAAS 2001*, (ExCel, London, 2001), pp. 693–696
11. Northrop Grumman Corporation, APH668 Product data sheet - Dec 2013. https://www.northropgrumman.com/BusinessVentures/Microelectronics/Products/Documents/pageDocs/APH668.pdf
12. A. Brown et al., High power, high efficiency E-band GaN amplifier MMICs, in *IEEE ICWITS, 11–16 Nov. 2012*, (IEEE, Piscataway, 2012)

13. M. Micovic et al., 92-96 GHz GaN power amplifiers, in *IEEE/MTT-S International Microwave Symposium Digest*, (IEEE, Piscataway, 2012)
14. A. Brown et al., W-band GaN power amplifier MMICs, in *IEEE MTT-S International Microwave Symposium Digest*, (IEEE, Piscataway, 2011)
15. D.F. Brown et al., W-band power performance of AlGaN/GaN DHFETs with regrown n+ GaN Ohmic contacts by MBE, in *IEEE IEDM Dig.*, (IEEE, Piscataway, 2011)
16. U.J. Lewark, J. Antes, J. Walheim, J. Timmermann, T. Zwick, I. Kalfass, Link budget analysis for future E-band gigabit satellite communication links. CEAS Space J **4**, 41–46 (2013)

# Chapter 15
# Accuracy: The Next Opportunity for MEMS

**Andrea Onetti**

## 15.1 Introduction

Thanks to their low cost, small footprint and reduced power consumption, MEMS-based devices have been used massively in personal electronics. Well-known products using these devices are smartphones, tablets, smartwatches, AR/VR headsets, and drones; but they are also present in niche applications like shock detection in sport environment.

STMicroelectronics manufactures its MEMS using a proprietary micromachining process, called ThELMA (Thick Epitaxial Layer for Micro-gyroscopes and Accelerometers). It was the first major manufacturer to start high-volume production on 200 mm wafers in 2006 of motion sensors for the consumer market. Since then, STMicroelectronics has continued to grow and diversify its portfolio into new types of devices and application segments; today, in addition to the personal electronics segment, the company features products dedicated to automotive and industrial markets.

Targeting these different markets implies different requirements and trade-offs in the design of products. As an example, power consumption has always been a key parameter in the field of personal electronics, to maximize the battery duration in portable devices like smartphones; conversely, this parameter is less important when dealing with automotive and industrial applications, where reliability and longevity are the primary focus. Figure 15.1 summarizes the relative importance of main requirements for the three market segments in the case of MEMS accelerometer.

This paper describes how STMicroelectronics is leveraging on its extensive knowledge coming from personal electronics to address the requirements of new

A. Onetti (✉)
AMS MEMS Sensors Division, STMicroelectronics, Milan, Italy
e-mail: andrea.onetti@st.com

© Springer Nature Switzerland AG 2020

A. Baschirotto et al. (eds.), *Next-Generation ADCs, High-Performance Power Management, and Technology Considerations for Advanced Integrated Circuits*,
https://doi.org/10.1007/978-3-030-25267-0_15

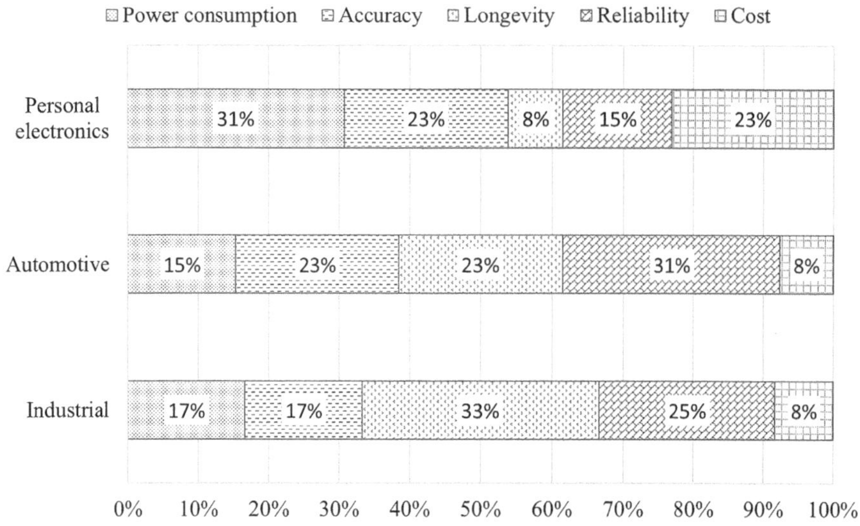


**Fig. 15.1** Relative importance of main requirements for the three market segments of personal electronics, automotive and industrial

market segments. It underlines how the final performance depends from the design of the device and from its packaging, testing and calibration. In particular, accuracy, that is a key to enable the pervasion of MEMS devices in new markets, depends strongly from the complete production chain.

As a design example, the paper provides details about the design of the new ASM330LHH six axes Inertial Module Unit (IMU) that targets automotive non-safety applications.

## 15.2 From Personal Electronics to New Markets

To address the requirements of the automotive and industrial markets a convenient approach is to exploit the extensive knowledge and experience in the design of products for the personal electronics. For example, by properly adjusting the balance of key trade-offs it is possible to address different applications but working on the same system architecture. Figure 15.2 shows an example of this approach used for the gyroscope design of two STMicroelectronics IMUs, the LSM6DSO and the ASM330LHH [1, 2]. The first product targets the personal electronics segment, and therefore the focus has been on current and noise reduction: for this reason, the mechanical element has a high sensitivity; the latter, targeting automotive non-safety applications, has been designed to provide high stability over time and temperature, and therefore has a lower mechanical sensitivity.

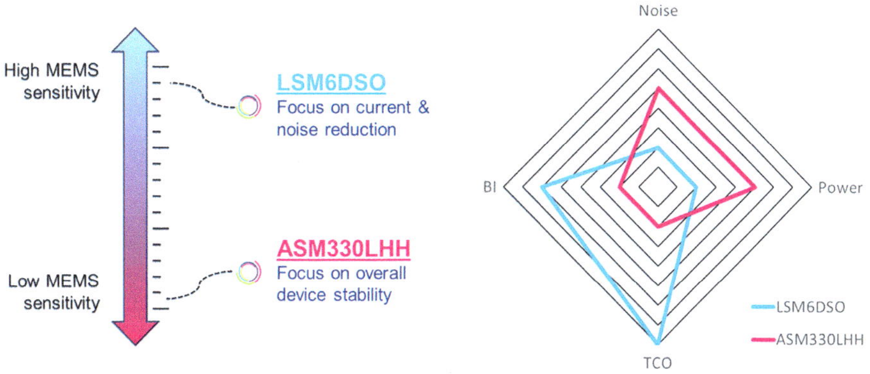


**Fig. 15.2** An example of how different requirements can drive the balance in the trade-offs during product design to address properly the end market segment. LSM6DSO and ASM330LHH gyroscopes are based on the same architecture

In addition, an already existing design can be used directly in other applications by adjusting the testing and calibration phases. This is convenient to streamline the overall supply chain by reducing the number of ASIC and MEMS to be manufactured and differentiating the product at the final production stages. It also provides high flexibility with the possibility to meet aggressive time-to-market requirements.

The waterproof pressure sensors and the ultra-low power accelerometers provide good examples of this concept. Both devices were designed to target personal electronics applications, like e-cigarettes and wearable devices for activity monitoring. However, with proper testing and calibration they can be exploited in other segments:

– The waterproof pressure sensor can be used in industrial applications for flow metering;
– The ultra-low power accelerometer can be used in automotive key fob applications, to monitor the free fall without affecting the battery lifetime.

## 15.3 MEMS for Smart Industry and Smart Driving

New market opportunities for MEMS devices are opening in the fields of smart industry and smart driving. Thanks to performance improvement driven by demanding applications in personal electronics in the last years, the accuracy, reliability and robustness of MEMS sensors improved considerably and are now aligned with the tight requirements of industrial and automotive applications.

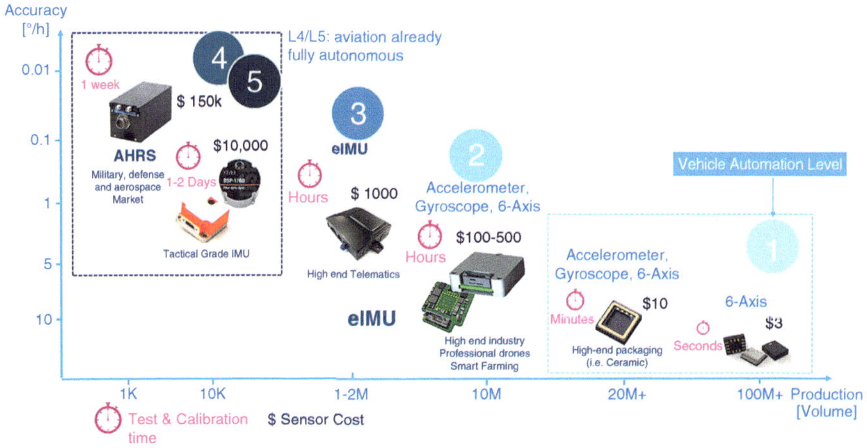


**Fig. 15.3** Accuracy requirements for inertial navigation systems according to different Vehicle Automation Levels

In the field of smart industry, sensors are used:

- To enable predictive maintenance, with real-time monitoring of factory equipment;
- For asset tracking along the whole supply chain, including the possibility to check proper handling of material and of environmental conditions;
- For shock detection, such that proper actions can be taken to avoid major damages or injuries.

Sensors can either be mounted on existing equipment, using wireless connection for data transmission, or be pre-installed in new machinery, which can rely on wired connections. Instead of sending out raw data, new MEMS devices from STMicroelectronics allow for local processing and fusion of the information from different devices. This approach embeds the intelligence directly in the device, that becomes a smart sensor, and relaxes the requirements for the central processing.

For what concerns the automotive non-safety, an increasing number of sensors is used with a particular focus on inertial navigation systems. In this respect, current MEMS devices are able to deliver an accuracy level in line with the requirements of Vehicle Automation Level 1, as shown in Fig. 15.3.

The key parameters for sensor accuracy in this field are:

- Noise, both at low and high frequencies and including vibration rejection;
- Stability over time and temperature;
- Repeatability of performance;
- Tolerance to external non-idealities, such that offset and sensitivity are not affected.

## 15.4 The ASM330LHH: Design Choices for High Stability

The ASM330LHH is the new six axes IMU from STMicroelectronics designed to target high stability applications in the automotive non-safety segment. As discussed in the Introduction, the design is based on the same architecture of the LSM6DSO with a different balance for the trade-offs. This paragraph provides details on the design of the three axes gyroscope to achieve the required stability and accuracy.

### *15.4.1 Design for High Stability Over Temperature*

One of the main environmental parameters that affects MEMS gyroscopes is temperature, which induces a drift of the zero rate level (ZRL). To best address this aspect, STMicroelectronics has developed a predictive model that accounts for the combined effect of the MEMS and the electronics from which solid design guidelines are derived.

This model has been used to design the ASM330LHH gyroscope. As discussed in [3], the drift of the ZRL over temperature is dominated by the variation of the phase delays of the different elements over temperature. A first remark is then linked to the choice of the working point of the mechanical element: to ensure a stable phase delay, it is necessary to work with relatively large frequency mismatches and this implies a reduced sensitivity. In turn, to maintain an acceptable level for the noise it is necessary to invest more current in the front-end stages. This is the opposite situation with respect to the LSM6DSO, where a large mechanical sensitivity is one of the key factors to have state-of-the-art current consumption and noise.

For what concerns the phase delays in the ASIC, the following considerations are made:

- Stable delays usually require larger current consumption, to move non-dominant poles far from the unitary gain bandwidth; again, a different trade-off on the current is present;
- When acting on the poles is not sufficient, it is important to use architectures that provide a stable bandwidth over temperature, like constant-gm topologies;
- If using a continuous time approach, stable components must be selected to obtain stable RC time constants. In general, this implies using a larger area as high temperature stability is linked to a reduced resistivity per square.

The following Table 15.1 summarizes the typical values of the main parameters for the gyroscope of the ASM330LHH compared to the LSM6DSO. Despite having the same system architecture, it is evident how the two products focus on different requirements.

**Table 15.1** Performance comparison between the ASM330LHH and the LSM6DSO

| | ASM330LHH | LSM6DSO |
|---|---|---|
| Frequency mismatch | Large | Small |
| Mechanical sensitivity | Small | Large |
| Current consumption (combo, high performance mode) | 1.3 mA | 0.55 mA |
| Rate noise density | 5 mdps/$\sqrt{\text{Hz}}$ | 3.8 mdps/$\sqrt{\text{Hz}}$ |
| TCO | ±0.005 dps/°C | ±0.010 dps/°C |

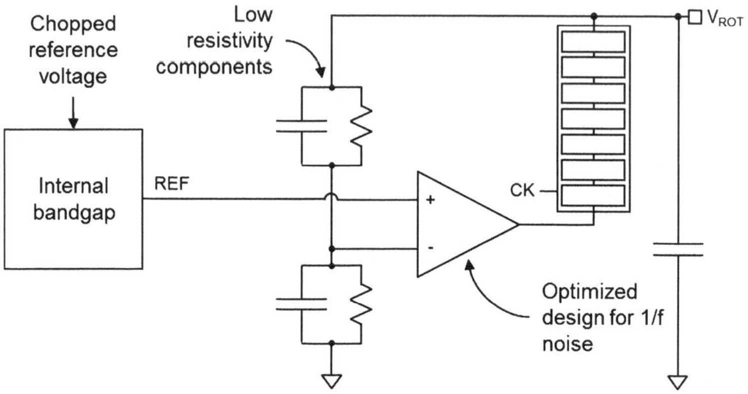


**Fig. 15.4** Schematic of the regulated charge pump to generate the rotor voltage for the rotor of the mechanical element

## 15.4.2 Design for High Stability Over Time

Stability of the stability over time is also crucial to achieve high accuracy in the final application. Low frequency noise in critical blocks must be addressed properly to ensure high bias stability and avoid errors in the estimated angle that increases with time.

The rotor voltage is particularly important for this aspect. Usually, a charge pump is required to bias the mechanical structure at suitable levels above the available power supply. In addition, it is important to regulate this voltage as it directly determines the overall charge sensitivity, according to the following formula:

$$\frac{q_c}{\Omega_c} = (V_{\text{rot}} - V_{\text{stat}}) * S_{\text{MEMS}}$$

where $q_c$ is the charge flowing to the front-end (in $C$), $\Omega_c$ is the input angular velocity to be detected (in dps), $S_{\text{MEMS}}$ is the sensitivity of the MEMS (in fF/dps), $V_{\text{rot}}$ and $V_{\text{stat}}$ are the biasing voltages of the rotor and the stator (in $V$), respectively. For this reason, the charge pump is usually regulated by means of a local feedback, as shown in Fig. 15.4.

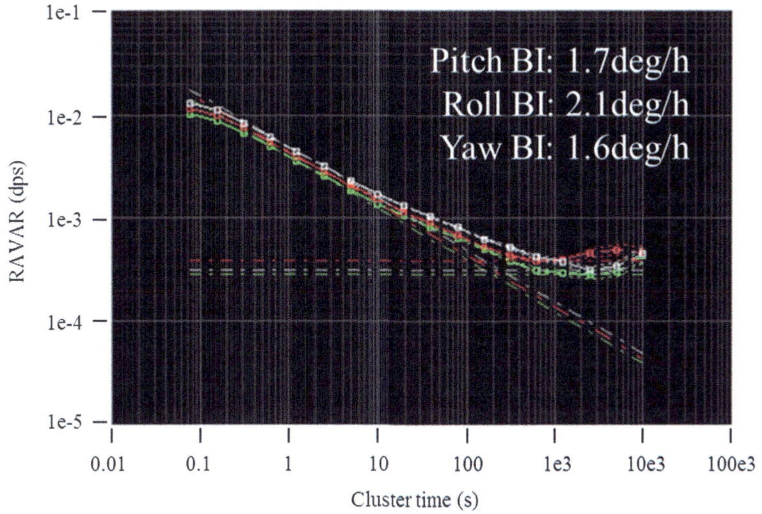


**Fig. 15.5** Typical Allan deviation plot for the gyroscope of the ASM330LHH

Low frequency noise on the rotor directly affects the stability over time of the gyroscope ZRL. To ensure low noise at the output it is crucial to optimize the overall chain. To this end, the following design choices are made:

- The internal bandgap of the device is chopped: low frequency noise components are moved to higher frequencies and filtered in the digital domain such that they do not appear in the output of the device; this is beneficial in general for the overall device;
- The operational amplifier in the charge pump is designed to minimize the 1/f component; in particular, this entails using donut devices as input pair to exploit their intrinsic good performance along with long channel transistors. Chopping of operational amplifier is not used since the inputs are connected to high impedance nodes;
- Feedback resistors have a low resistivity per square, which minimizes their low frequency noise components at the expense of large area occupation.

This design optimization, along with a general reduction of 1/f noise in critical blocks along the readout chain, allows the ASM330LHH to reach extremely good levels of bias stability, as shown in Fig. 15.5.

### *15.4.3 Sensor Accuracy: Measurements*

The accuracy of the ASM330LHH has been evaluated using a tunnel test in which the absence of GNSS signal is simulated to estimate the position error obtained

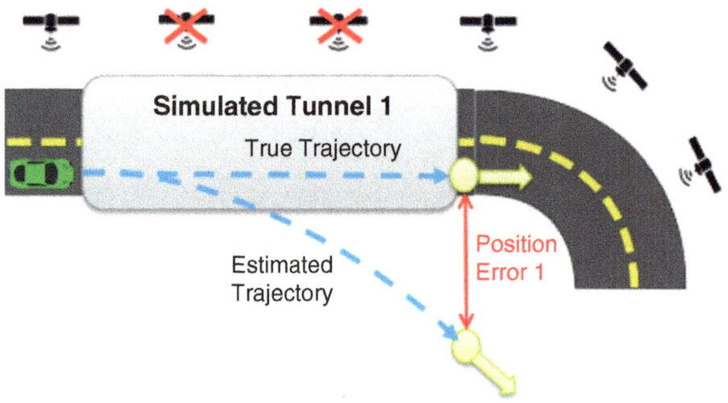


**Fig. 15.6** Tunnel test environment to evaluate the accuracy of ASM330LHH in the absence of GNSS signal

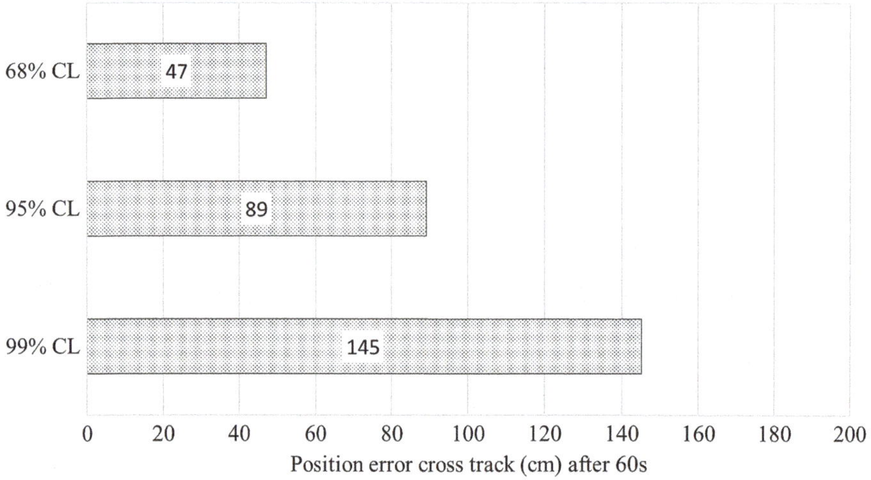


**Fig. 15.7** Estimated position error at the exit of the tunnel for three different confidence levels

using pure inertial navigation when exiting the tunnel. In the test, the IMU is travelling at 15 m/s in a straight tunnel of 900 m, therefore requiring 60 s from input to output (see Fig. 15.6).

The test has been run for 200 times to extract the position error statistics, and the results are shown in Fig. 15.7 for three different confidence levels. The estimated position is within 1 m of the true one with a confidence level above 95%.

Another way to characterize the accuracy of the ASM330LHH gyroscope is to measure the heading error after integrating the outputs over 1 min. In fact, due to noise the heading error grows over time:

- When white noise is dominant, integrating the outputs results in zero-mean random walk with a standard deviation that grows with the square root of time;
- For larger integration time, flicker noise becomes relevant and the variance of the integrated outputs grows over time.

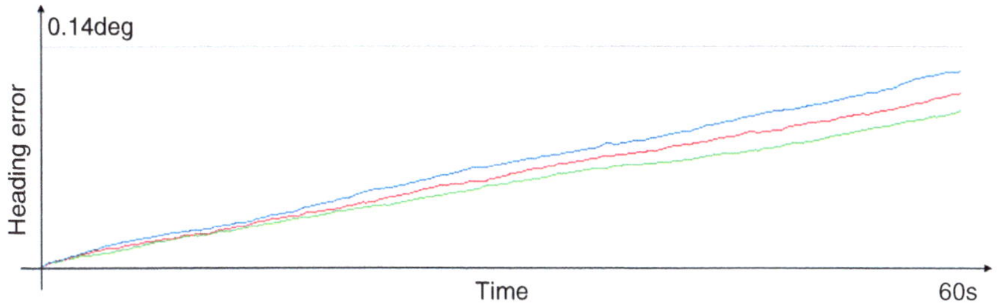


**Fig. 15.8** Heading error at one sigma measured over a population of 50 devices

Figure 15.8 shows the heading error at 1σ for measurements over a population of 50 ASM330LHH. After 60 s, the error is below 0.14 ° for the three directions.

## 15.5 Conclusions

Today, MEMS devices have the possibility to meet the requirements of different market segments beyond personal electronics, including industrial and automotive applications. A particular focus is placed on accuracy of data from MEMS sensors that are fundamental to enable high-performance applications.

Testing and calibrations, on top of design, play a fundamental role to this end and give the flexibility to address different needs by ensuring the product specifications are aligned to market requirements. By adjusting the whole supply chain, STMicroelectronics has already in place the possibility to address emerging market trends exploiting its expertise in the field of MEMS devices.

## References

1. ASM330LHH datasheet. https://www.st.com/resource/en/datasheet/asm330lhh.pdf
2. LSM6DSO datasheet. https://www.st.com/resource/en/datasheet/lsm6dso.pdf
3. S. Facchinetti et al., Development of a complete model to evaluate the zero rate level drift over temperature in Coriolis vibrating gyroscopes, in *2017 IEEE Inertial Sensors & Systems, Conference Proceeding* (IEEE, Piscataway, 2017), pp. 125–128

# Chapter 16
# Robustness, Reliability and Diagnostic Aspects in Sensors for Automotive Applications: The Magnetic Sensors Case

**A. Laville, M. Pardoen, G. Close, M. Poezart, and D. Gerna**

## 16.1 Introduction

Position sensors applications include pedal, steering wheel, throttle valve, and numerous others [1]. These sensors perform various magnetic sensing functions: sensing the angle or linear position of a magnet, or directly sensing the magnetic field. Position sensors can be classified according to the measured variable (see Fig. 16.1): proximity, linear displacement sensors, or angle sensors.

Low cost non-rare earth magnets have typical field strengths in the range of 10 mT and show a high negative temperature coefficient. Hall-based integrated sensors exploit the Hall effect [2] to generate a signal proportional to the magnetic field. The resulting single-digit mV range input signal imposes stringent requirements on the accuracy and sensitivity of the low-voltage electronics. The readout electronics is designed to be limited by the thermal noise of the resistive Hall elements (HEs). Another factor that impacts the precision are the residual offsets of the HEs. Offsets, referred to the HE terminals, are to be kept below 10–20 $\mu$V over temperature and lifetime.

The high-voltage regime is also a key factor for the design: the sensor has to face transient operating voltages >40 V due to the extreme noisy automotive environment. Although the device has a typical supply voltage of 5 V it has been designed using 40 V CMOS technology. All these interferences are described in several EMC automotive compliance standards; this paper describes few examples on how the device achieves both low emissions and good immunity.

A. Laville (✉) · M. Pardoen · G. Close · M. Poezart · D. Gerna
Melexis Technologies S.A., Bevaix, Switzerland
e-mail: alv@melexis.com

© Springer Nature Switzerland AG 2020

A. Baschirotto et al. (eds.), *Next-Generation ADCs, High-Performance Power Management, and Technology Considerations for Advanced Integrated Circuits*,
https://doi.org/10.1007/978-3-030-25267-0_16

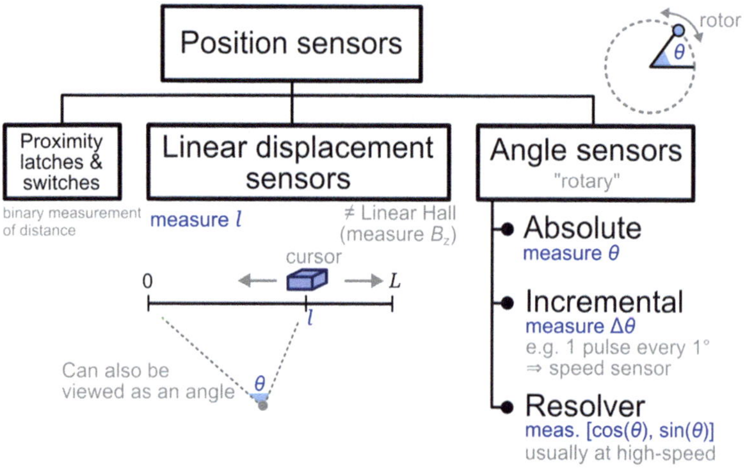


**Fig. 16.1** Classification of magnetic position sensors

## 16.2 Architecture

### *16.2.1 Sensor Configurations*

The chip integrates twelve horizontal HEs, realized as diffusion resistors. To sense in-plane fields, a combination of horizontal and vertical Hall transducers can be used [3]. Vertical Hall transducers are typically less sensitive than their horizontal counterparts. In this design, we use integrated magneto-concentrator (IMC) post-processed on the chip surface to redirect the in-plane field lines. This allows to sense the magnetic field in oblique directions. Full 3D sensing is achieved by combining several sensed signals together [4]. Figure 16.2 illustrates how the combination of HEs and IMC geometries allows to sense multiple field components and serves various magnetic applications [5].

### *16.2.2 Magnetic Signal Path*

Figure 16.3 depicts the block diagram of the complete magnetic signal path. Depending upon the magnetic mode, a subset of the HEs is selected. The polarity and the bias current direction are programmable, allowing a large choice of applications. The current spinning technique is used to rotate the bias current of the HEs such that four phases are generated. The offsets of the HEs (larger than the signal, in the order of several mV) appear then as DC component. The Hall voltage, on the contrary, is modulated, thereby also rejecting the $1/f$ noise of the amplifiers. The HE outputs are combined in the current domain for easy summation. A feedback

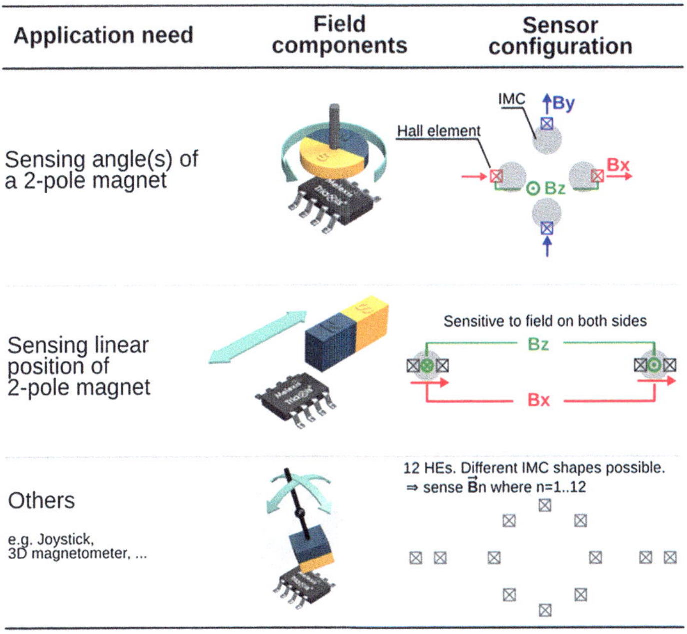


**Fig. 16.2** Magnetic sensing modes and corresponding applications

loop adjusts a rough offset current to center the resulting combination. The centered signal is then amplified without risk of early saturation. The amplifier chain supports two decades of programmable gain to cover a wide range of applications and also to compensate for the magnet field change and Hall sensor sensitivity change over temperature. The resulting voltage is then converted to the digital domain. In angular applications, two components *cosine* and *sine* are acquired and processed with four spinning phases each.

In addition, a temperature and a mechanical stress sensor [6] are also included. The temperature information is critical to perform fine offset drift corrections while the inclusion of the stress sensor allows the compensation of piezoresistive and piezo-Hall effects.

### 16.2.3 Measurement Sequencing and Digital Corrections

The amplified analog signals are converted by a 15-bit ADC. Due to circuit imperfections, these signals are in general neither of the same amplitude nor in perfect quadrature (see the inset *Before correction* in Fig. 16.3). Fine corrections, offsets and cross sensitivities, are calculated by an embedded software based on adjustable coefficients. The coefficients are factory calibrated on each individual

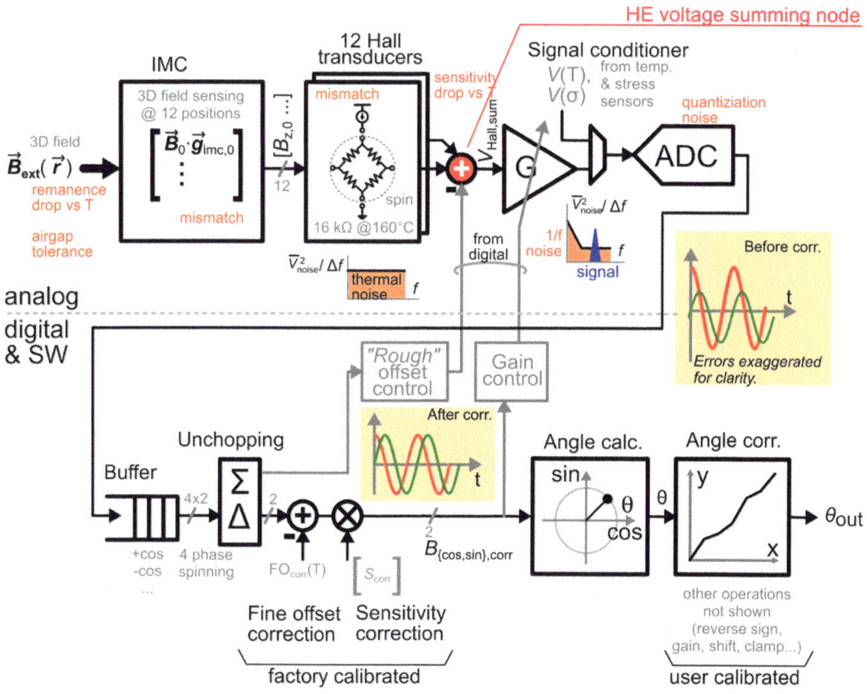


**Fig. 16.3** Block diagram of the magnetic signal path

sensor using on-chip non-volatile memory. After the software corrections, the resulting signals are then a pair of ideal cosine and sine signals (see the inset *After correction* in Fig. 16.3).

In angle sensing applications, the corrected magnetic signals are used for angle calculation via the arctangent function. The calculated angle is reshaped and corrected via a collection of user-calibrated signal transformations. The embedded software also adjusts the feedback signals, setting the gain and rough offset of the analog amplifier chain.

As this sensor targets safety-relevant applications (x-by-wire), the embedded software is also continuously running multiple self-diagnostic functions (e.g., ADC self-tests). Finally, the angle output is either transmitted as an analog voltage, or encoded and transmitted into digital frames of standard automotive protocols.

## 16.2.4 Safety Concept

The circuit was developed according to the automotive functional safety standard ISO26262:2011 [7]. It requires a rigorous development process: management of

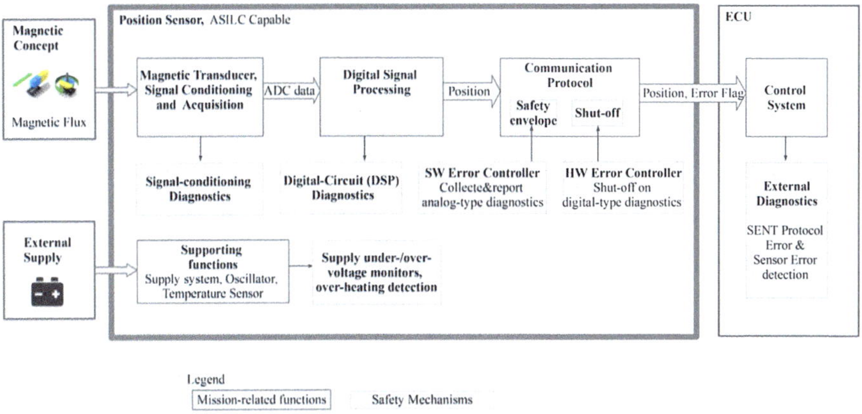


**Fig. 16.4** Safety concept block diagram

requirements, issues, documentation, configurations; strict adherence to the internal processes. This will objectively reduce the risk of systematic failures down to an acceptable level from Automotive Safety Integrated Level (ASIL) A up to ASIL D.

The safety standard requires that random defects (typically caused by aging) are detected and controlled through adequate diagnostic functions within the so-called fault handling time interval.

The safety concept block diagram shown in Fig. 16.4 describes how both the systematic errors and the random hardware faults are monitored and communicated to the electrical control unit (ECU).

During the design phase, the sensing function and the monitoring functions are separated and isolated to reduce the common cause of failures.

The fault detection mechanisms of the analog circuit comprise:

- Voltage test patterns of the amplification chain and of the A/D converter
- Valid range monitoring of
  - the rough offset code
  - the gain code
  - the temperature value
  - the mechanical stress sensor value
  - the Hall signals themselves
- 6 supply over voltage and under voltage monitors
- 2 bias current monitors

The ASIL C/D development rigor calls for a verification of the diagnostic coverage by means of fault injection simulations, which was performed on the HEs connections to the analog frontend [8].

The ISO26262 single-point failure metric (SPFM) shows the fraction of single-point random hardware faults, which are neither safe nor detected. The ASIL C target of the SPFM (97%) was achieved.

## 16.3 Magnetic Amplification Chain

Figure 16.5 shows the circuit schematic of the HE biasing and first amplification stage. The biasing circuit can be configured to supply either voltage or individually trimmed current to up to 4 different HEs. When voltage biasing is applied, the HEs are connected to the regulated 3.3 V supply voltage, and maximum sensitivity is achieved at the expense of mismatch. In case of current biasing, a reference Hall element (HE_Ref) is used to generate a reference current. The current is then mirrored and supplied to the HEs with individually trimming adjustments. This allows to correct sensitivity mismatch while still operating close to the maximum supply voltage. Sixteen switches around each HE support the spinning current technique for offset cancellation. The first amplifying stage exhibits a noise of 4 nV/$\sqrt{}$(Hz) with each of the input transistors biased at 70 μA. It is summing four HE output voltages. The circuit also removes offset of the complete front-end chain up to ±24 mV, per each measurement. The amplifier is split in a transconductance and a transimpedance stage. In between these stages a differential current, $I_{\text{offset}}$, can be injected in order to correct the offset without introducing non-linearity. The second stage provides a programmable gain, which in combination with additional two amplifying stages spans over two decades allowing measurements of magnetic fields between typically 10 and 70 mT.

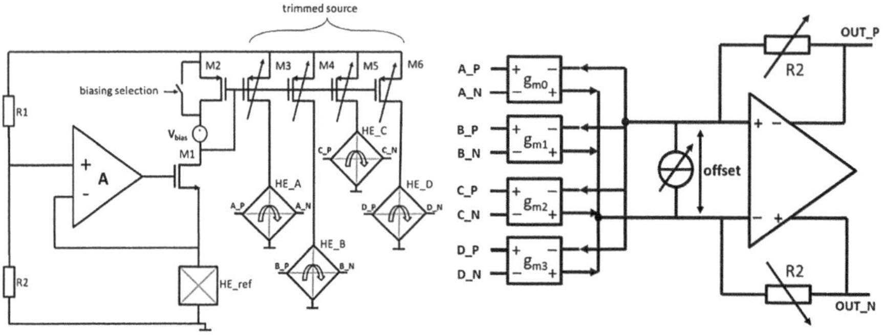


**Fig. 16.5** Front-end circuit schematics: (**a**) Biasing circuit. (**b**) HE outputs sum and first amplifier

## 16.4 Interfacing with the Outside World

Sensors for automotive applications need to be designed to have certain robustness when the system is aggressed by electrostatic discharge (ESD) pulses, conducted radio frequency (RF)-disturbances or environmental stress affecting the supply voltage stability.

In order to quantify the robustness of systems, standards describing test setups and measurements have been developed. Examples are the ISO-16750-2 [9] which defines a significant number of supply voltage variation setups and the IEC 62132-4 [10], that defines a direct power injection (DPI) setups applied to signal and supply pins.

There are different status of robustness identified by a functional status classification (ISO 7637-1 [11])

Status I: The function performs as designed during and after the test

Status II: The function does not perform as designed during the test, but returns to normal operation after the test

Status III: The function does not perform as designed during the test and does not return to normal operation without a simple driver/passenger invention, such as turning off/on the DUT, or cycling the ignition switch after the disturbance is removed.

Status IV: The function does not perform as designed during and after the test and cannot be returned to proper operation without more extensive intervention, such as disconnecting and reconnecting the battery or power feed. The function shall not have sustained any permanent damage as a result of the testing.

For obvious reasons the ultimate robustness goal is to have the systems reach "Status I." Some design examples on how we get these results will be presented.

### *16.4.1 Supply System Design*

An interesting test is the handling of the RI-130 signal [12]. As Fig. 16.6 shows one can observe the input voltage, even with a 100 nF capacitor attached to it, to swing between 30 V and −20 V (blue trace).

This means that a supply system must include bidirectional ESD protection structures and a reverse voltage protection feature. As the sensor has analog circuitry operating at 3V3 and digital circuitry running on a 1V8 regulator, we've actually implemented an analog 3V3 low drop out (LDO) regulator having a "reverse current protect" feature: the pass transistor is switched off when its output voltage is higher than the supply voltage. The external load capacitance reduces the output noise, improves load transients, and is used to supply the complete sensor circuit when the supply disappears for a while. This is the "autarky" operation mode.

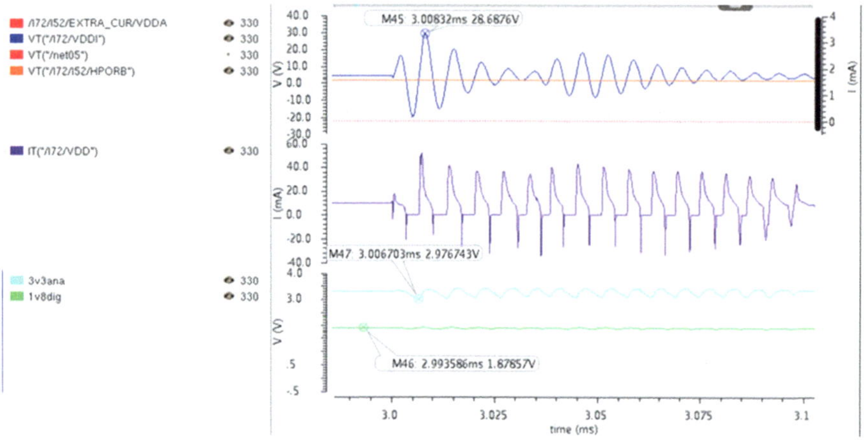


**Fig. 16.6** Supply system under RI130 signal

The purple trace above shows the LDO output current and one can observe that the external load capacitance needs to be recharged during some us. This test essentially sets the LDO output current limit and the regulator's unity gain bandwidth (GBW), which are chosen to be around 40 mA and 1 MHz. The cyan curve shows the ripple of the analog regulator output, which, thanks to the autarky operation, stays high enough to correctly feed the digital LDO, leading to no reset event.

## *16.4.2 ESD Structure Design*

Besides the usual human body model (HBM) and contact discharge (CD) robustness, there are specific test pulses that automotive sensors attached to the wire harness need to comply to. One of them, probably the most challenging one and described in the GMW3097 (GM) [13] is the 85 V slow direct capacitive coupling (DCC) pulse. It is actually applied in a worst case 150 °C scenario. The potential energy that a DCC voltage generator can unleash into the integrated circuit using 0.5*C*V**2 equals to roughly 0.5*100 nF*85 V**2 = 360 μJ. Comparing this with the energy absorbed by a 8 kV HBM of around 100 ns into a SCR with a hold voltage of 5 V and an Ron of 1 Ω around 100 ns*5.3 A*10 V = 5.3 μJ, one can appreciate that the DCC energy pulse unleashes energy an order of magnitude higher than what a typical 8 kV HBM ESD-SCR structure can handle.

In order to avoid large ESD structures, a bi-directional silicon controlled rectifier (SCR) structure has been integrated having a low hold voltage for the physical layer (PHY) outputs and a bi-directional high hold voltage SCR with about 50 V trigger voltage for the supply pin. The supply pin high hold voltage is chosen to avoid supply latch-up and profit from the fact that 100 nF capacitor is attached to the supply pin; as the DCC generator has a 100 nF feed capacitance, using the same load capacitance will divide the maximum voltage seen at the supply pin by 2. Using 50 V trigger voltage allows us to deal with at least a 100 V slow DCC pulse without triggering the SCR.

The PHY output, however, doesn't have the luxury of seeing a large capacitive load as the SENT protocol shows 6 μs 5 V to 0 V falling and 18 μs or so rising edges and does not allow supply ripple currents to be above 9 mA. This is the main reason why one uses a load capacitance in the order of 2–10 nF. With such a low load capacitance, the DCC source voltage will not undergo a division and is seen on the PHY out pin in full force. However, as latch-up is not a threat, one can now use low hold voltage SCRs. This way, one can benefit from the fact that the DCC source resistance is specified to be 2 Ω, e.g., using a 1 Ω SCR hold resistance with 1 Ω routing resistance, the injected energy is actually divided by $(2 + 1 + 1) = 4$. Figure 16.7 shows a simulation of the current through the device taken into account routing resistance. One can observe that the ESD structure absorbs "just" around (5 A*5.6 V*1.5 μs) ~40 μJ for a 85 V slow DCC pulse. (Red trace is current, red trace in second strip is voltage across the ESD structure, purple is the energy absorbed).

The failure energy of typical ESD protection SCRs injected using a 1600 ns long duration transmission line pulse (LD-TLP) measurement technique is about three times higher than the failure energy injected with a 100 ns TLP, thanks to the heat spreading. After applying the right surface sizing, we implemented a "reasonably"

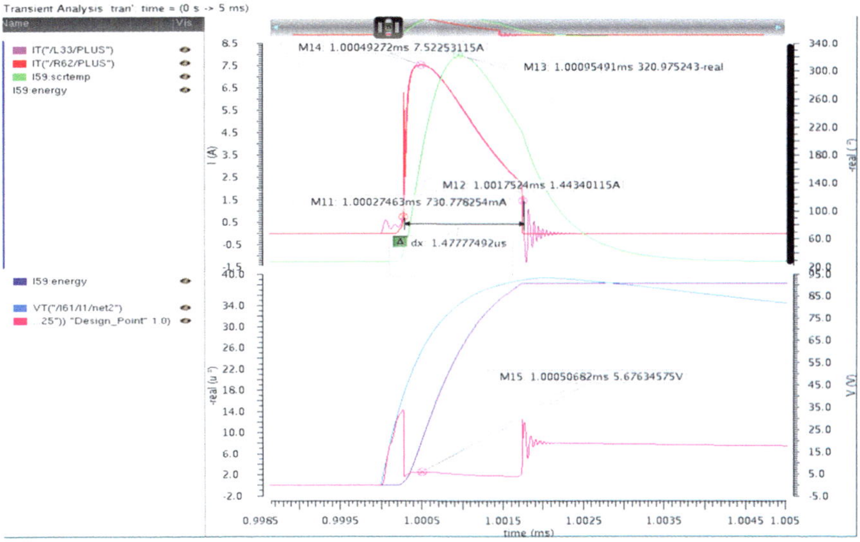


**Fig. 16.7** Voltage and current of the ESD protection under slow 85 V DCC discharge

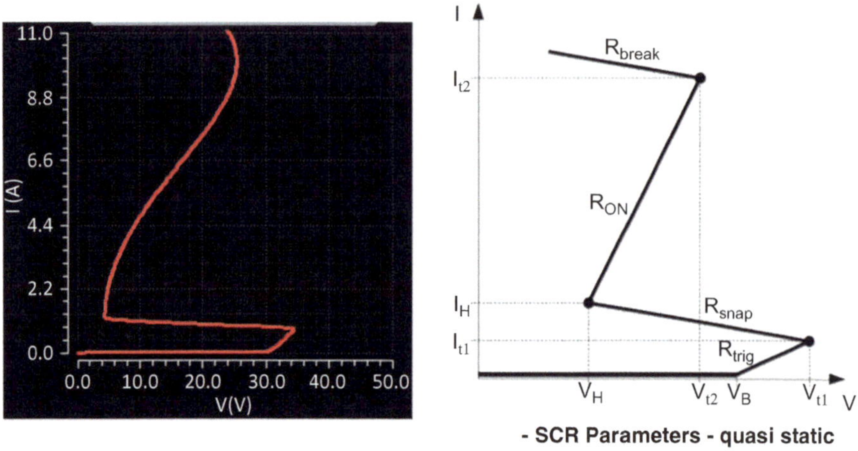


**Fig. 16.8** SCR VI characteristics: model vs. quasi-static parameters

large 140 × 300 um ESD protection structure for the PHY out pin. Measurement data showed structure surviving a 110 V GM-DCC pulse at the PHY out pin.

Designing for the highest functional class forced us to simulate complete circuit behavior while applying pulses on the supply and PHY out pins. It is obvious that this can only be achieved when the ESD protection structure is correctly modeled. Therefore a verilog-A model has been developed, which models the SCR behavior. Spline functions were used to create a continuous and smooth function (such that derivatives are continuous too) to avoid convergence problems during transient simulation. The resulting current vs. voltage plot can be seen in Fig. 16.8.

### *16.4.3 Analog/PWM/SENT PHY Design*

Sensors that have a physical distance from ECUs use PHY interfaces to communicate.

The PHY technology typically used for sensor communication used today is analog, pulse width modulation (PWM) or the single edge nibble transmission (SENT) [14] protocol; SENT can be used both with or without the Short PWM Code (SPC) [15] extension allowing bidirectional communication. The first two are obvious. SENT data is transmitted using time intervals between two falling flanks representing a 4-bit data word (Nibble).

The PHY interface (the circuitry directly connected to the connection wires) is normally supplied by the application voltage and therefore does not really benefit from the 3V3 LDO autarky capacitor. The reason for this is that the voltage levels specified at the PHY output are ratio-metric for the analog mode and the levels are absolute voltages for the SENT mode.

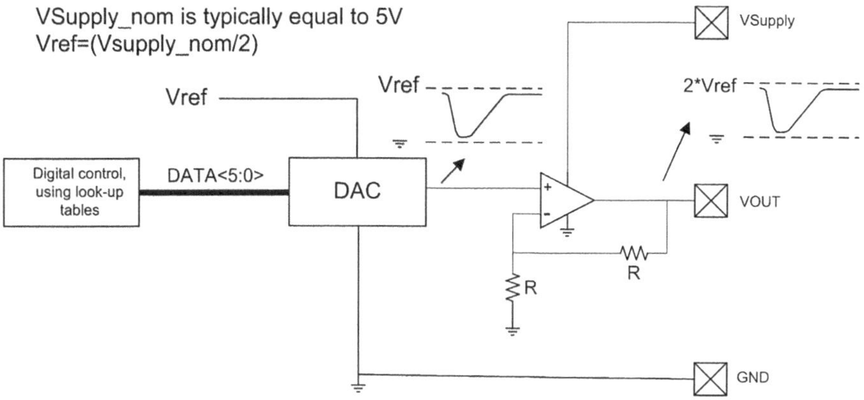


**Fig. 16.9** SENT PHY architecture: a low resolution DAC followed by an output buffer

That means that when a supply micro-cut for example arrives at the moment of a SENT negative flank generation, the significant signal distortion leads immediately to a Nibble bit detection error at the ECU receiver side. Those errors manifest themselves as cyclic redundancy check (CRC) or checksum errors and in order to pass functional Status I, just a couple of those are allowed in a certain time frame.

In order to strive for functional Status I for SENT mode of operation, one needs to avoid distortion of the falling flanks. Figure 16.9 shows a typical state of the art PHY architecture used for the generation of SENT signals.

A digital signal is generated using for example look-up table techniques and is followed by a digital-to-analog converter (DAC). The signal at the output of the DAC looks as shown in Fig. 16.9. The full range of the DAC would be then between $V_{GND}$ (typically 0 V) and $V_{ref}$. For a decent DAC operation, $V_{ref}$ is lower than the applied supply voltage ($V_{Supply}$). A good choice would be $V_{ref} = 2.5$ V. This way, amplifying the signal by 2 generates the correct SENT signal levels. A low resolution DAC complies with emission EMC requirements.

A simple solution to allow for analog ratio-metric operation and SENT operation would then be to use a simple thermometer DAC architecture, where for SENT operation an internal stable reference voltage will be imposed by implementing a dedicated buffer amplifier and for analog operation the supply voltage itself is used as a voltage reference. This architecture is shown in Fig. 16.10.

This way the SENT signal generation is quite independent of $V_{supply}$ and any $V_{supply}$ variation, as long as it is above the 4.5 V, it will not be noticed at the output.

**Improving the Output Amplifier Precision During the Micro-Cut Event**
The LDMOS MN2 and HV-PMOS MP2 implement a low drop diode to protect the low side MN1 driver transistor against negative voltages at the output. The bulk switch protects against reverse currents into the supply voltage when the output voltage rises above the supply voltage.

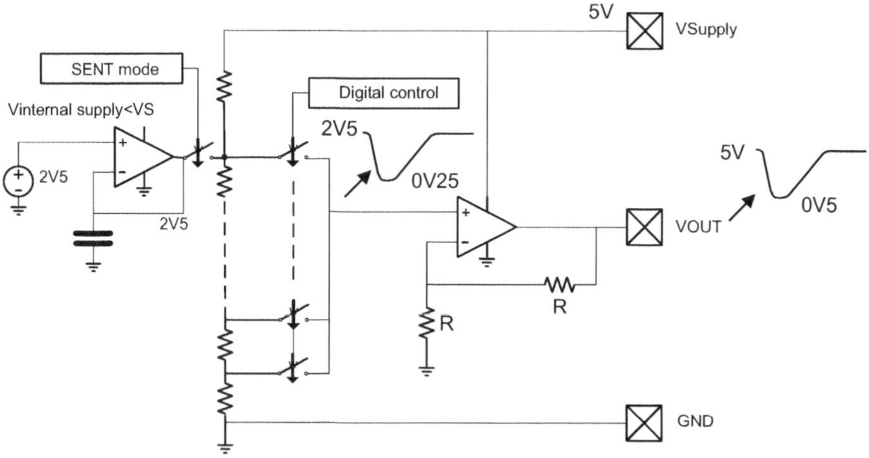


**Fig. 16.10** SENT architecture

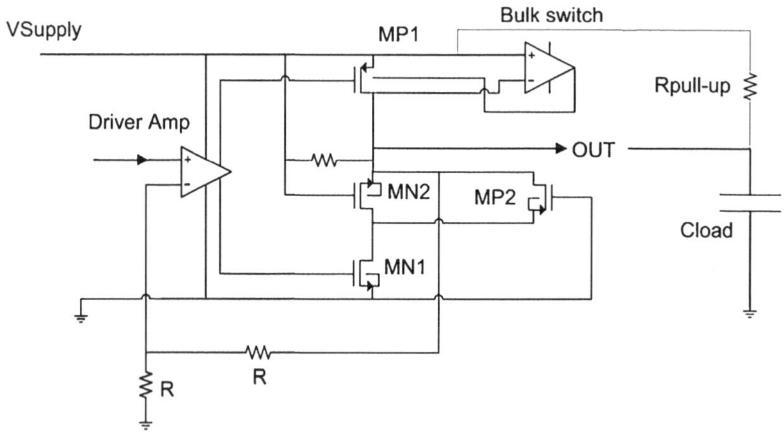


**Fig. 16.11** Circuit diagram of a typical PHY output stage

As can be seen in Fig. 16.11, the driver amplifier is supplied by $V_{\mathrm{Supply}}$ allowing the implementation of class-AB quiescent current control of the output transistors MN1 and MP1, however, when the supply disappears during a micro-cut event, MN1 cannot be switched on and the negative flank generation is effectively halted. Implementing a "supply switch" switching the supply of the Driver Amp to the maximum available voltage, the negative flank will not be interrupted. Mathematically:

$$V_{\mathrm{supply}}\ \mathrm{driver\ amp} = \max\left\{V_{\mathrm{out}}, V_{\mathrm{supply}}, V_{\mathrm{int\ supply}}\right\}$$

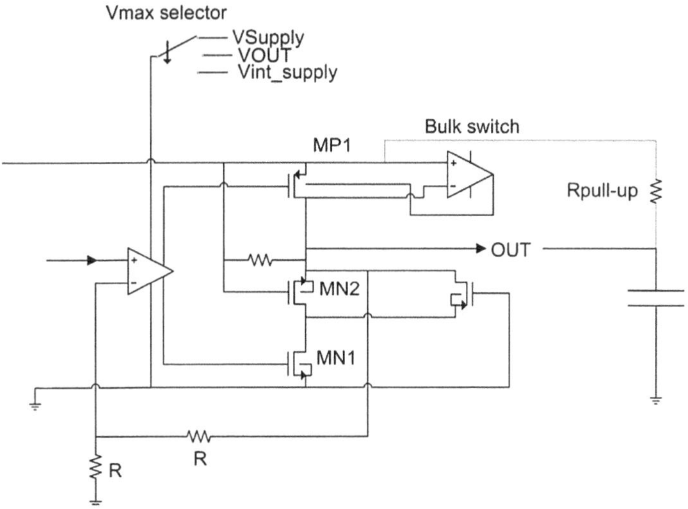


**Fig. 16.12** Max available voltage selection architecture for PHY

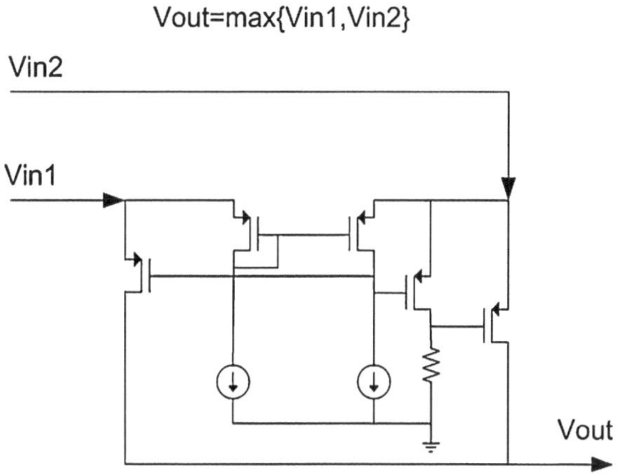


**Fig. 16.13** A simple two input voltage, maximum voltage selector

Figure 16.12 shows the basic principle of this idea.

Just by cascading two maximum selectors like the one shown in Fig. 16.13, a three input max voltage selector is designed.

Figure 16.14 shows how the PHY output signal (red) is not disturbed, even though the supply voltage (blue) is practically absent.

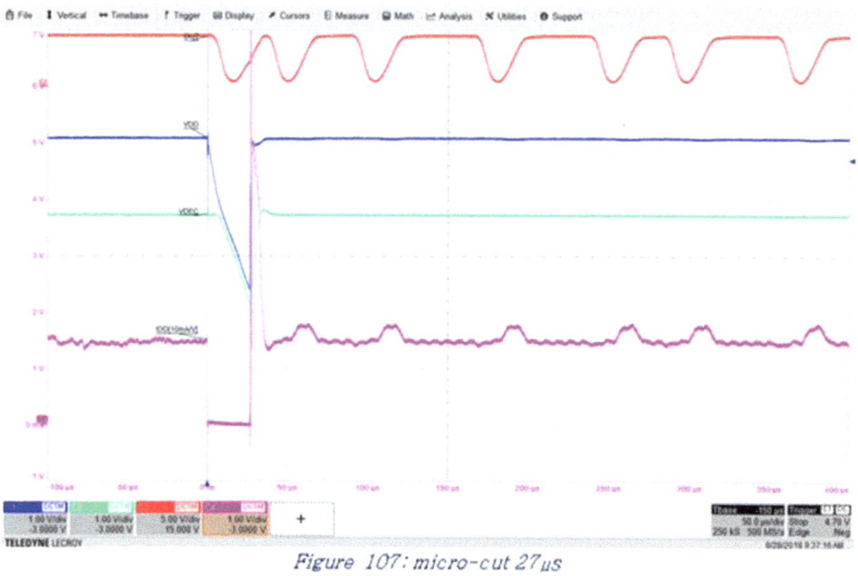


**Fig. 16.14** PHY micro-cut behavior measurement. $V_{supply}$ is blue, PHY out pin is red, current consumption (10 mA/V) is purple and the 3V3 LDO output is green

### *16.4.4 Emission Performance of PHY Design*

The use of a digital signal generation of a smooth SENT signal is able to lower the emission profile significantly compared to previous design based on slew rate control even with external pi-filters (Fig. 16.15).

## 16.5 Experimental Magnetic Results

### *16.5.1 Offsets*

Among all parameters that can impact the precision the most critical one is the residual Hall offset. Physically, the residual Hall offset is caused by piezoresistive effects, geometrical imperfections, and temperature gradients [16, 17]. An example of geometrical imperfection is the mismatch in the lead bringing the bias current to the HE. This causes a net average self-generated magnetic field over the HE—naturally this self-generated field is also sensed, yielding a self-induced Hall voltage. The total residual offset is typically in the order of 10–100 uV. It is compensated by the embedded software via the fine offset correction. There we seek a perfect cancellation—which we can never reach because of low-frequency noise, drift, and imperfect calibration. The plots below illustrate the residual offsets after fine offset correction in the embedded software, referred to the HE terminals (Fig. 16.16).

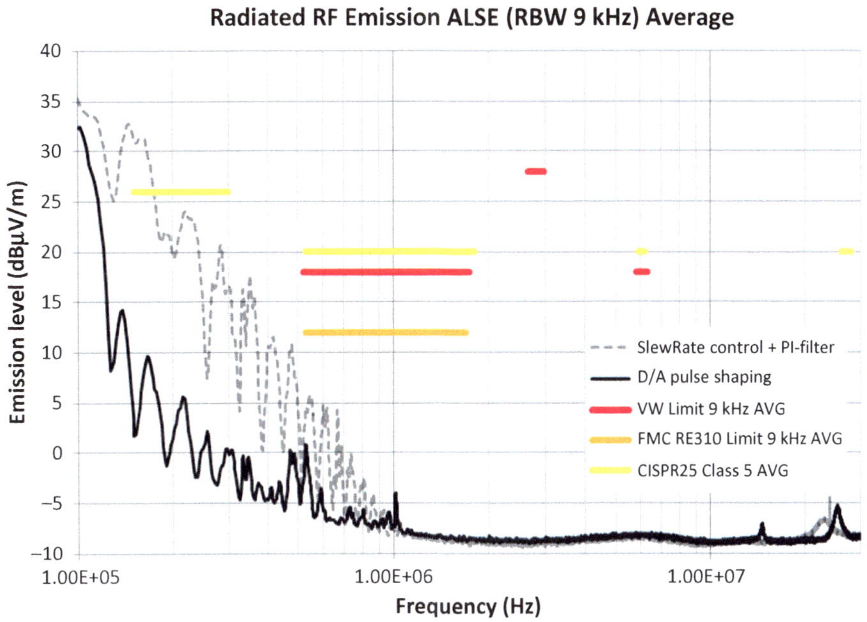


**Fig. 16.15** PHY emission performance

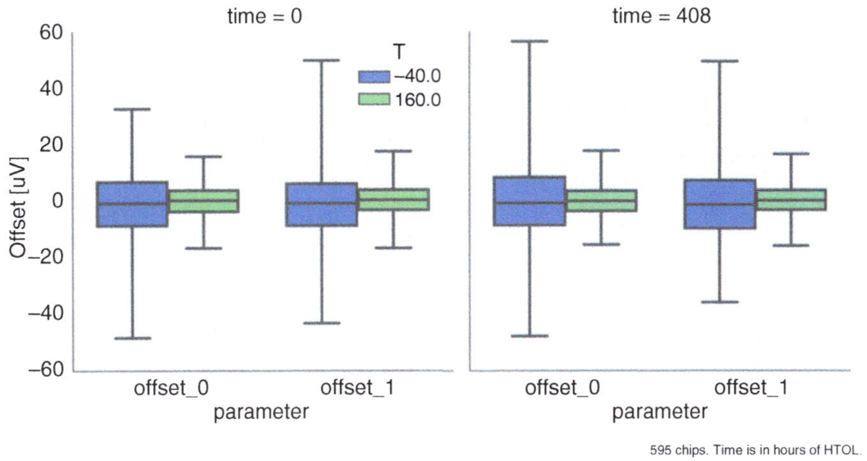


**Fig. 16.16** Offset component of the angle drift, expressed in uV referred to the HE terminals. The left panel shows the initial situation after factory trimming. The right panel shows the same offsets at the end of operating life, emulated by an accelerated test (408 h at 175 °C). The suffixes (_0 and _1) correspond to the cosine and sine components

### 16.5.2 Angle Accuracy

To quantify the accuracy of the sensor, we characterized a batch of 130 sensors in terms of their angle non-linearity error. A rotating field gradient of 10 mT/mm was applied. In this mode, the sensor measures the angle of the field gradients. Taking into account the multiple HEs and the distances, the total sensed field is 30 mT. At the highest temperature of 160 °C this yields $V_{\text{Hall}} = 2.3$ mV of Hall voltage signal. The results are shown in Fig. 16.17. The non-linearity error reaches 0.4° at 160 °C after lifetime (automotive qualification according to AEC-Q100).

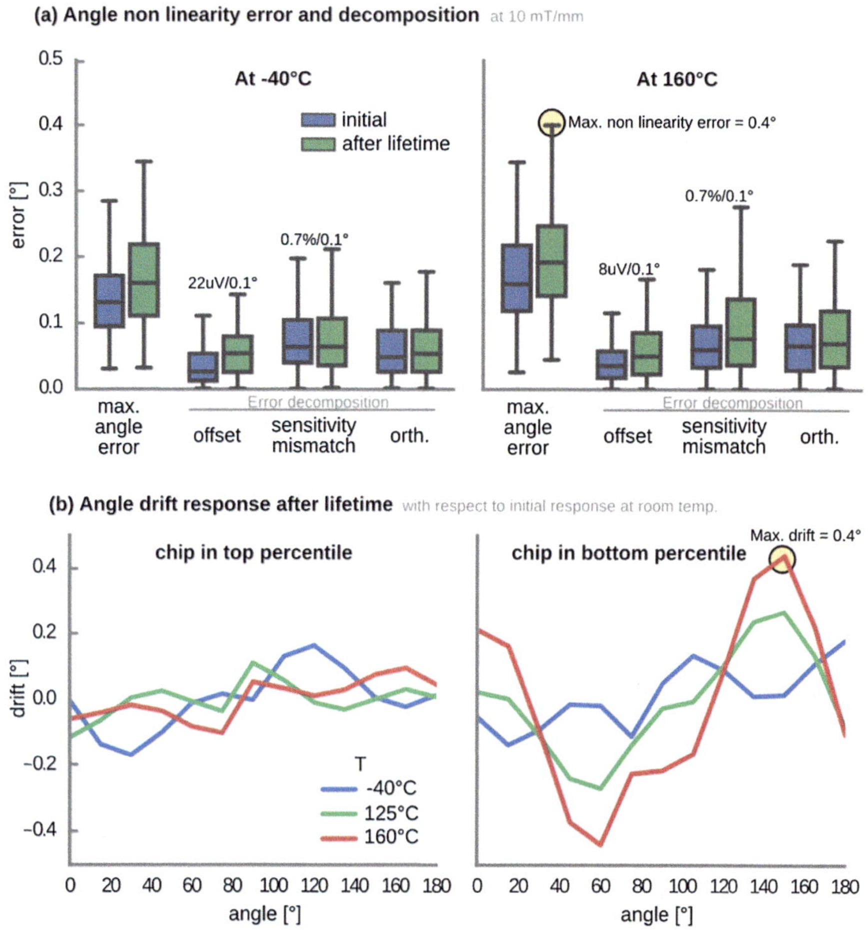


**Fig. 16.17** Accuracy measurements over the temperature range and lifetime with 10 mT of signal. (**a**) Distributions of angle non-linearity error and its components for 130 sensors. (**b**) Angle error curve for the extreme chips

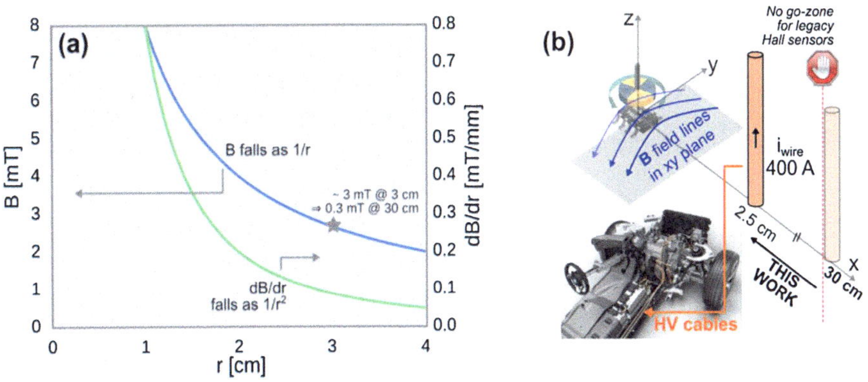


**Fig. 16.18** (**a**) Magnetic field and field gradient as a function of distance r to a wire carrying *i* wire = 400 A. (**b**) In legacy Hall sensors, this generates an exclusion zone around the high-voltage cables in electrified cars. Photo is from BMW i8

### *16.5.3 Stray-Field Robustness*

Magnetic stray fields are naturally present in modern cars, especially electrified cars. Large fields in the order of 1 mT have been measured in real cars. The most critical location is naturally close to the high-power cables, typically marked in orange, in hybrid and electric cars see Fig. 16.18 (photo is from BMW i8). The automotive industry has therefore normalized stray-field immunity tests [18].

To put the above requirement in perspective, consider what happens in legacy Hall sensors. They typically operate with 40 mT. The 5 mT stray field induces arctan(5/40) = 7° of error. This highlights the severity of the problem. Legacy sensors are an order of magnitude out of specification with respect to the stray-field immunity required in modern cars. In legacy sensors, the options are very limited. This is a fundamental issue with magnetometer-based sensing. One approach is to increase the signal with just stronger magnet (and higher cost), or with special magnetization concentrating the 360° angle range into a smaller linear stroke [19]. This is not a sustainable strategy. Anyway it still falls short as a general solution as it is not feasible to reach 0.4° of error. The final option is to increase the minimum clearance with respect to localized stray-field sources. This is not a practical solution: it would require more than 30 cm distance for legacy sensors. A final option is to use shielding. Again, it just shifts the problem to the applications, and it is not sustainable.

Our design uses instead gradient-based measurement. The gradiometric concept is implemented in a CMOS chip [20] with an appropriate integrated magneto-concentrator (IMC). The sensor is sensitive to magnetic field gradients in the plane of the chip. In production tests, the sensors are trimmed for optimum stray-field rejection by adjusting the sensitivities of the individual Hall elements. The trimming circuitry is based on fine tuning the bias current of each HE. A reference Hall

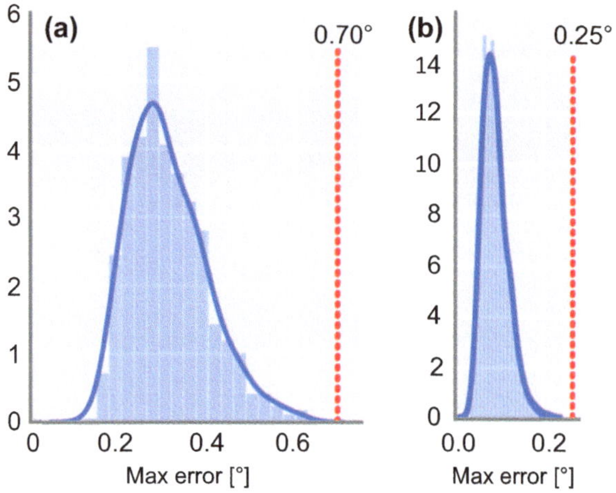


**Fig. 16.19** (**a**) Angle thermal and lifetime drift in the absence of stray field. (**b**) Extra worst case stray-field induced error in the presence of a 5 mT stray field. Sample size = 600 chips

element *HE_ref* is used to generate a reference current. The current is then mirrored and supplied to the HEs with individually trimmable adjustments (step size around 0.7%). This allows to correct the sensitivity mismatch while still operating close to the maximum supply voltage.

The stray-field rejection is thus limited by residual mismatch [21]. Figure 16.19a shows that the baseline accuracy, limited by thermal drift, is 0.7° in the absence of stray field for about 600 chips. Figure 16.19b shows the impact, for same 600 chips, of a 5 mT uniform stray field. Such a stray field induces an extra error of at most 0.2°.

## 16.6 Implementation

The device has been implemented in XFAB 0.18 um HV bulk technology. Figure 16.20 shows the die photo.

## 16.7 Conclusions

This paper has shown design techniques to address the challenges of a fully integrated Hall-based position sensor. XFAB's 0.18 um HV CMOS technology allows the integration of Hall elements, IMC, high and low voltage components, and a complete microcontroller with non-volatile memory. This allows a significant reduction of external component count. The sensor is able to sustain high operating temperatures in harsh environments while achieving angle precision better than 1° over lifetime. Parasitic electromagnetic emissions are also limited well below

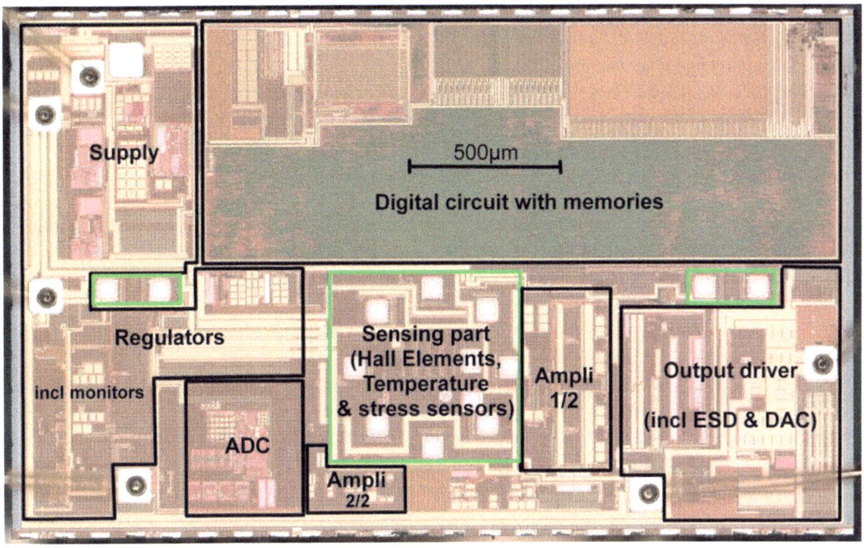


**Fig. 16.20** Die photo

tolerated level. Overall, the robust single-chip sensor can be deployed in safety-critical ASIL-C automotive applications.

## References


1. W. Granig, S. Hartmann, B. Köppl, Performance and technology comparison of GMR versus commonly used angle sensor principles for automotive applications, in *SAE Technical Paper Series, volume 1 of SAE Technical Paper Series*. 400 Commonwealth Drive, Warrendale, PA, United States (SAE International, 2007)
2. R.S. Popovic, *Hall Effect Devices* (CRC Press, Boca Raton, 2003)
3. E. Kejik F.B. Schurig, R.S. Popovic, First fully CMOS-integrated 3D Hall probe, in *The 13th International Conference on Solid-State Sensors, Actuators and Microsystems, 2005. Digest of Technical Papers. Transducers '05*, vol 1 (2005), pp. 317–320
4. V. Hiligsmann, P. Riendeau, Monolithic 360 degrees rotary position sensor IC, in *Proceedings of IEEE Sensors*, vol 3 (2004), pp. 1137–1142
5. S. Leroy, S. Rigert, A. Laville, A. Ajbl, G.F. Close, Integrated Hall-based magnetic platform for position sensing, in *ESSCIRC 2017 - 43rd IEEE European Solid State Circuits Conference*, Leuven, Belgium (2017), pp. 360–363
6. C.S. Huber, O. Paul, Package stress monitor to compensate for the Piezo-Hall effect in CMOS hall sensors. IEEE Sens. J. **13**(8), 2890–2898 (2013)
7. ISO 26262:2011, Road vehicles — functional safety
8. N. Chaabani, A methodology for analog fault injection at primitive level, in *Proceedings of the Cadence User Conference 2017*, EMEA - Munich, Germany, May 15–17, Session MS14
9. ISO-16750-2:2012, Road vehicles — environmental conditions and testing for electrical and electronic equipment Part 2: electrical loads, in *International standard 2012* (International Organization for Standardization, 2012)

10. IEC 62132-4:2006, Integrated circuits - measurement of electromagnetic immunity 150 kHz to 1 GHz - part 4: direct RF power injection method, in *International Standard 2006* (International Electrotechnical Commission, 2006)
11. ISO 7637-1:2015, Road vehicles — electrical disturbances from conduction and coupling — part 1: definitions and general Considerations (International Organization for Standardization, 2015)
12. FMC1278, Electromagnetic compatibility specification for electrical/electronic components and subsystems, in *Ford Motor Company*, 2016
13. GMW3097, General specification for electrical/electronic components and subsystems, in *Electromagnetic Compatibility*, GM, 2015
14. SAE International Surface Vehicle Information Report, in *SENT – Single Edge Nibble Transmission for Automotive Applications* J2716, Rev. 2008
15. L. Beaurenaut, Short PWM code: a step towards smarter automotive sensors. in *Advanced Microsystems for Automotive Applications* (2009), pp. 383–395
16. P. Ruther, U. Schiller, W. Buesser, R. Janke, O. Paul, Thermomagnetic residual offset in integrated hall plates. Proc. IEEE Sens. **2**, 763–766 (2002)
17. A.A. Bellekom, Origins of offset in conventional and spinning-current Hall plates. PhD thesis, TU Delft, Delft University of Technology, 1998
18. ISO 11452-8:2015, Road vehicles, component test methods for electrical disturbances from narrowband radiated electromagnetic energy: part 8, immunity to magnetic fields
19. M. Delbaere, Robust magnetosensitive position sensor for demanding applications. ATZ World **119**(4), 54–57 (2017)
20. S. Huber, J.-W. Burssens, N. Dupré, O. Dubrulle, Y. Bidaux, G. Close, C. Schott, A gradiometric magnetic sensor system for stray-field-immune rotary position sensing in harsh environment. Proc. Eurosens. **2**(13), 809 (2018)
21. N. Dupré, O. Dubrulle, S. Huber, J.W. Burssens, S. Christian, G.F. Close, Experimental demonstration of stray-field immunity beyond 5 mt for an automotive-grade rotary position sensor. Proc. Eurosens. **2**(13), 763 (2018)

# Chapter 17
# Rad-Hard Mixed-Signal IC Design, Theory and Implementation

C. Calligaro and U. Gatti

## 17.1 Introduction

Even if it is a common belief that space is a niche market, it is estimated that 1450 satellites with a mass greater than 50 kg will be launched on average every year by 2025. They will represent a market value of $250 billion for the space industry. Even the small satellite market projections for the 2021 estimate a value of $5.32 billion. Moreover, on the medium-to long-term space robots will replace men in extravehicular (EV) maintenance. The success of space missions depends on performing, monitoring, and controlling an extensive amount of functions on-board and in robot. All these systems contain a lot of electronics components, so that for academia and companies it is worth to be active in this field. To take benefit from innovative technologies developed for terrestrial applications, a technology transfer from terrestrial to space is envisaged. Unfortunately, the interactions of space electronics with space environment factors like solar flares, ionizing radiation, and extremes of temperature can cause degradation of materials in ICs. This chapter focuses on the use of standard sub-micrometric CMOS processes for the development of mixed-signal circuits (i.e., data-converters, sensor interfaces, opamp, etc.) widely used in electronics and addresses the methodology and solutions adopted to overcome extreme environments effects.

C. Calligaro (✉) · U. Gatti
RedCat Devices, Milan, Italy
e-mail: c.calligaro@redcatdevices.it; u.gatti@redcatdevices.it

© Springer Nature Switzerland AG 2020

A. Baschirotto et al. (eds.), *Next-Generation ADCs, High-Performance Power Management, and Technology Considerations for Advanced Integrated Circuits*,
https://doi.org/10.1007/978-3-030-25267-0_17

## 17.2 Extreme Environments

There are several degrees of "extreme," some of which are more challenging than other. Indeed, extreme environments are diverse, but include operation in radiation rich environments, operation in very high/low temperatures, in vibrational intense ambient, in chemically corrosive systems, etc. The focus here will be on radiation effects.

The Earth and its environment are protected by the atmosphere, which acts as a semi-permeable "screen," to let throughout light and heat, while stopping radiations and UVs. However, microelectronics circuits that are employed in space applications do not benefit from the protection of Earth atmosphere [1].

Ionizing radiation is ubiquitous in space, in the form of charged particles, coming mainly from the Sun, supernovae explosions, and other galaxies, and which can be trapped and concentrated by planetary magnetic fields. When striking electronic components, these energetic particles cause both cumulative and single event effects. The former induces a degradation in performance, such as increased leakage currents. The latter are generally non-destructive, and in memory components, may induce bit flips. Thus, in addition to satisfy regular specifications, space qualified circuits must maintain their performance under increased level of radiations and temperatures [2].

Analog and mixed-signal circuits (i.e., voltage regulators, A/D and D/A converters, opamp, etc.) are widely used in almost all systems intended for space application. They usually need very low sensitivity to external conditions, so they shall be designed to guarantee an adequate level of immunity to radiations (radiation-hard or radiation-tolerant).

### *17.2.1 Space Environment*

In the vacuum of space, energetic particles can travel over great distances without energy loss. In outer space, away from any radiation sources (i.e., stars), the amount of particle radiation is very small. However, in proximity of the Sun and the Earth, the particle radiation flux is significant. The particle radiation from the Sun to the upper atmosphere of the Earth can be divided into three different categories: (1) particles generated during solar events, (2) particles trapped in the Earth's magnetic field, and (3) galactic cosmic rays (GCR). The main sources of natural space radiation are summarized in Table 17.1.

Radiation in space is variable. Satellite systems in low earth orbit, such as Iridium, may receive 30 krad (Si) over a 10-year mission, whereas Globalstar can receive ten times that amount [4]. The requirements on TID and SEE depend on the final application, i.e., for an interplanetary mission the TID required for the components is in the range of MRad, while for low orbital satellites the TID is $\sim$100 kRad. Commercial integrated technologies do not have sufficient radiation

**Table 17.1** Main sources of natural space radiation environment [3]

| | | |
|---|---|---|
| Radiation belts | Electrons | eV—10 MeV |
| | Protons | keV—500 MeV |
| Solar flares | Protons | keV—500 MeV |
| | Ions | 1 to few 10 MeV/n |
| Galactic cosmic rays | Protons and ions | Max flux at about 300 MeV/n |

**Table 17.2** Main sources of natural space radiation environment

| Dose | Application |
|---|---|
| 0.3 rad | Natural (annual at ground) |
| 0.3 rad | X-ray exam |
| 1 rad | TAC (computed axial tomography) |
| 400 rad | Whole body median lethal dose for acute radiation syndrome |
| 1 krad | Typical radiation tolerance of ordinary ICs (accumulated) |
| 4–8 krad | Typical radiotherapy dose, locally applied |
| 100 k–1 Mrad | Typical tolerance of radiation-hardened ICs (accumulated) |
| 10 Mrad | Long distance mission |

robustness to guarantee a good reliability for all these environments, leading to the need of finding new solutions. At the same time, a correct functionality must be guaranteed in an extended temperature range (−55 to 125 °C), and in some cases at more extreme temperatures. Avionic, automotive, oil drilling, industrial applications, etc. share with space some of these requirements.

### *17.2.2 Other Harsh Environments*

Particle radiation is not restricted only to space. Natural and artificial sources of particle radiation may be found on Earth at high altitude, on the ground and underground. Sources could be either natural or artificial. The former can be cosmic rays, radionuclides in the soil, and thermal neutrons. Instead, artificial radiation sources can be accelerators (e.g., cyclotrons, synchrotrons), lasers, nuclear power plant, medical equipment (X-ray, Tomography, etc.).

Radiation is also present at avionics altitudes. When the protons enter the Earth atmosphere, some of them undergo nuclear spallation reaction to produce several particles including neutrons, which in turn release cascades of spallation reactions in the atmosphere that reach the Earth surface. The neutron flux at avionics altitude can be higher by a factor of 100 than at ground [5].

As an idea of the involved magnitudes, Table 17.2 summarizes some dose examples.

Radiation is not the only “hostile” condition in space. Temperature is another issue that shall be faced. For example, the ambient temperature on Moon ranges from +120 °C (in the sunshine) to −230 °C (in the shadowed polar craters). If the electronic circuits can be designed a priori to operate reliably in the space environment, the need for protection would be eliminated. By getting rid of the need for shielded "warm" boxes (the current practice) results in dramatic reductions in system weight and volume, and thus cost for each mission.

## 17.3 Radiation Effect on CMOS Technologies

The impact of the various types of radiations on electronic devices and systems consists of ionization effects (TID) and a variety of single event effect (SEE). TID effects are due to electron-hole pairs created in normally insulating layers and separated by an electric field, thus leading to unwanted charge build-up and insulator degradation. SEE usually arises from the impact of an energetic heavy ion or proton and, depending on the energy, leads to single event upsets (SEU) or latch-up (SEL). SEU are mainly pertinent to digital circuits where the state of a bit can be flipped.

### *17.3.1 Cumulative Effects*

Standard CMOS transistors (both nMOS and pMOS) widely use silicon dioxide. It is employed to isolate polysilicon gate from transistors channel and to isolate contiguous differently biased transistor. Silicon dioxide manufacturing is one of the key processes in a silicon foundry CMOS flow and relevant efforts making pure oxides have been done in the past to obtain a low number of defects both in thin (channel) and thick (isolation). With modern technologies, a technique, called shallow trench isolation (STI), makes more scaled and efficient isolation between contiguous transistors than in the past. Charge trap phenomena are almost the same in both gate oxides and isolation oxide. Thick oxide means a higher probability to have defects and, consequently, to give rise to charge traps in case of holes generated by charged particles. The same may occur in thin gate oxide, but considering that the oxidation is thin, a lower number of defects are expected, thus decreasing the probability to have charge traps. Deep submicron CMOS processes (below 100 nm) are regarded by the space community safe enough to consider any TID effect negligible at transistor channel level [6]. Of course, the same cannot be said for STI; considering the thickness, they still have to guarantee isolation between neighbor transistors. Since charge traps are responsible for leakage paths, we divide TID effects into two:

- Intra-device leakage.
- Inter-device leakage (Fig. 17.1).

Intra-device leakage affects the single transistor only shaping its width and length and modifying its general performance; inter-device leakage affects different

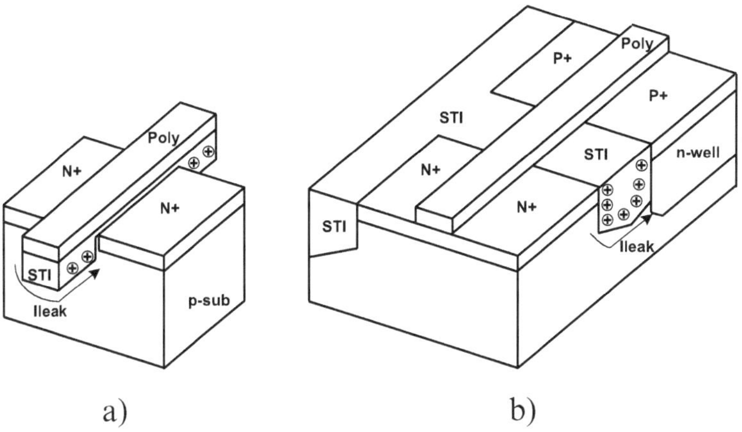


**Fig. 17.1** Simplified view of intra-device (**a**) and inter-device (**b**) leakage phenomena

transistors with a biasing of the first affecting the performance of the second (and vice versa). Inter- and Intra-device leakage may also combine their effects leading to a further degradation that is not recoverable.

The progressive effects of electron-hole pairs (HEPs) generation with time lead to a collection of positive traps (holes trapped by defects in STI) in proximity of the interface between silicon dioxides and active areas. This effect is negligible in all drain and source regions except those corresponding to channel transistor borders. In the middle of the channel, there are not relevant effects, but where thin channel oxide becomes thick isolation oxide (the two edges), a conductive path can bring charges from source to drain overriding the role of transistor gate. From the electrical point of view, the overall effect is a deviation of the nMOS threshold voltage whose value will be the medium value of the three contributes.

Once the leakage paths on border will be strong enough, the transistor will not be able to switch on and off, becoming a leaky transistor. For pMOS transistor, the behavior is the same in terms of charge trapping but opposite in terms of threshold shift.

Inter-device leakage involves neighbor transistors, and the effect is the biasing between different terminals. In the case of Fig. 17.1b, the leakage path can make a connection between the n+ diffusion of one of the nMOS transistor terminals and the n-well containing the pMOS transistor. Inter-device leakage may be critical in analog design where few millivolts can affect the working conditions.

For a MOS transistor the transconductance is proportional to the carrier mobility at the proximity of the oxide/silicon interface. Mobility is strongly influenced by the carrier scattering. As radiation dose increases, the lattice damage due to the oxide and interface traps increases, as well, consisting in a degradation of the average collision time and, thus, of the mobility [7], which is a critical factor in analog ($N_{\mathrm{ot}}$ and $N_{\mathrm{it}}$ are oxide and interface trap density, respectively):

$$\overline{\mu} = \frac{\mu_0}{1 + \alpha_{ot} N_{ot} + \alpha_{it} N_{it}}$$

This is a factor which is particularly critical for analog application.

### 17.3.2 Single Event Effects

Single event effect is caused by the passage of a highly ionized particle (i.e., heavy ion) through sensitive regions of a microelectronic device, for example through a reverse-biased p-n junction in a CMOS. The junction may be part of a MOS transistor (drain-body or source-body) or may be a well-substrate junction. SEE are categorized based on the consequences that they have on a device (Table 17.3).

A single event upset occurs when a SEE changes the logic value of a memory cell. If a SEU affects two or more memory cells, a multiple bit upset occurs.

Another well-known problem is the latch-up induced by particles colliding with very high energy (single event latch-up, SEL). This issue typically arises when nMOS and pMOS transistors are faced one to each other (e.g., building single inverter) so presenting wells at different voltage level (n-well of pMOS transistors connected to power supply and p-substrate of nMOS transistor connected to ground). Due to silicon fabrication, different doped areas may be connected forming a p-n-p-n junction (known as thyristor or SCR) with an intrinsic positive feedback path (Fig. 17.2).

**Table 17.3** Taxonomy of the SEE and their consequences

| | | |
|---|---|---|
| SEU | Single event upset | Change of bit in memory element |
| SET | Single event transient | Temporary variation of a bit or voltage |
| MBU | Multiple bit upset | Change of several bits coming from SET/SEU |
| SEFI | Single event functional interrupt | Corruption of data path in state machines |
| SEL | Single event latch-up | High current conditions leading to hard error |
| SEGR | Single event gate rupture | Rupture of the gate coming from high field |

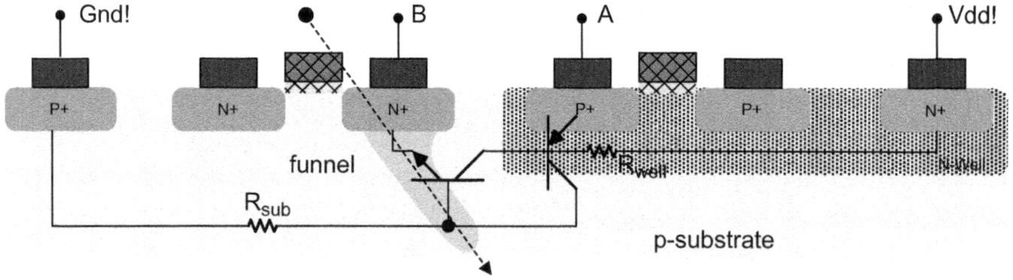


**Fig. 17.2** Cross-section of the p-n-p-n junction forming a SCR [6]

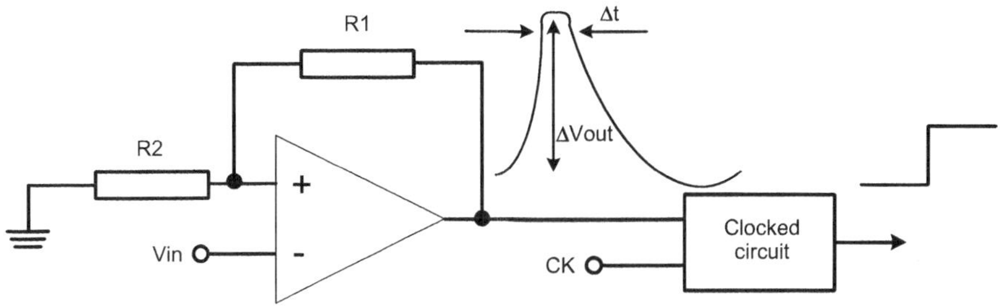


**Fig. 17.3** Example of ASET [8]

SCR is normally in off condition, and current flowing is negligible. However, the energy released by the particle may be enough to activate this positive feedback loop. In this case, a single current spike overriding the hold current of the parasitic SCR takes the region in the latched state and current from power supply takes advantage of the positive feedback of the SCR causing the latch-up.

Single event transients (SETs) occur in digital and analog circuits, called DSET and ASET, respectively. In the digital world, a SET is a transient glitch which affects the voltage of a node in combinational logic. Transients are temporary; however, they may propagate to adjacent nodes where the effect of other SETs can be added. SETs can lead to SEUs if in the path there are registers, flip-flop or SRAM cells.

In analog, a SET in a linear device is caused by the generation of charge by a single particle (proton or heavy ion) passing through a sensitive node in the linear circuit.

The ASET consists of a transient voltage pulse generated at that node that propagates to the device output (Fig. 17.3), where it appears as the same voltage transient, an amplified version of this transient, or a change in the logical output (e.g., in an opamp). It can cause a change in a circulating current, as well. SETs have been observed in many different types of linear microcircuits such as opamp, voltage references, comparators, ADCs. ASETs in linear devices differ significantly from other types of digital SEE.

Bias conditions significantly impact both the device SET sensitivity and the SET characteristics. A mitigation could be envisaged by means of low-pass filtering (impacting on performance), with redundancy techniques or novel architectures.

## 17.4 Radiation Hardening Techniques

Radiation hardening-by-design (RHBD) may be considered as a set of techniques used to mitigate radiation effects on silicon devices. Since semiconductor components may be different according to the technology used (CMOS, BiCMOS) and the application they target (analog, digital, memories), RHBD techniques may be adapted and applied to fit requirements coming from the space domain. Generally, three main approaches have been considered:

- process enhancements (Radiation Hardening-By-Process, RHBP)
- design enhancement (Radiation Hardening-By-Design, RHBD)
- shielded packages (Radiation Hardening-By-Shielding, RHBS).

RHBP uses solutions involving dedicated fabrication processes or specific technologies (i.e., SOI, SiGe, etc.). RHBP shows several disadvantages:

- generally dedicated processes are expensive
- the maturity of such processes is not comparable with standard ones
- specific manufacturing variations (extra masks, specific doping profiles, etc.) lead to low yields and low repeatability.

Moreover, these special processes are usually under national rules, such as US ITAR (International Traffic Arms Regulation) restrictions.

In general, RHBP represents an effort that can be supported only by large silicon foundries, but rad-hard market represents a niche (because of low volumes).

RHBS uses customized packages (e.g., Rad-Pak® [9]). The effectiveness of shielding depends on the radiation environment. For this reason, custom solution should be studied case by case and typically a shielding package is heavier than a comparable ceramic part. Their effectiveness against high energy particles is also a concern. An estimate of trapped proton fluxes transmitted by plane aluminum shields of various thicknesses (from 0.5 g/cm$^2$ to 2 g/cm$^2$ Al) for a low earth polar orbit during minimum solar activity showed a residual transmitted flux for proton energies ranging from few MeV to $\sim$500 MeV [10].

While RHBP and RHBS have been already employed for the development of commercial rad-hard digital and analog components, very few examples exist of RHBD utilization, especially for mixed-signals. In the pure RHBD methodology, the fabrication process is a standard process (i.e., CMOS or BiCMOS) and the radiation hardness is achieved considering only the circuit design at different levels, as we'll seen later.

A good and robust architecture is the starting point for almost every rad-hard component [6]. The right macro-blocks positioning, redundancy, and a preliminary floor-plan is the main brick for making easier any successive step both from the topological point of view (the circuits) and the physical layout point of view. This is called RHBD at Architectural Level (RHBD-AL).

In the pyramidal top-down approach the next step is represented by the circuit design where topological solutions focused on minimizing the impact of charged particles are considered to avoid hard- and soft error. This is called RHBD at Circuit Level (RHBD-CL).

The final step is represented by the physical layout design. A suitable architectural approach combined with robust circuit solution paves the road to rad-hard devices, but the layout plays a crucial role since the main hard errors coming from TID and SEL can be solved with a good and reliable design. This is called RHBD at Layout Level (RHBD-LL).

RHBD on standard CMOS technology may count on the following advantages:

- maturity of the process and very high repeatability;
- availability from many silicon foundries;
- low integration cost;
- MPW (multi project wafer) approach available;
- RHBD technical solutions easily portable.

In the next paragraphs, we'll describe the RHBD techniques adopted for the digital and analog domains, respectively.

### *17.4.1 Digital Domain*

The RHBD-AL can have an impact on area and performance of the component we are going to design, and it's always the outcome of a compromise. For example, in a generic semiconductor memory at architectural level radiation hardness can be improved by using redundant logic, such as ECC (error correcting codes), which depending on its complexity can require a significant area overhead (i.e., Hamming (12.8) algorithm demands a 50% area increase in the memory array). Another example is the "scrambling" in a memory array: the physical location of bits does not correspond to the logical bit position, to avoid logical MBU. A further improvement can be obtained by storing each bit of a byte into a different memory sub-array, and by providing each memory sub-array with separate bit-line and word-line decoders, to avoid MBUs due to address upset (Fig. 17.4). An alternative to the multiplication of decoders is the introduction of redundancy in decoders or DEMUXs using voting schemes.

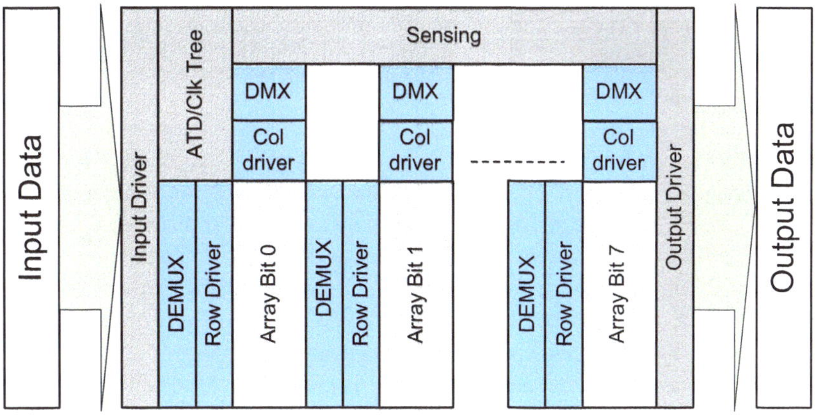


**Fig. 17.4** RHBD-AL with arrays, drivers, and DEMUXs split applied to a generic SRAM

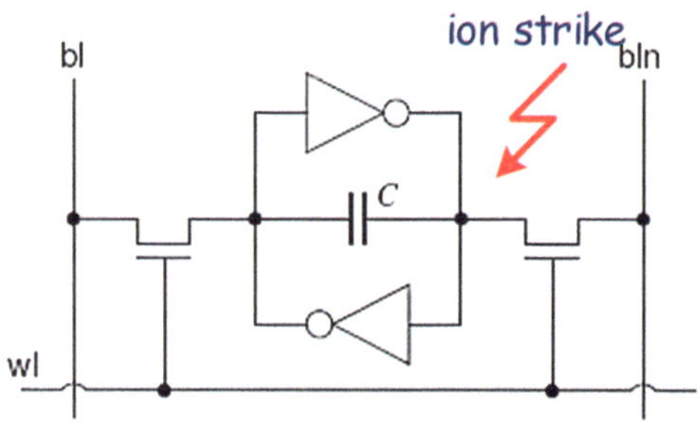


**Fig. 17.5** 6T SRAM bit-cell with fringe capacitor against SEU

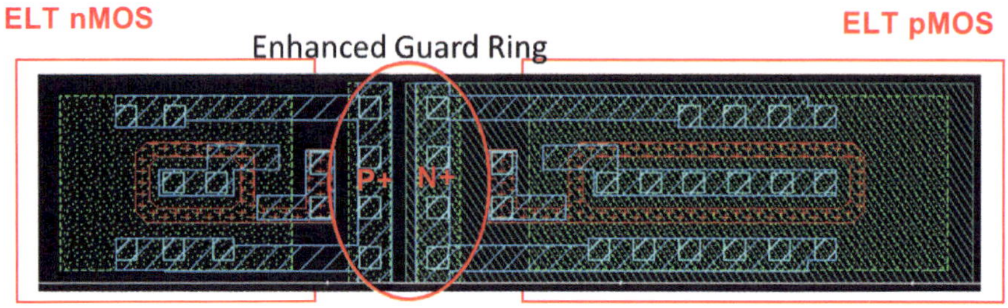


**Fig. 17.6** ELT transistors and guard-ring protection

RHBD-CL is the typical domain of analog designers. Single blocks such as sense amplifiers, digital port, I/O buffers, latches, SRAM, and analog cells are considered. For example, in a 6T SRAM cell, adding Miller capacitor enhances the critical charge, $Q_{cr}$, (i.e., the minimum charge necessary to trigger a SEE) in internal nodes, thus reducing SEU: a higher linear energy transfer, LET (i.e., a measure of the energy released per unit length) is required to upset (Fig. 17.5).

A possible solution to minimize SET propagation in a logic port consists in reducing the number of transistors not directly connected to supply and ground. In this case, if SET does not exceed threshold voltage, it cannot propagate towards other logic gates. This approach can be applied to analog stacked transistor, as well, for example when choosing an opamp topology.

At layout level (RHBD-LL) some special design techniques can be adopted to improve device tolerance to radiation both in terms of TID and SEE. To overcome TID leakage effects on nMOS, edge-less transistors, ELT (or annular transistor) physical design have been adopted [11]. They are MOS transistors with annular gate shape. This geometry was proved to reduce current leakage due to cumulative effects in nMOS transistors, even at very high total doses, at the expense of a larger area, as shown in Fig. 17.6. Holes trapped in STI do not have effect on pMOS since the conduction between its drain and source is made by holes. Even if it not strictly necessary, construct pMOS devices with ELT shapes allows to mitigate drive-strength unbalancing.

The use of enhanced guard rings (EGR) around p-well and n-well, biased to constant voltages, prevents SELs. Moreover, the use of guard rings also around transistors of the same type biased at different voltages reduces inter-device leakage, since positive charges trapped in the STI oxide cannot induce a parasitic channel between n-type diffusions at different voltages.

### *17.4.2 Analog Domain*

Analog and mixed-signal circuits suffer from the same issues when coping with radiations. TID effect causes variations in transistors parameters, i.e., increase of parasitic currents, increase or decrease of threshold voltages, and decrease of transconductance, $g_m$. The latter consequences could have a strong impact on analog circuits performance, since the accuracy of analog functions depends on transconductance and output resistance of transistors. SEUs are mainly pertinent to digital circuits where the state of a bit can be flipped. In analog circuits an ASET can arise but its impact shall be evaluated case by case: sometimes it subsides sometimes it can induce biasing variation that are not easily recoverable. SELs can destroy both digital and analog circuit.

In order to make analog circuits more robust to radiations, two ways can be followed. The easiest foresees the adaptation in analog of RHBD techniques (architecture, ELT, guard-ring, etc.) that already proved their effectiveness with digital circuits. A smarter solution focuses on identification of novel circuital solutions able to overcome radiation issues. Obviously, the literal translation of digital RHDB technique in analog is not straightforward and can bring to problems or to trade-off. Usually regular shaped transistors can sustain up to 10–30 krad (Si) radiation without degradation. The minimum provision to avoid TID effects is to choose L of MOS transistors greater than a certain value. However, in analog circuits even a small change in the threshold voltages can be relevant and ELT becomes a need. However, the ELT shape puts several concerns in analog design:

- Modeling of the ELT transistor not easy
- Limitation in the W/L viable ratio
- Asymmetry
- Mismatch
- Poor density
- Reliability

In analog technology, scaling not always can be a help. It causes the gate oxide to become thinner leading to a reduction of the radiation-induced charge trapping. Thus, TID effects result mainly related to thick lateral isolation. Low-voltage, which means lower electric field, helps to furtherly reduce TID. On the contrary, owing to the decrease in supply voltages, in node capacitance and the higher density, the sensitivity to ASET could be larger.

Let's have a look on how different levels of RHBD apply to an analog design with an example of ADC converter [12], which has been used inside the dosimeter test-case. RHBD-AL has an impact on the choice of the type of converter. Traditional

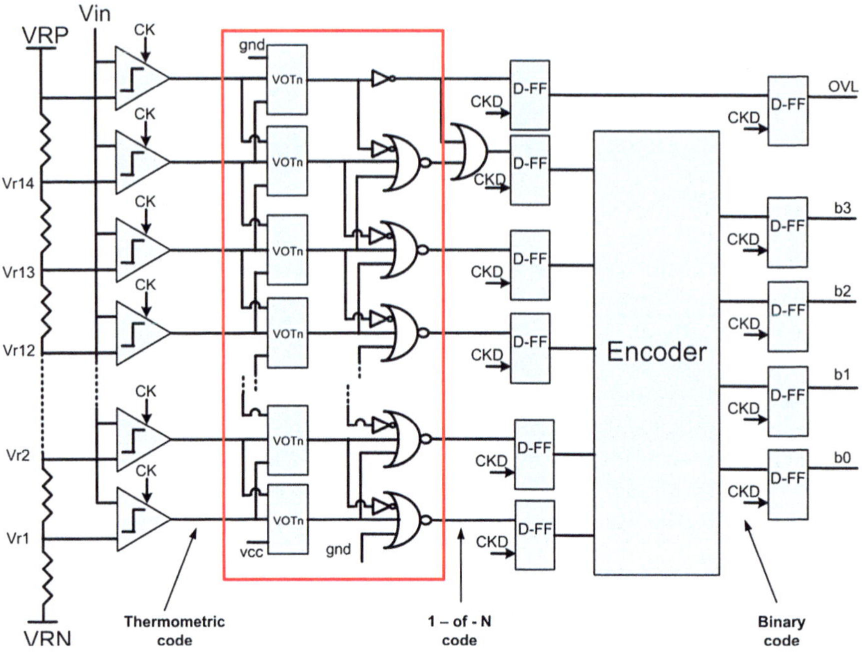


Fig. 17.7 Simplified schematic of the flash ADC [12]

ADC architectures (flash, SAR, sigma-delta, pipelined) are the starting point. Pipelined architectures include a lot of analog stuff and are highly prone to variations due to radiation effects. Amplifiers and comparators suffer from offset, gain variations, and inaccuracy in current mirrors due to TID, so that special custom techniques must be introduced to overcome such limitations [13]. SAR suffers of offset issues, as well, and, if implemented with charge redistribution techniques a second effect of TID arises, that is the decrease of the nMOS threshold which increases the leakage onto the top of the capacitor array [14]. The SAR architecture is particularly prone to SEU. If an upset occurs in the control logic a bit-flip error results during the conversion. Since the state of the switches in the subsequent steps depends on the value of the former bits, a single error in one bit may propagate to the remaining conversion bringing to multiple bit errors. Finally, sigma-delta constitutes a potentially interesting option for rad-hard ADC. The analog part of the circuit should be less immune to radiations effects, because ideally all faults introduced by extra-charges in integrators capacitors are transformed in out-of-band noise by the modulator [15]. In Ref. [12] an ADC with a conventional flash topology has been chosen (Fig. 17.7). It is based on a resistive string, which generates a series of reference voltages that are compared to the input signal by means of $2^N$ comparators. A custom two-stage error correction logic (red square) and an encoder 1-of-$N$ follow the comparators. Finally, a binary encoder provides the output bits.

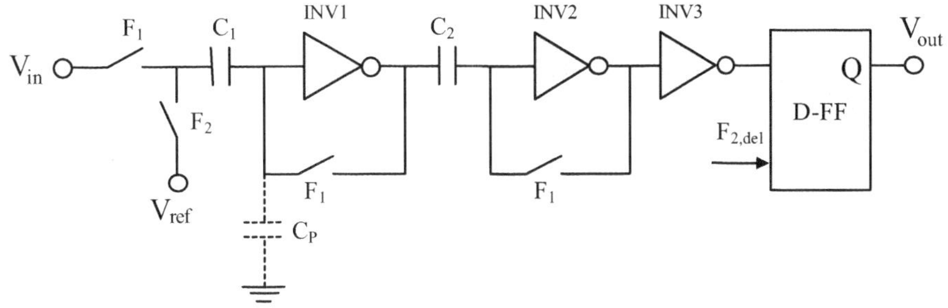


**Fig. 17.8** Schematic diagram of the auto-zeroed comparator

From a rad-hard point of view, this architecture has some intrinsic advantages:

- The converter has a "digital design" style
- The converter has parallel signal processing
- The comparator has a distributed S&H

The flash converter is mainly based on digital ports, including the comparator. The digital cells have been made resilient to radiations with RHBD techniques. The flash topology allows a parallel processing of the input signal, so that an error on one row does not propagate to entire bit word and can be corrected using conventional techniques (or more effective) for flash converters.

At circuit level it is possible to operate both on analog and digital sections to make it robust against radiations. The comparator (Fig. 17.8) is based on the so-called chain of two stages auto-zeroed inverters [16, 17]. It is just the auto-zero feature of the CMOS inverter that makes this comparator ideal for operation in an environment where the MOS threshold voltages can shift dynamically.

After that the device has been exposed to radiation, because of threshold voltages variations, the transfer curve of the inverter simply shifts. However, the inverter is still operating along its maximum gain region. Someway the increase in nMOS $g_m$ and decrease in pMOS $g_m$ tend to compensate and the gain stays almost constant. Thus, the auto-zeroing technique automatically compensates for dynamic threshold variations. The inverters do not need any bias circuits which can be prone to TID. A fortuitous consequence is that the auto-zeroing also improves the ASET behavior. If the MOS in the inverter is hit by a single event, the transient output error is immediately corrected during next auto-zero phase, so that the upset limit duration is limited to a single-clock phase.

Specific countermeasures have been proposed in the ADC at circuit level:

- Selection of proper passives (polysilicon resistors insensitive to TID)
- Quite high sampling capacitance to avoid SEE (critical nodes shall be enough "capacitive" to be resilient to the $Q_{cr}$ released by a hitting ion)
- Voter for bubble error correction based on "democracy" plus second correction stage based on NOR3

And at layout level:

- Switches based on ELT (leakage is a concern)
- Binary decoder implemented with ELT shape (n- and p-MOS)
- Isolation techniques for latch-up prevention

However, ELT topology imposes constraints in the choice of MOS W/L. In particular the annular shape fixes a limit on the minimum *W* which is possible to draw. For example, in an analog design the sizes of switches are always the results of a trade-off. Two main phenomena shall be considered:

- The $R_{on}$ of the CMOS shall allow enough speed and linearity
- The clock feedthrough shall be kept at minimum

The first condition pushes for using complementary switches with higher aspect ratio, while the clock feedthrough pulls for minimum MOS dimensions. But radiation robustness introduces additional constraints, so that the switches size does not correspond to minimum lithographic.

As concerns SELs in analog, a special approach has been proposed in [18], which consists in a custom protection circuitry. However, since the operating and latch-up currents vary for different devices, each device needs to be characterized for these parameters before the protection circuit can be designed. To maintain the design as more portable as possible this technique has not been applied to the Flash ADC design.

The converter has been integrated in a standard 180 nm process from two different foundries (TowerJazz and X-Fab). The measured prototype of the Flash ADC (Fig. 17.9) showed a TID immunity up to 300 krad (Si) and the characterization under single event (Xe ions, energy: 1217 MeV, LET: 60 MeV cm$^2$/mg) indicates that neither latch-up nor errors in the output code arise. Achieved results are quite similar for the prototypes of the two foundries demonstrating the quality and portability of the proposed RHBD techniques.

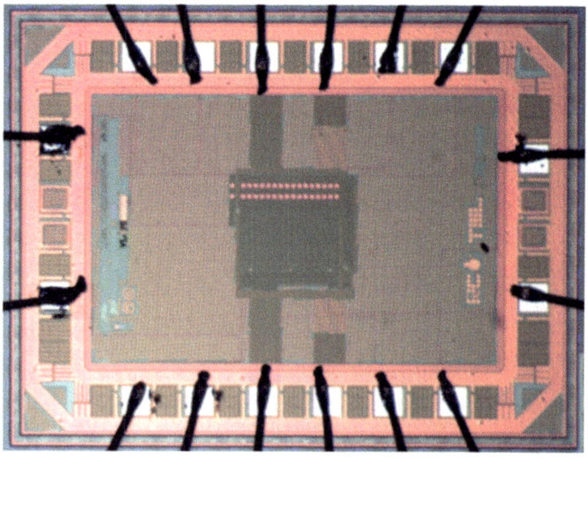

**Fig. 17.9** Microphotograph of the rad-hard Flash ADC

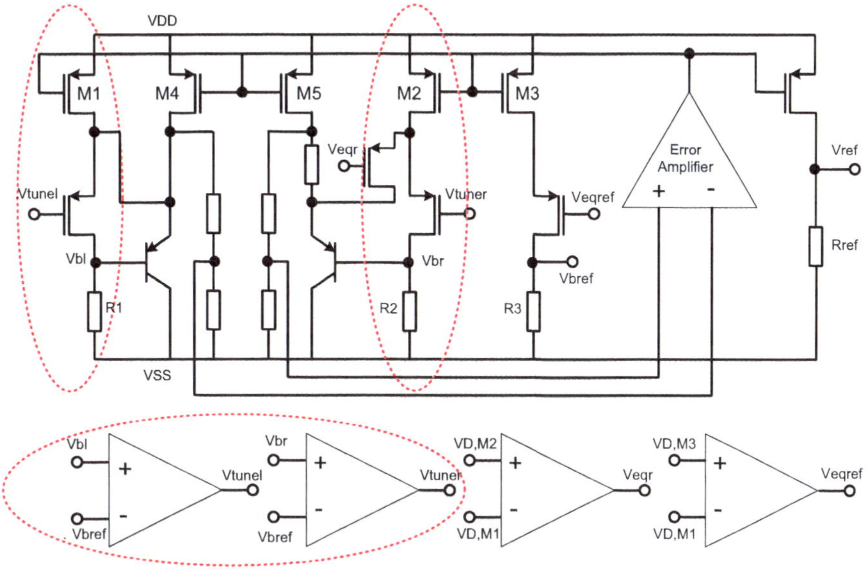


**Fig. 17.10** Schematic of the rad-hard bandgap using DBLC technique [19]

**Table 17.4** Main measured performance of DBLC bandgap [19]

| Feature | Value |
|---|---|
| Technology | 130 nm CMOS |
| Reference voltage | 0.6 V |
| Output voltage spread | ±15 mV (before trimming) |
| | ±4 mV (after trimming) |
| Temperature coefficient | 15 ppm/°C |
| Output voltage variation after 4.5 MGy TID (measured on ten samples) | 17% (without DBLC) |
| | ±3% (with DBLC) |

Another example of circuital solution adopted to make an analog circuit more robust to radiation can be find in the bandgap of [19]. One of the main issues in a bandgap circuit based on diode-BJT is related to radiation effects which arise due to TID in the diodes. This is because irradiation induces holes trapping near $SiO_2$-Si interface, thus increasing base leakage current and decreasing BJT current gain. Under dose the reference output voltage, which is derived from BJT base-emitter voltage and collector current, changes accordingly.

The proposed solution consists in a dynamic base leakage compensation (DBLC). It is based on a feedback compensation circuit (Fig. 17.10), whose purpose is that all the base leakage currents induced by radiation flows only into the compensation network, keeping constant the bandgap current $I_{D,\,\mathrm{M4}}$ and $I_{D,\,\mathrm{M5}}$. And so the output voltage. The benefits of the solution are shown in the Table 17.4. By applying the DBLC the bandgap output voltage variation with dose drops down from 17% to 3%, as measured by authors over ten samples (5 of conventional bandgaps and 5 of DBLC ones) at 4.5 MGy.

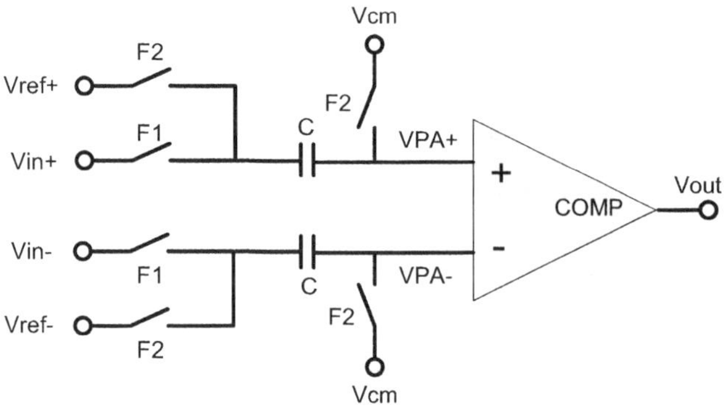


**Fig. 17.11** Switched-capacitor comparator [20]

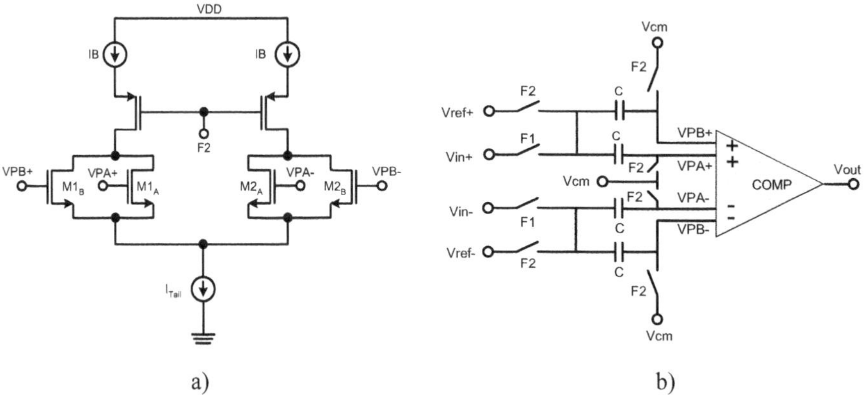


**Fig. 17.12** Differential amplifier with split MOS (**a**) and SC comparator (**b**) [20]

Another example of RHBD-CL applied to analog is given by the switched-capacitor (SC) comparator presented in [20]. In general, switched-capacitor circuit topologies exhibit much greater single-event sensitivity than their continuous-time counterpart due to the presence of floating nodes in the signal path. These nodes do have any discharge path, so charge deposited on a floating capacitor by a single-event strike will persist until the next clock phase. As shown in Fig. 17.11 the input network of the switched-capacitor comparator is particularly vulnerable during the evaluation phase ($F_1$).

A "brute force" solution is to increase the capacitor size. A smarter solution is based on the so-called dual path design technique. The technique can dramatically reduce the vulnerability of floating nodes with a 2× capacitor area penalty. Figure 17.12a shows the comparator input circuit with dual inputs implemented in the differential input stage. Input transistors $M1$ and $M2$ have been split into two identical transistors connected in parallel, $M1_A//M1_B$ and $M2_A//M2_B$, such that the *W/L* ratio of each parallel device is half the *W/L* ratio of the original transistor. When the gates of $M1_A$ and $M1_B$ are shorted together, this configuration is identical to a standard differential amplifier. However, the mitigation technique requires the gates to maintain isolated signal paths. Therefore, the external switched-capacitor differential network must also be duplicated as shown in Fig. 17.12b. However, the capacitors are not halved in size, since they are already sized to a minimum value to prevent mismatches.

A single-event hit on the floating node $V_{PA-/A+}$ will perturb the gate voltage of $M1_A$ in Fig. 17.12a. If the gate voltage of $M1_A$ decreases, the drain current $I_{M1A}$ through transistor $M_{1A}$ will decrease, as well. For a large enough perturbation, transistor $M_{1A}$ will turn-off completely. In a standard differential amplifier, the decrease in current would cause an increase in the current through the $M2_A//M2_B$ branch and a consequent error voltage at the amplifier output. However, thanks to the dual input transistors structure, MOS $M1_B$ provides an alternative signal path for the current. The parallel not-affected floating node for input transistor $M1_B$ keeps the correct voltage to be compared, and the comparator therefore operates correctly.

## 17.5 A Case Study: Circuits for Dosimetry

Radiation sensors based on CMOS technology and implemented monolithically on a single chip with the read-out circuitry are promising candidates in the design of low-power re-usable dosimeters. They are small, low cost and find application in several fields, such as in vivo dosimetry in radiation therapy [21]. The floating gate (FG) MOSFETs as dosimeter for ionizing radiation has been investigated in several studies [22]. Out of the previously reported designs, the one proposed in [23, 24] is attractive for low-cost and compact micro-system solutions. Such sensors can be fabricated in any standard CMOS technology allowing sufficient isolation of the FG by the used dielectrics. Such devices have good sensitivity to gamma or X-ray radiation. The floating gate is pre-charged before irradiation, while the resulting FG discharge (threshold $V_t$ decrease) during irradiation is determined by injection of charge carriers to and from the FG and generation of electron-hole pairs in the device dielectrics. So, the absorbed dose is related from the change in $V_t$, as preliminary reported in [25]. A precise $V_t$ extraction circuit is necessary to build-up a complete dosimeter. The used sensing device is based on an FG "C-sensor." "C" relates to the read-out that uses a CMOS inverter principle and shown in Fig. 17.13a. In the solution proposed in [24], the $V_t$ is derived from the drain current, $I_d$, flowing through nMOS only, while pMOS is grounded. The FG of the C-sensor is capacitively coupled to the control capacitor, CG, tunneling capacitor (for pre-charge), TG, and the two nMOS and pMOS transistors. The tunneling capacitor and the transistors have ~110 Å gate oxide, and the control capacitor uses ~3500 Å STI oxide (Fig. 17.13b).

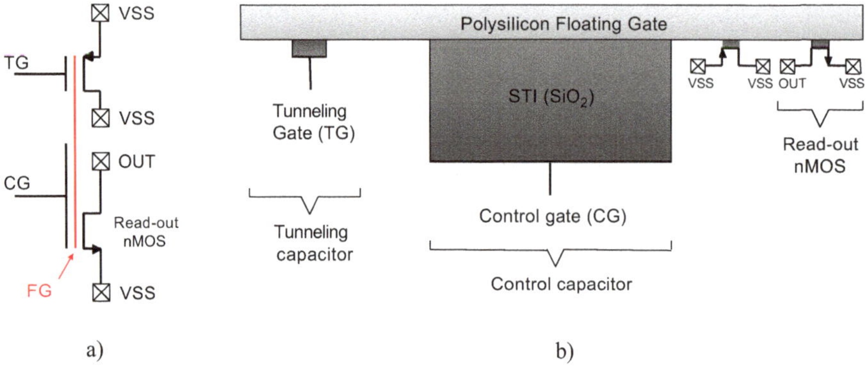


**Fig. 17.13** FG C-sensor scheme (**a**) and cross-section (**b**) [24]

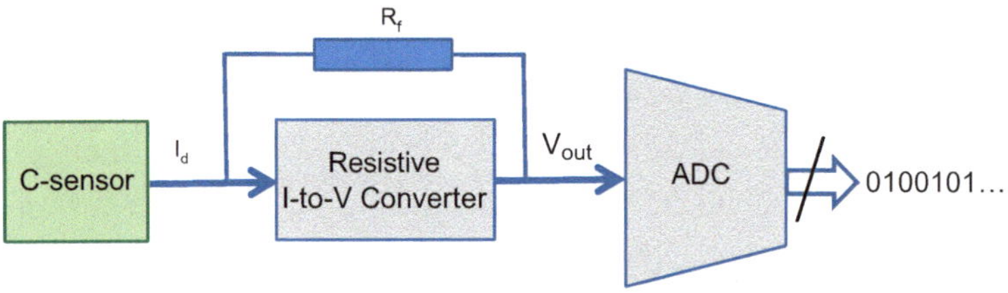


**Fig. 17.14** Block diagram of the monolithic dosimeter (C-sensor + interface)

The pre-charging is performed by applying 8.5 V to the CG and −4.5 V to the TG, while drain is grounded. The $V_t$ shifts towards higher values proportionally to the charge stored in the FG. The read-out circuit for $V_t$ extraction is designed in a standard 180 nm CMOS technology (Vcc=5V). RHBD techniques have been applied since even if the typical absorbed dose for each use is low (~1 krad) the dosimeter is re-usable several times and electronics could be subject to degradation. Current $I_d$ is transformed into voltage by a current-to-voltage (*I*-to-*V*) interface based on a resistive detection, while a 5-bit flash ADC similar to that one presented above has been cascaded to get digital data (Fig. 17.14). Its speed was limited to 1 MS/s since this speed in not really needed for the application. The *I*-to-*V* output dynamic is $2V_{pp}$.

It was observed on wafer stand-alone FG sensors that low-noise operation could be achieved in the linear regime with $I_d$ current ranging from 25 μA to ~100 μA.

The *I*-to-*V* interface processes the measured current, while the required voltage span for $V_{out}$ is $2V_{pp}$. The ADC sampling frequency of $V_{out}$ is limited to 1 MS/s. The simulation results, performed in various PVT conditions, show a complete compatibility of the sensor interface with the operation requirements.

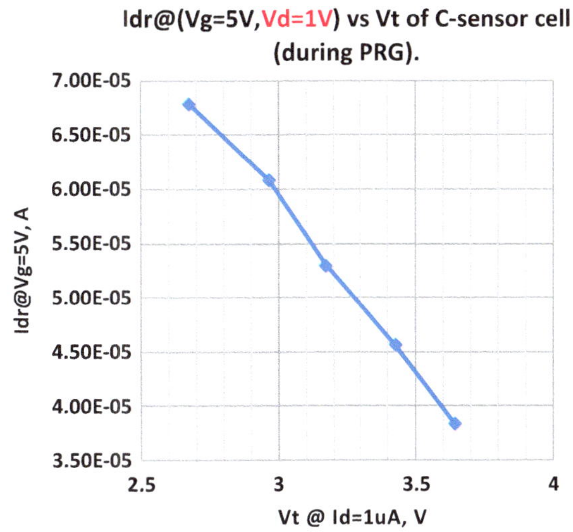


**Fig. 17.15** Drain current as function of $V_t$ [24]

### 17.5.1 C-Sensor Characteristics

The preliminary measurements performed on the stand-alone FG test sensors showed that with the read-out from nMOS in the linear mode the sensor operation is less noisy in contrast to sub-threshold and saturation modes. Charging of the FG is done by applying short voltage pulses and measuring $I_d$ after each pulse until the desired level of $I_d$ is reached. Experimentally, it was found that suitable drain voltage $V_{ds}$ in read should be $\sim$1 V. The gate voltage must be at least 1.5 V higher than the obtained $V_t$ of the charged sensor. Figure 17.15 shows the measured drain current during pre-charge as a function of the $V_t$ when $V_{ds} = 1$ V and $V_{gs} = 5$ V ($V_t$ is defined as the $V_{gs}$ at which the transistor drain current is equal to 1 μA).

During characterization shown in [24] at the end of the pre-charge, $V_t = 3.63$ V and $I_d = 38$ μA. Then, the sensor was exposed to different doses of ionizing radiation. The absorbed radiation dose is measured in terms of Gray (Gy = 100 rad), which corresponds to the deposit of a joule of radiation energy in a kg of matter. After absorption of 10 Gy (Cobalt-60 source), the $I_d$-vs-$V_{gs}$ curve of the sensor is shifted by about 1 V which corresponds to a $V_t$ equal to 2.67 V and $I_d = 68$ μA. In turn, this corresponds to linear response of 3 μA/Gy or 30 nA/rad.

### 17.5.2 Circuital Solutions

Sensor reading can be performed in different ways. One of them is to exploit the propagation delay (which depends on $V_t$) of the C-sensor used as an inverter. Another solution makes use of a VCO-based ADC conversion [21]. A simpler way

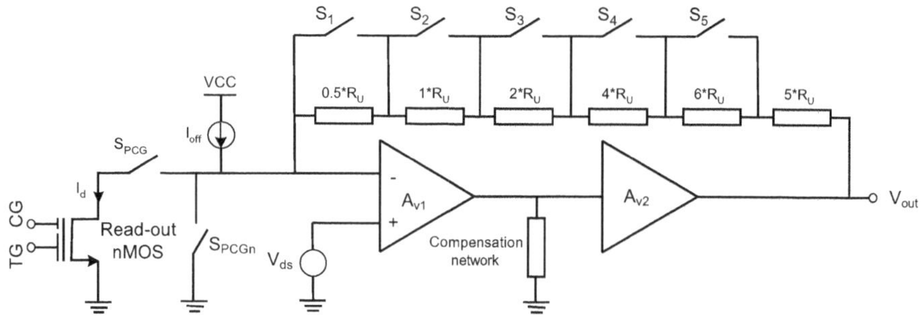


**Fig. 17.16** Schematic of the proposed current-to-voltage interface [24]

to deduce the $V_t$ value is to perform a $I$-to-$V$ conversion of nMOS drain current. The common-gate detection and the resistive detection are between the used techniques for the $I$-to-$V$ conversion [26]. In the former solution, the nMOS sensor current flows into the source of a common-gate stage. In order to have a good SNR, a relatively large bias current is required. The $V_{ds}$ of the C-sensor must be kept fixed. The resistive detection shown in Fig. 17.16 uses resistances in feedback to the operational amplifier to convert the sensor current into voltage.

In this case, the feedback resistance itself could be the main noise source. Nevertheless because of the relatively modest accuracy required, high feedback resistance is not an issue even for resistances in the tens of kohm range.

Besides the first opamp, a second low-output impedance stage is needed since the $I$-to-$V$ interface gain is given by:

$$V_{out}/I_d = R_f - R_{out}/A_v \tag{17.1}$$

where $R_f$ is the overall feedback resistor, $R_{out}$ is the second stage output resistance and $A_v >> 1$. The first stage was designed having high $A_{v1}$ gain and output impedance, while the second has $A_{v2} \approx 1$ and low output impedance.

$R_f$ has been implemented as a programmable resistor to consider the process spread and operating temperature span. The switches $S_1$–$S_5$ are selectable with logic signals. Process spread can cause $I_d$ variations of ±25% and resistor variation of ±12%, while temperature can give additional ±10% variation in 0 °C to 85 °C range. To be compliant with the ADC specifications the $I$-to-$V$ best output range was set from 1.5 to 3.5 V. Considering that the expected sensor currents span (30 μA) corresponding to 1 V shift in $V_t$, nominal $R_f$ must be equal to 2 V/30 μA = 66 kohm to exploit all the 2 $V_{pp}$ range. Since $V_{out}$ varies between 2.5 V (corresponding to $I_d$ = 38 μA) to 4.5 V ($I_d$ = 68 μA), an offset current $I_{off}$ has been added to provide a suitable voltage shift. Both switches configuration and $I_{off}$ are determined during a calibration phase after pre-charge and before radiation exposure. During the pre-charge, the nMOS is disconnected from the interface by means of $S_{PCG}$ switch and grounded by means of $S_{PCGn}$ switch.

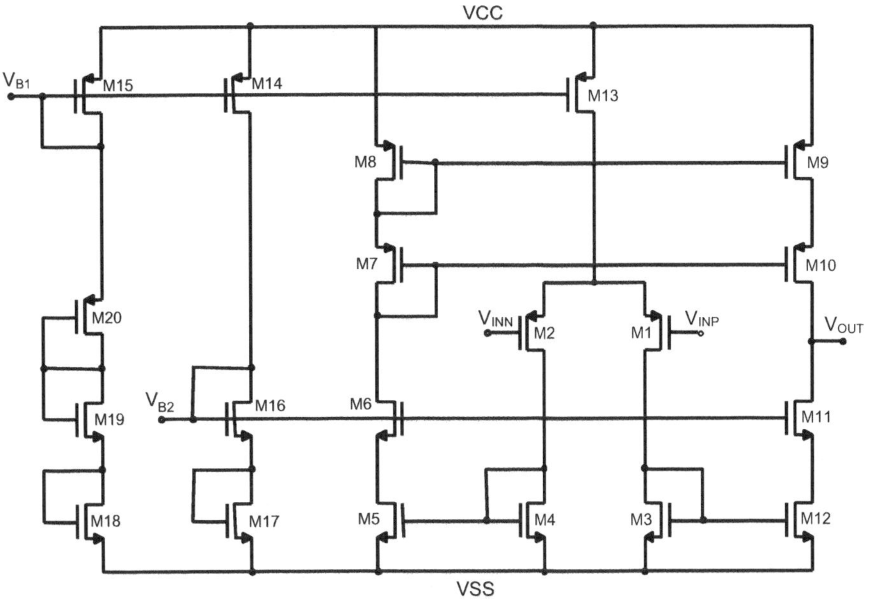


**Fig. 17.17** Schematic of the mirrored cascode opamp [24]

The cascade of the two amplifiers satisfies the following requirements:

- 2 $V_{pp}$ output swing around $V_{cc}/2$
- Input stage able to manage $V_{ds} = 1$ V
- $A_{v1} > 60$ dB
- Low output impedance and capability to drive ADC input capacitance (12 pF in this case)

Bandwidth is not an issue in this case, since the sampling is slow. The first stage was implemented as a mirrored cascode opamp (Fig. 17.17). This design guarantees optimum input common mode range and high-gain (>78 dB with 14 MHz bandwidth) with phase margin >80°. The input transistors are pMOS because they are less noisy and work well at 1 V input.

The second stage has the gain $A_{v2} \sim 1$, high input impedance and low output impedance ($R_{out} = 1/g_m = 500\ \Omega$). It is based on the so-called battery level-shifter topology (shown in Fig. 17.18). In our implementation, the "battery" shift is fixed to ~0 V ($V_{in} - V_{out}$ is the equal to $V_{gs,M1} - V_{gs,M2}$). A compensation network (series of $R$ and $C$) has been added to introduce a zero in the frequency response. Overall power consumption is lower than 2 mW.

A 5-bit flash ADC was used in the conversion of $V_{out}$ into a digital word to be used for further calibration and averaging. We selected this ADC since it has already proven to be rad-hard up to 300 krad TID and is not sensitive to single events and SEL [12], even if it is surely not the optimum choice from the power consumption point of view.

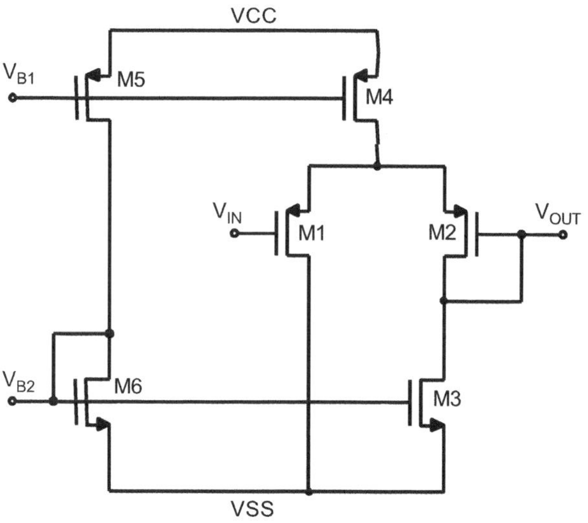


**Fig. 17.18** Schematic of the "battery" level shifter [24]

As said, besides the sensor itself, all the electronics must be radiation-hard up to the maximum absorbed doses during whole dosimeter life. This is why we applied RHBD techniques in the design. In particular, the RHBD used here is focused on TID effects mitigation and SEL countermeasures in layout. This is done by using ELT transistors and surrounding each of them by its own guard-ring. Interdigitated structures, which improve the matching, are implemented by alternating ELT transistors. Passive elements have been also chosen according to rad-hard requirements. The resistors are made with heavily doped polysilicon, which is insensitive to TID [2], while the capacitors are metal-insulator-metal (MIM) devices.

### *17.5.3 Measurement Results Under Total Dose*

The chip contains one C-sensor, the *I*-to-*V* interface, and the flash ADC and has been implemented in the 180 nm standard CMOS technology by TowerJazz. The C-sensor is as small as 20 μm × 20 μm (Fig. 17.19a). The core area is less than 0.25 $mm^2$ (Fig. 17.19b).

Figure 17.20 shows the sensor current as a function of the $V_t$ for three different $V_{gs}$ before programming. The current values and behavior are in fair agreement with that one expected.

Six samples of the designed chip have been radiated with Cobalt-60, with three doses: 300 rad, 600 rad, and 1 krad, to verify linearity. As can be seen in Fig. 17.21 the samples achieve excellent linearity. The flash ADC performance is similar to the one already illustrated above, featuring one more bit of resolution.

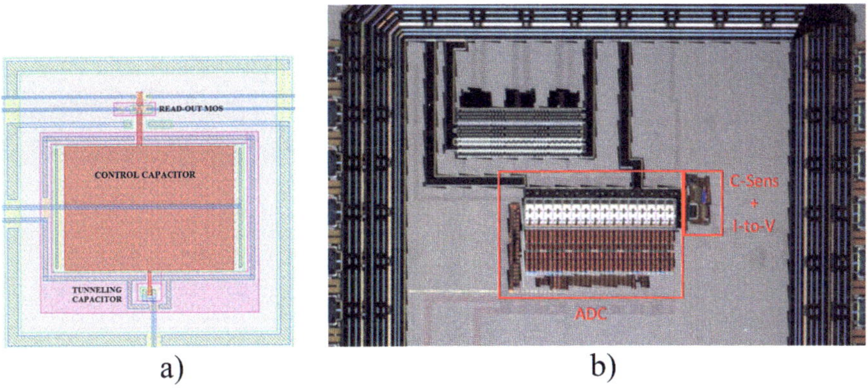


**Fig. 17.19** C-sensor layout (**a**) and dosimeter chip photo (**b**)

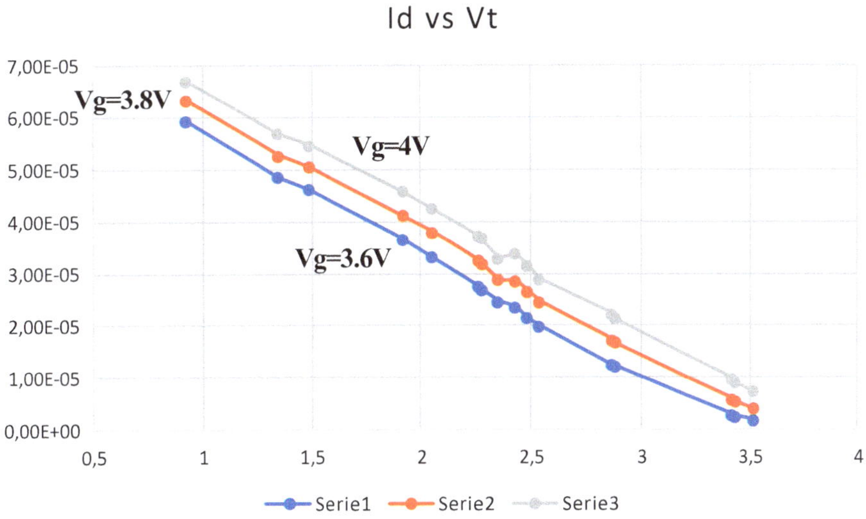


**Fig. 17.20** Measured current sensor as a function of the $V_t$ for three $V_{gs}$ before programming

In order to evaluate the robustness of the adopted RHBD, the dosimeter samples have been pre-charged and irradiated (discharged) several times, with any deviation from the correct operation.

## 17.6 Conclusions

In this chapter, we learned why harsh environments are an issue in ICs and which design approach is the best compromise between performance and resiliency. In particular, this is true for analog and mixed-signal circuits, where accuracy in every

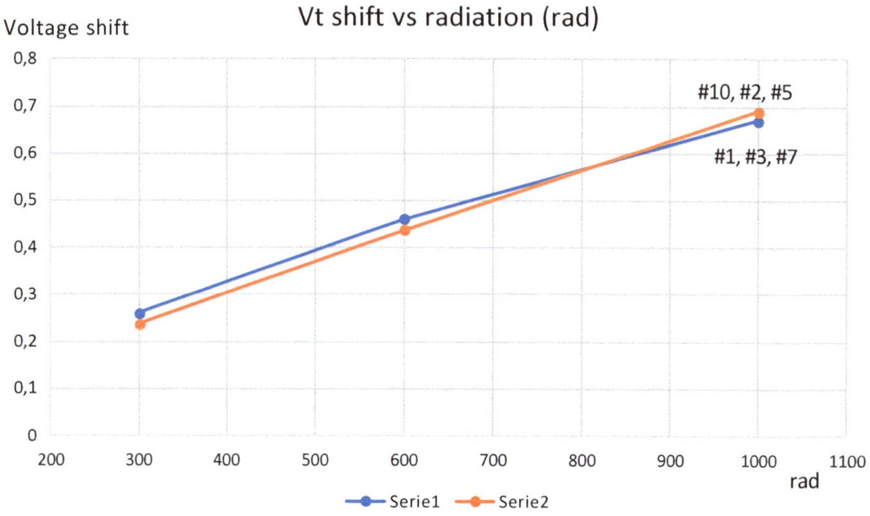


**Fig. 17.21** $V_t$ shift as a function of radiation (from 300 rad to 1 Mrad) measured over six samples

operation condition is a need. This has been investigated in several circuits, mainly from the academia and research, and the solutions adopted have been proven with concrete examples. Between them, a novel dosimeter able to operate under several radiation cycles featuring low cost with smaller area than previous implementations [27, 28] has been presented. It confirms the suitability of the RHBD methodology applied to standard CMOS processes and represents a promising solution for re-usable medical equipment.

## References


1. Space Faring – The Radiation Challenge, Educational Serie. NASA (2008)
2. J. Cressler, *Extreme Environment Electronics* (CRC Press, Boca Raton, 2013)
3. R. Velazco et al., *Radiation Effects on Embedded Systems* (Springer, Berlin, 2007)
4. J.M. Benedetto, Economy-class ion-defying ICs in orbit. IEEE Spectr. **35**(3), 36 (1998)
5. E. Ibe, *Terrestrial Radiation Effects in ULSI Devices and Electronic Systems* (Wiley, Hoboken, 2015)
6. C. Calligaro, U. Gatti, *Rad-Hard Semiconductor Memories* (River Publisher, Delft, 2018)
7. J.R. Schwank et al., Radiation effects in MOS oxides. IEEE Trans. Nucl. Sci. **55**, 1833–1853 (2008)
8. S. Buchner et al., Single event transient in linear ICs, in *Proceeding of NSREC*, Seattle, 2005
9. D.G. Millward et al., The Effectiveness of Rad-Pak ICs for Space-Radiation Hardening, in *Electronic Components and Technology Conference* 1990, pp. 913–916
10. T. Bion et al., A model for proton-induced SEU. IEEE Trans. Nucl. Sci. **36**, 2281–2286 (1989)
11. G. Anelli, NSS-MIC Short Course. 2006
12. U. Gatti, C. Calligaro, et al., Radiation-hardening methodologies for flash ADC. Analog Integrated Circuits Signal Proc J **87**, 141–154 (2015)

13. A. Sternberg et al., Single-event sensitivity and hardening of a pipelined analog-to-digital converter. IEEE Trans. Nucl. Sci. **53**, 3532–3538 (2006)
14. A. Lanot et al., Analysis of the Effects of Single Event Transients on an SAR-ADC Based on Charge Redistribution Proceedings of LATW, Los Angeles, 2015
15. K. Makris et al., Very High-Resolution Analog-to-Digital Converter at 1 kHz for Space Application, *in Workshop AMICSA* 2014
16. A. Dingwall et al., An 8-MHz CMOS Subranging8-bit A/D converter. IEEE J. Solid State Circuits (JSSC) **20**, 1138–1143 (1985)
17. E. Mikkola et al., SET tolerant CMOS comparator. IEEE Trans. Nucl. Sci. **51**, 3609–3614 (2004)
18. P.J. Layton, et al., Single Event Latch-Up Protection of Integrated Circuits, in *Proceedings of RADECS* 1997, Cannes, pp. 327–331
19. Y. Cao et al., A 4.5 MGy TID-tolerant CMOS bandgap reference circuit using a dynamic base leakage compensation technique. IEEE Trans. Nucl. Sci. **60**, 2819–2824 (2013)
20. B.D. Olson et al., Single-event effect mitigation in switched-capacitor comparator design. IEEE Trans. Nucl. Sci. **55**, 3440–3446 (2008)
21. E.G. Villani et al., Monolithic 180 nm CMOS dosimeter for in vivo medical applications. IEEE Trans. Nucl. Sci. **60**, 843–849 (2013)
22. Y. Wang et al., A Novel Fully Integrated Floating-Gate MOSFET Radiation Dosimeter Using VTH Extractor, in *Proceedings of NEWCAS* 2004
23. E. Pikhay et al., Characterization of single poly radiation sensors. IEEE Electr. Dev. Lett. **36**, 618–620 (2015)
24. U. Gatti, C. Calligaro, E. Pikhay, Y. Roizin, Re-usable 180nm CMOS dosimeter based on a floating gate device. Proceedings ICECS, Montecarlo, 2016
25. E. Pikhay et al., Degradation study of single poly radiation sensors by monitoring charge trapping. Microelectron. Reliab. J. **59**, 18–25 (2016)
26. T.A. Roessig, R.T. Howe, A.P. Pisano, J.H. Smith, Surface Micromachined 1 MHz Oscillator with Low-Noise Pierce Configuration, in *Proc. IEEE Solid-State Sensor Actuator Workshop*, 1998
27. M. Arsalan et al., Ultra low power CMOS-based sensor for on-body radiation dose measurements. IEEE J. Emerging Sel. Top. Circuits Syst. **2**, 34–41 (2012)
28. E.G. Moreno et al., Temperature compensated floating gate MOS radiation sensor with current output. IEEE Trans. Nucl. Sci. **60**, 4026–4030 (2013)

# Chapter 18
# 1-GRad-TID Effects in 28-nm Device Study for Rad-Hard Analog Design

**Marcello De Matteis, F. Resta, A. Pipino, F. Fary, S. Mattiazzo, C. Enz, and A. Baschirotto**

## 18.1 Introduction

The study of radiation interactions with materials and more specifically with MOS transistors is one of the major aspects in development of reliable integrated circuits for space and research applications [1, 2]. Radiations interact with silicon circuits in several ways. Incident high-energy particles generate electron/holes pairs on silicon, transiently modifying the local charge concentration across the several layers building active and passive devices.

Oxide layers are typically more sensitive due to the several charge trapping phenomena that can critically change the MOST structures build-up electrical fields and hence the transistors static and dynamic electrical features. Gate oxide (GOX), shallow-trench-isolation (STI) islands, and finally spacer oxides (over the lightly doped-drain (LDD)) regions are all affected by charge trapping phenomena with complex transient behaviors that depend on several parameters regarding radiations (deposited dose, particles typology (electron, protons, alpha, etc.), time lapse from initial particles bunch shot) and silicon devices (MOST channel doping typology, density, impurities, MOST geometry (i.e., *W*/*L* aspect ratio), biasing point, etc.).

M. De Matteis (✉) · F. Resta · A. Pipino · F. Fary · A. Baschirotto
Department of Physics and INFN, University of Milano Bicocca, Milan, Italy
e-mail: marcello.dematteis@unimib.it; federica.resta@unimib.it; alessandra.pipino@unimb.it; f.fary@campus.unimib.it; andrea.baschirotto@unimib.it

S. Mattiazzo
INFN Sezione di Padova, Padova, Italy
e-mail: serena.mattiazzo@pd.infn.it

C. Enz
STI IMT ICLAB, Ècole Polytechnique Fédèrale de Lausanne, Lausanne, Switzerland
e-mail: christian.enz@epfl.ch

© Springer Nature Switzerland AG 2020

A. Baschirotto et al. (eds.), *Next-Generation ADCs, High-Performance Power Management, and Technology Considerations for Advanced Integrated Circuits*,
https://doi.org/10.1007/978-3-030-25267-0_18

The advent of nm-range technologies has probably improved the hardness of both analog and digital circuits in front of radiations damages thanks to the reduced GOX thickness achieved by advanced high-k oxides [3]. On the other hand, deeply nm technologies have strict layout rules that practically prevent classical hardware-by-design solutions (i.e., edgeless/enclosed layout [4, 5]) to mitigate leakage currents and threshold voltage variations in radiation exposed analog and digital circuits. This paper reviews some of the most relevant results achieved by the Scaltech28 project (founded by the Italian National Institute for Nuclear Physics [6, 7]) that has the main aim to study and characterize the impact of high dose levels (up to 1 Grad TID) on commercial CMOS 28 nm integrated circuits. The proposed study is organized in two main parts. Section 18.2 presents the complete device characterization vs. dose of single-finger structures p/nMOST, highlighting the most relevant electrical characteristics like leakage current, maximum ON current, and threshold voltage [7].

The sub-threshold slope vs. dose will be also analyzed and the consequent effect on MOST efficiency ($g_m$/I) will be commented, since adoption of 28 nm technology often operates analog circuits in weak and/or moderate inversion region due to the reduced voltage headroom imposed by the higher threshold voltage in turn forced by high-k GOX. In Sect. 18.3 a complete design of a read-out analog front-end for HEP pixel detectors will be presented. The analog channel performs both charge-to-voltage and charge-to-time conversion at 30 mV/fC and arises 25 ns time delay at 1.8 μW static power [6]. The effects of 1 Grad TID will be presented in terms of time and frequency domain electrical measurements (i.e., comparing pre- and postradiation damages). At the end of this paper, conclusions will be drawn.

## 18.2 1 Grad Total-Ionizing-Dose Effects on SP nMOST

The irradiated test structures consist of two sets of ten standard-process (SP) single-finger MOS transistors (n-channel and p-channel), with different geometries and diode-based ESD protection.

The p/nMOST share the same substrate, bulk-source are shorted whereas drain and gate contacts are accessible and used to bias the devices. The available sizes are reported in Table 18.1. The TID tests have been carried out at the X-ray facility installed at the Physics and Astronomy Department of the University of Padova [8]. A photo of the silicon die is reported in Fig. 18.1. It was mounted on a semi-automatic 4-in. wafer probe station within the X-ray irradiation cabinet and contacted by a custom probe card. The probe card features 32 probe tips (two columns of 16). A Keithley 707 switching matrix connects the four single-measuring-units (SMUs) of the semiconductor device analyzer (HP4156) and the voltage supply to specific probe tips. In this way, the selected set of structures could be properly biased during irradiation and measurements. Given the limitation in the

**Table 18.1** Device-under-test geometry

| MOST typology | Width ($W$) (μm) | Length ($L$) |
|---|---|---|
| nMOS/pMOS (single-finger) | 3 | 1 μm |
| | 3 | 30 nm |
| | 1 | 30 nm |
| | 1 | 60 nm |
| | 1 | 90 nm |
| | 0.1 | 1 μm |
| | 0.2 | 1 μm |
| | 0.4 | 1 μm |
| | 0.1 | 30 nm |
| | 0.3 | 30 nm |

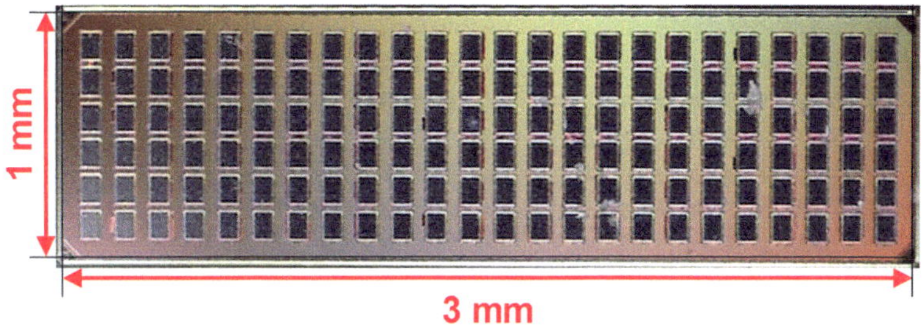


**Fig. 18.1** Devices under tests—chip photo

number of connections available in the switching matrix, only one set of transistors could be biased and irradiated at a time. This meant that the irradiation study was performed on different dies for nMOSFETs and pMOSFETs. The chip was placed as close as possible to the tube exit in order to maximize the dose rate, which could be set at 8MRad/h ($SiO_2$). During irradiations, the devices were kept with both drain and gate biased, while source and bulk were shorted to ground [5]. The $I_{DS}$-$V_{GS}$ characteristics for both MOST sets are taken in:

- Saturation region, with $|V_{DS}| = 1.1$ V (approximately assuming 0.5 V threshold voltage ($V_{TH}$) for SP MOST in 28 nm CMOS node);
- Linear region, with $|V_{DS}| = 0.1$ V in order to extract the threshold voltage by adopting maximum transconductance method in linear region [9].

Irradiations were performed at room temperature up to 1 GRad, in steps. Measurements were taken before irradiation and after each dose step. The whole setup was remotely controlled, so that the sequence of irradiation and measurements could be performed automatically. No annealing was performed afterwards, given limitations to the available setup.

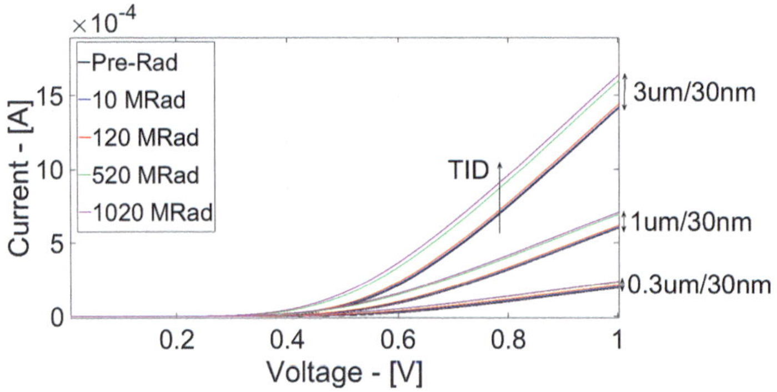


**Fig. 18.2** nMOST $I_{DS}$–$V_{GS}$ for $W/L = 3$ μm/30 nm, 1 μm/30 nm and 0.1 μm/30 nm

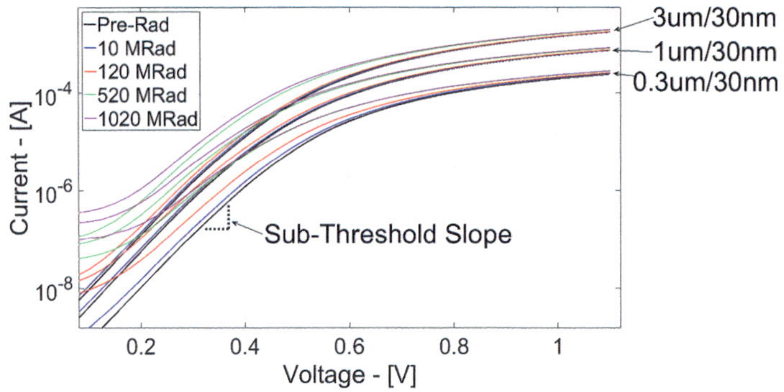


**Fig. 18.3** nMOST $I_{DS}$–$V_{GS}$ for $W/L = 3$ μm/30 nm, 1 μm/30 nm and 0.1 μm/30 nm—Log Scale

## *18.2.1 nMOST Electrical Characterization*

Figs. 18.2 and 18.3 show the drain-source current ($I_{DS}$) vs. gate-source voltage ($V_{GS}$) in natural units and log scale, respectively, of MOST whose $W/L = 3$ μm/30 nm, 1 μm/30 nm, and 0.3 μm/30 nm (gray background in Table 18.1).

The maximum allowed finger width in CMOS 28 nm process node is 3 μm and the minimum 0.1 μm. Unfortunately the 0.1 μm width irradiations tests are not available (most of analyzed data are meaningless) and hence the selected minimum width MOST for this study is 0.3 μm.

Thus, the three nMOST samples, whose $I_{DS}$–$V_{GS}$ characteristic is shown in Figs. 18.2 and 18.3, allow a comprehension of the radiation damages at minimum gate length and vs. width (maximum, medium, and minimum).

The $I_{DS}$–$V_{GS}$ curves are plotted at different TID levels (from 10 MRad up to 1 GRad), in order to evaluate the variations of the electrical characteristics vs. dose. Figs. 18.2 and 18.3 allow deducing some relevant points:

- Threshold voltage ($V_{TH}$) decreases with increasing radiations dose. This effect is mainly due to the charges trapping at both GOX and STI level. Positive trapped charges attract negative charges. They favorite the channel spatial charge creation just beneath the gate, reducing the external gate voltage needed to create the channel [10]. Nonetheless, the positive interface traps in nMOST could also increase the leakage current leading to potential issues in both analog and digital circuits.
- The maximum ON current increases with dose.
- Sub-threshold slope (SubThS, i.e., the $I_{DS}$–$V_{GS}$ slope in the 0.3 V – 0.4 V regions in Fig. 18.3) decreases with dose and this generally leads to lower MOST efficiency (lower $g_m$ for a given current).

The SubThS measures the MOST current switch-on/off rate as a function of a certain gate-source voltage variation. It is a popular parameter involved in several small-signal characteristics of the MOST in terms of electrical ($I_{DS}$ rate vs. $V_{GS}$) and physical (depletion region vs. GOX capacitance) aspects. It can be expressed as [11]:

$$\text{SubThS} = \frac{1}{n \bullet \ln(10) \bullet V_{\text{Thermal}}} \tag{18.1}$$

where $V_{\text{Thermal}}$ is 25 mV and n is the slope factor, given by the ratio between the depletion region capacitance ($C_{\text{DEP}}$, depending in turn on biasing) and GOX capacitance ($C_{\text{OX}}$):

$$n = 1 + \frac{C_{\text{DEP}}}{C_{\text{OX}}} = \frac{\gamma}{\sqrt{|2 \bullet \Phi_F| + V_{\text{BS}}}} \tag{18.2}$$

$\Phi_F$ is the Fermi level, $V_{\text{SB}}$ is the source-bulk voltage, $\gamma$ is the body effect constant [12], and it is proportional to the square substrate doping density $\sqrt{N_A}$ (and related with electron charge q, silicon and vacuum permittivity ($\varepsilon_{\text{SI}}$ and $\varepsilon_0$) and GOX capacitance, $C_{\text{OX}}$):

$$\gamma = \frac{\sqrt{2 \bullet q \bullet N_A \bullet (\varepsilon_{\text{SI}} \bullet \varepsilon_0)}}{C_{\text{OX}}} \tag{18.3}$$

Basically, it can be concluded that SubThS reduction depends on changes in doping concentration profile due to the charge trapping after radiations.

The same physical phenomena that reduce $V_{TH}$ lead to SubThS reduction inducing lower MOST $g_m/I_{DS}$ efficiency, that is given by the well-known equation:

$$\frac{g_m}{I_{\text{DS}}} = \frac{\frac{\partial I_{\text{DS}}}{\partial V_{\text{GS}}}}{I_{\text{DS}}} = \frac{\frac{I_{\text{DS}}}{n \bullet V_{\text{Thermal}}}}{I_{\text{DS}}} = \frac{1}{n \bullet V_{\text{Thermal}}} = \text{SubThS} \bullet \ln(10) \tag{18.4}$$

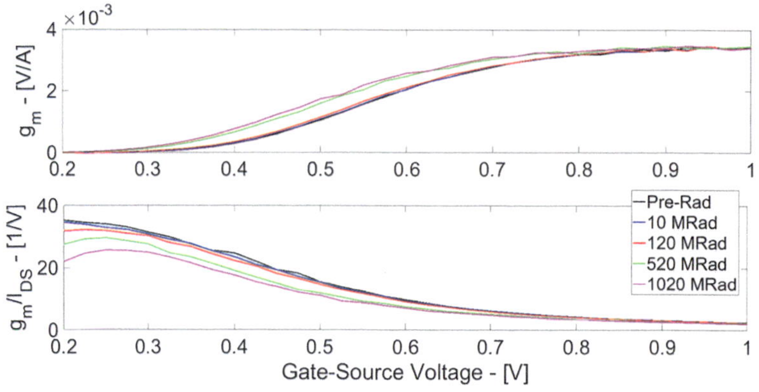


**Fig. 18.4** nMOST $g_m$ and $g_m/I_{DS}$ vs. $V_{GS}$ for $W/L = 3$ μm/30 nm

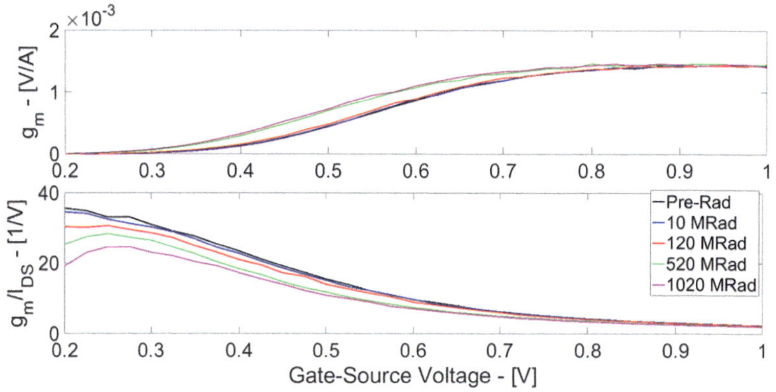


**Fig. 18.5** nMOST $g_m$ and $g_m/I_{DS}$ vs. $V_{GS}$ for $W/L = 1$ μm/30 nm

This is demonstrated by Figs. 18.4, 18.5, and 18.6 that show the transconductance ($g_m$) and the $g_m/I_{DS}$ curves vs. dose for $W/L = 3$ μm/30 nm, 1 μm/30 nm, and 0.3 μm/30 nm MOST, respectively.

As a conclusion, Fig. 18.7 shows the percent deviation of the maximum achievable $g_m/I_{DS}$ vs. dose that for narrow and short nMOST ($W/L = 0.3$ μm/30 nm) can reach 30% loss.

Finally, it is interesting to analyze the OFF current ($I_{OFF}$) behavior in nMOST devices overexposed to radiations. Fig. 18.8 shows the evolution of the leakage current as a function of TID, for all tested transistors. Leakage current increases of 2/3 orders of magnitude for all nMOST. Therefore the degradation of $I_{OFF}$ should be carefully taken into account since it would cause a large increase of power consumption in complex circuits.

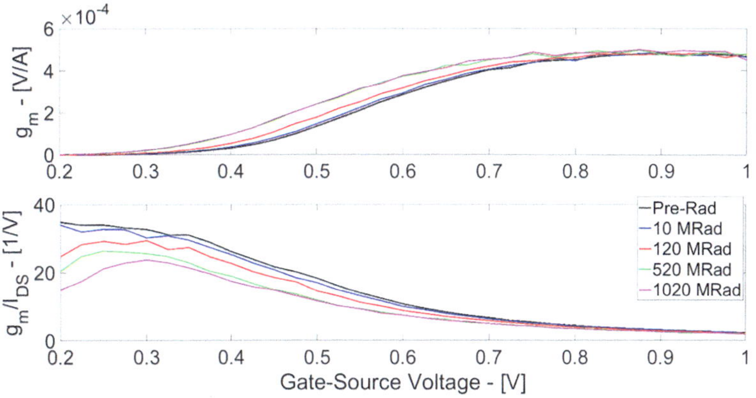


**Fig. 18.6** nMOST $g_m$ and $g_m/I_{DS}$ vs. $V_{GS}$ for $W/L = 0.3$ μm/30 nm

**Fig. 18.7** nMOST Max $g_m/I_{DS}$ vs. TID dose

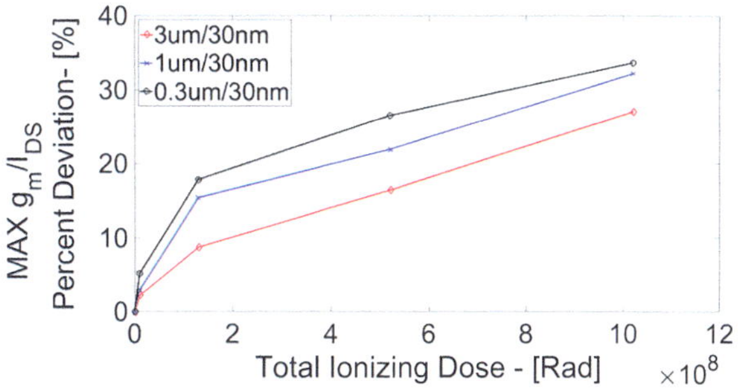


## 18.2.2 pMOST Electrical Characterization

For pMOSFETs there is no degradation in the off-state leakage current, since they are not subject to sidewall leakage current. The degradation of the sub-threshold slope, which is related to the accumulation of interface traps [10], is scarcely relevant (1/SubThS < 10 mV/dec for all tested transistors).

This might be a possible indication of a negligible accumulation of interface traps in the oxide. These considerations are confirmed by Figs. 18.9 and 18.10 where $I_{DS}$–$V_{GS}$ characteristic is plotted vs. dose and vs. pMOS W/L.

Moreover, the stability of SubThS vs. dose is also confirmed by Fig. 18.11 where both $g_m$ and $g_m/I_{DS}$ curves are plotted vs. $V_{GS}$ for those pMOST having $W/L = 3$ μm/30 nm, 1 μm/30 nm, and 0.3 μm/30 nm. While $g_m$ curves are clearly separated, the MOST efficiency curves are compacted in a small region, that is an evidence of small SubThS variations.

The parameter which is mostly degraded by TID for the pMOSTs is the saturation current ($I_{ON}$), i.e., the drain current for $|V_{GS}| = |V_{DS}| = 1.1$ V

Figure 18.12 plots the variation of the saturation current with the dose (expressed in percent variation with respect to the value before irradiation, $I_{ON}/I_{ON,PRE}$), for all

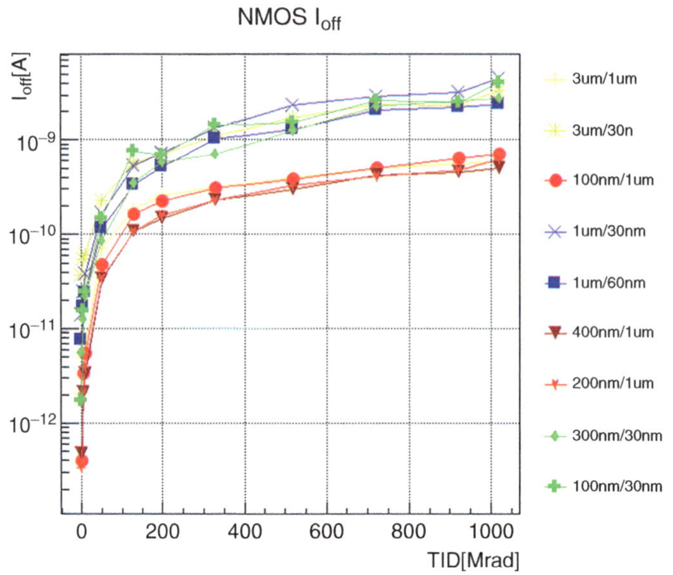


**Fig. 18.8** nMOST $I_{\mathrm{OFF}}$ current

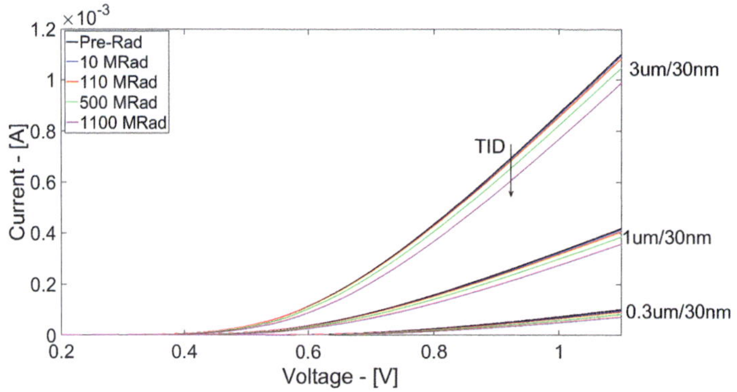


**Fig. 18.9** pMOST $I_{\mathrm{DS}}$–$V_{\mathrm{GS}}$ for $W/L = 3$ μm/30 nm, 1 μm/30 nm, and 0.3 μm/30 nm

tested devices. It is clear that $I_{ON}$ decreases for all the devices, but the degradation is limited for some of them, while it is more severe for others.

In order to better understand the dependence on the transistor geometry, notice that the four pMOST sharing long channel ($L = 1$ μm, the maximum available, so reasonably no short channel effects) and having a width W ranging from 3 μm down to 100 nm, the drain current loss is limited to 20% at the maximum dose for wider pMOST, while for the minimum width the loss of about 90%.

The degradation of the maximum drive current for narrow transistors is attributed to radiation induced narrow channel effect (RINCE) [5]. For these devices, the

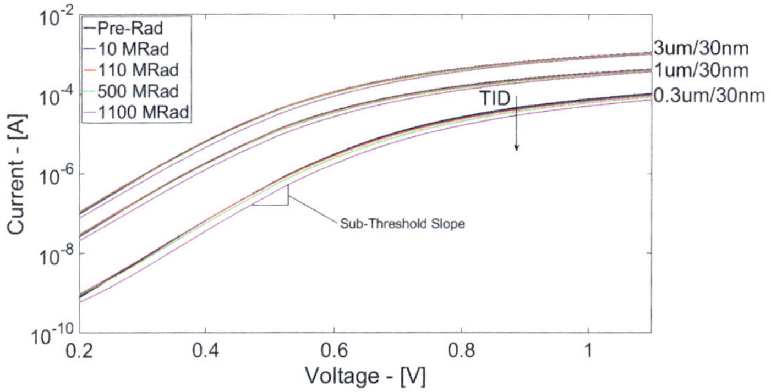


**Fig. 18.10** pMOST $I_{DS}$–$V_{GS}$ for $W/L = 3$ μm/30 nm, 1 μm/30 nm, and 0.3 μm/30 nm

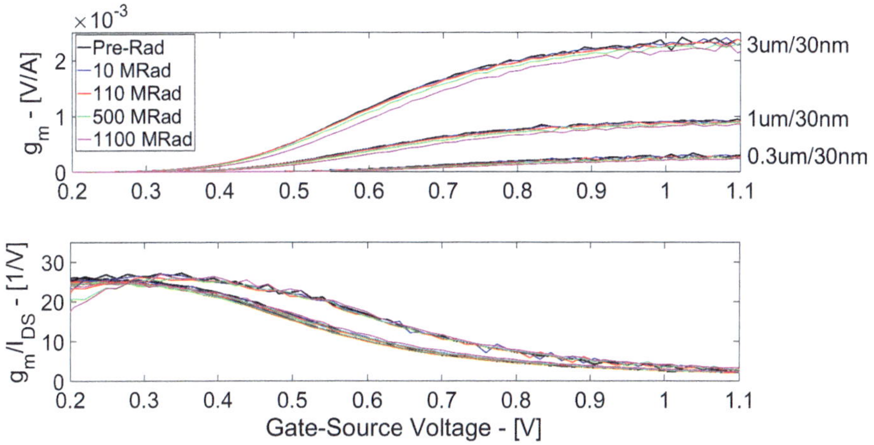


**Fig. 18.11** pMOST $g_m$ and $g_m/I_{DS}$ for $W/L = 3$ μm/30 nm, 1 μm/30 nm, and 0.3 μm/30 nm

radiation-induced positive charge trapped in the lateral STI influences the main transistor; it prevents the edge channel inversion, with the consequence of reducing the available channel width.

Interestingly there seems to be no (or very small) effect related to the channel length. Considering those pMOST in Fig. 18.12 sharing minimum channel length of 30 nm, the current loss is below 20% at 1 Grad. This is different from what has been measured for other technology nodes (like 65 nm [5]) and might have a strong impact on the design of digital circuits, where designers might have more flexibility in the choice of the device geometry (especially minimum L MOST).

In order to further investigate the origin of this dramatic current loss vs. $W$, both the threshold voltage ($V_{TH}$) and the transconductance ($g_m$) evolution with TID are plotted in Figs. 18.13 and 18.14, respectively.

The loss in the drain current can only partially be explained by a decrease of mobility, since the variation of $g_m$ with respect to the initial value is in the worst case

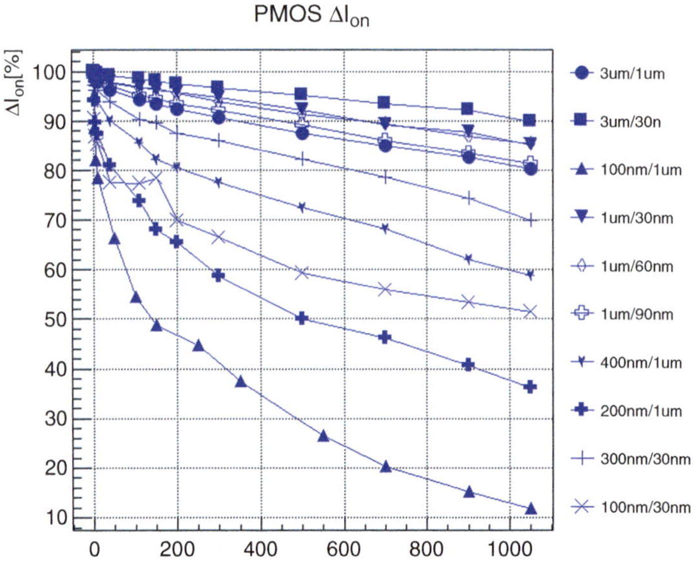


**Fig. 18.12** pMOST $I_{\mathrm{OFF}}$ current

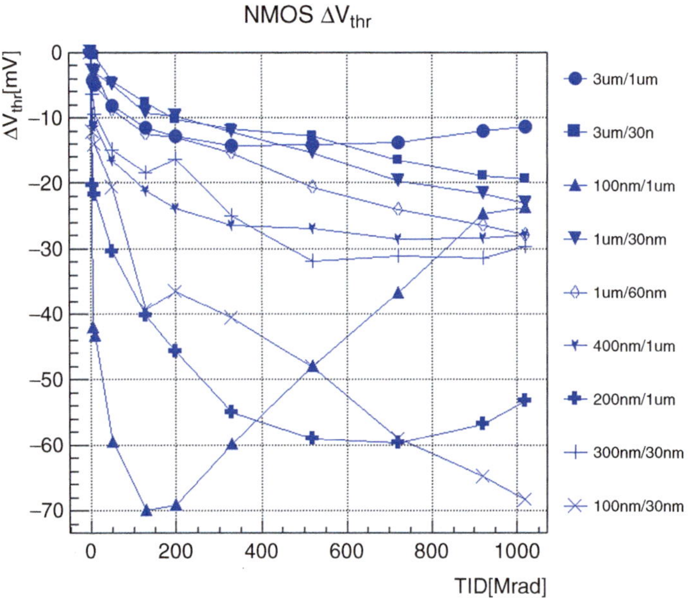


**Fig. 18.13** pMOST $\Delta V_{\mathrm{TH}}$ current

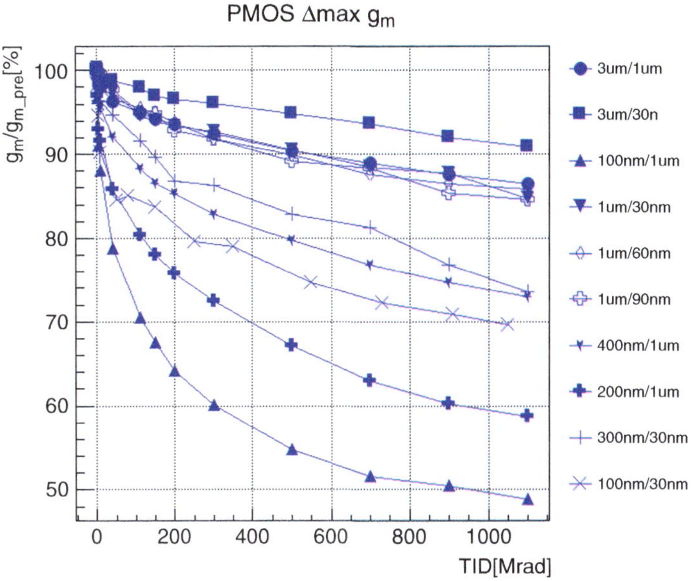


**Fig. 18.14** pMOST$g_{m,\mathrm{MAX}}$

(minimum $W$ device) of 50%. In Fig. 18.15 the free carrier mobility ($m$) variation, with respect to the pre-irradiation value, is shown.

The highest mobility reduction is in the narrowest/longest channel devices ($L = 1$ μm, $W = 100$ nm, 200 nm).

Another contribution comes from the shift in the threshold voltage (Fig. 18.13), which is limited to few tens of mV for wide devices, while it reaches hundreds of mV for narrower transistors.

## 18.3 Integrated Circuit for PIXel Detector in 28 nm CMOS Technology

Another relevant aspect for rad-hard design is to explore the behavior of SP MOST operating in analog signal processing stage. This section reports the most important measurements results of a single-channel analog read-out front-end for pixel detector in both pre- and postradiations conditions. The channel complies with the specifications of HEP pixel detector. The analog front-end is composed by the cascade of a charge-sensitive-preamplifier (CSPreamp, responsible of charge-to-voltage conversion [13]) and a switched-comparator for voltage-to-time conversion. Thus the front-end provides both charge arrival time information (the event) and time-over-threshold (the amount of charge).

The CSPreamp is based on a single-MOST (M1) gain stage with feedback capacitance ($C_F$). Krummenacher feedback [14] is used to manage the large leakage

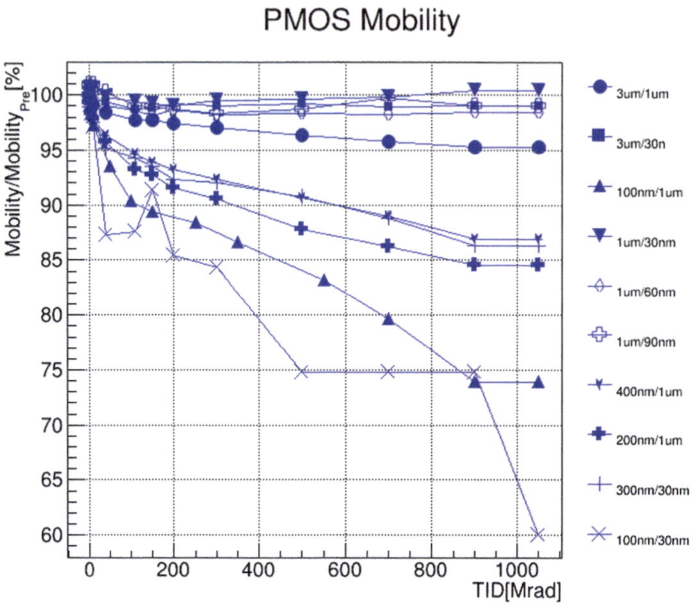


**Fig. 18.15** pMOST mobility

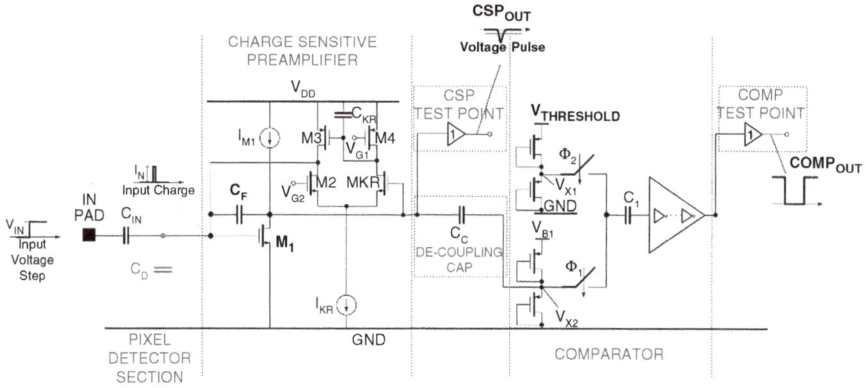


**Fig. 18.16** IC-PIX28 schematic

current coming from the detectors and that could increase after radiation damages. The feedback composed by MKR-M2-$C_F$ should synthesize an equivalent feedback resistance in the order of MΩ. In first approximation the equivalent feedback resistance is equal to $2/g_{m\mathrm{KR}}$ (where $g_{m\mathrm{KR}}$ is the MKR transconductance). Hence, the biasing current for M2-MKR-M3-M4 stage (50 nA) is very much lower than M1 biasing current (1.6 μA).

The developed circuit is called integrated circuit for PIXel detector in 28 nm technology (IC-PIX28) and its simplified scheme is shown in Fig. 18.16 [6, 7].

The channel has been tested using a proper printed circuit board (PCB) in presence of different levels of TID (i.e., 480 Mrad, 750 Mrad, 909 Mrad, 1 Grad). X-ray irradiation has been done at Physics and Astronomy Department of the University of Padova with the system reported in Fig. 18.17. With a dose rate of 6.64 MRad/h ($SiO_2$) at room temperature, the maximum 1 Grad-TID dose level has been reached in 152.02 h (about 6.3 days).

Looking at time (CSPreamp output voltage in Fig. 18.18) and frequency domain measurements (CSPreamp output signal and noise, Figs. 18.19 and 18.20) there are three main effects of radiation damages:

- The recovery time (the time lapse needed to the CSPreamp output voltage to recovery the baseline) after radiation exposure is dramatically increasing from 390 ns up to 1.8 μs.

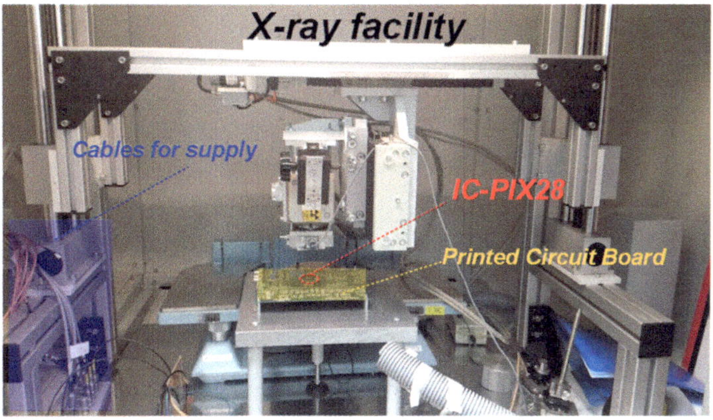


**Fig. 18.17** X-ray facility and IC-PIX28 irradiation setup

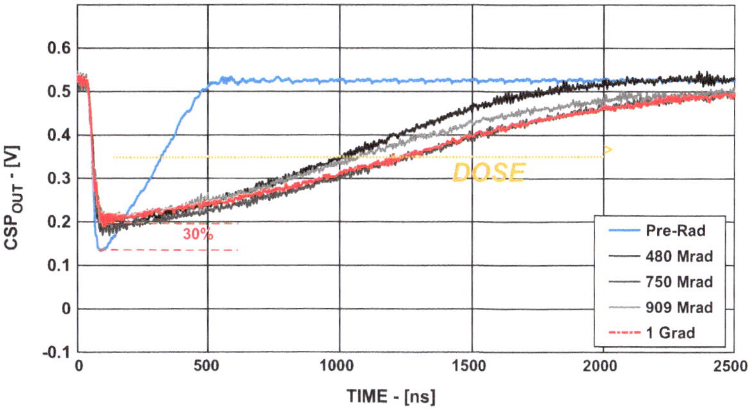


**Fig. 18.18** IC-PIX28 time-domain measurements (CSPreamp output voltage)

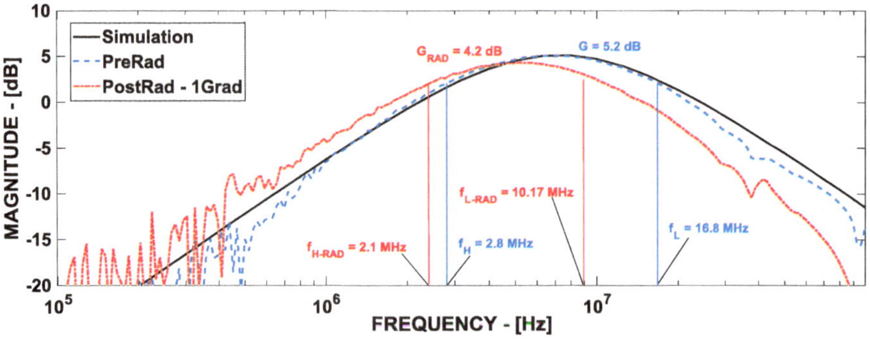


**Fig. 18.19** IC-PIX28 CSPreamp output vs. input frequency response

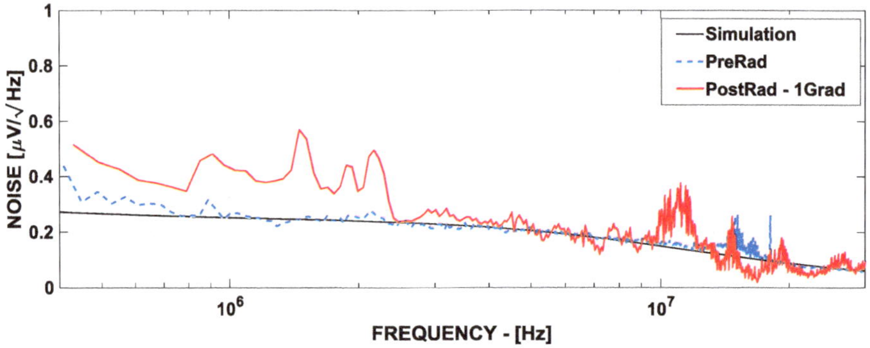


**Fig. 18.20** IC-PIX28 CSPreamp output noise power spectral density

- Both sensitivity (i.e., charge to voltage conversion rate) and peaking time delay feature a slight decreasing (sensitivity goes from 30 mV/fC down to 21 mV/fC at 1 Grad, whereas peaking time delay increases from 31 ns up to 90 ns, as reported in the performance resume in Table 18.2).

The recovery time increasing can be mainly attributed to the $I_{OFF}$ current increasing caused by both STI charge trapping and GOX leakage.

Several related works in literature [4] on less scaled technologies demonstrate that, for a given MOST finger, the $I_{OFF}$ current is proportional to both finger count and length. The IC-PIX28 adopts a four finger by 3 μm/100 nm each one, resulting in a MOST with an equivalent $W/L = 12$ μm/100 nm M1.

Measured leakage current for a 3 μm/30 nm W/L nMOST is about 0.8 nA (Fig. 18.8), leading to 0.0267 nA/nm $I_{OFF}$ density.

The estimated $I_{OFF}$ current for M1 is thus 4·(0.0267 nA/nm)·100 nm = 10 nA, more than one order of magnitude higher than preRad $I_{OFF}$.

This operates the Krummenacher amplifier out of its proper operating region. The leakage-compensating feedback is not anymore operating after overexposing

**Table 18.2** IC-PIX28 main performance

| Parameter | PreRad | PostRad |
|---|---|---|
| Detector capacitance | 100 fF | |
| Input range | 7fC–20 fC | |
| Output CSP voltage range—[mV] | 476–482.9 | 258.3–334 |
| Average sensitivity | 30 mV/fC | 21 mV/fC |
| Maximum peaking time delay | 31 ns | 96 ns |
| Minimum ToT @ $Q_{IN,MIN} = 7$ fC | 396 ns | 1849 ns |
| Maximum ToT @ $Q_{IN,MAX} = 20$ fC | 763 ns | 2717 ns |
| Maximum comparator delay | 5 ns | 20 ns |
| CSP pass band gain | 5.2 dB | 4.2 dB |
| CSP bandwidth—[MHz] | 2.8–16.8 | 2.1–10.17 |
| Integrated noise—[$\mu V_{RMS}$] | 806 | 960 |
| ENC—[fC] | 0.027 (165 e−) | 0.04 (247 e−) |
| SNR @ $Q_{IN,MIN} = 7$ fC | 47.6 dB | 45 dB |
| SNR @ $Q_{IN,MAX} = 20$ fC | 52.5 dB | 47.8 dB |

the CSPreamp to 1 Grad dose and thus the decay time in Fig. 18.18 dramatically increases.

The second aspect emphasized by both time and frequency domain measurements results is some losses in terms of both gain and peaking time delay. This is due to the efficiency loss explained in previous section (see Figs. 18.4 and 18.7). The efficiency reduction leads to lower $g_m$ at constant current and this reduces the CSPreamp loop gain with evident effects on closed-loop frequency response. Thus the loop gain decreases and the dominant pole time constant increases. This is confirmed from measured frequency response (pre- and postirradiation) shown in Fig. 18.19 (where the input signal has been provided by a pulse generator driving an integrated input capacitor).

Notice that the passband gain decreases of about 1 dB and the dominant pole right shifts of a factor two approximately (from 14 MHz preRad down to 8 MHz postRad).

The $g_m$ reduction is also confirmed by a slightly higher noise power in postRad noise power spectral density measurements. The main performance of IC-PIX28 pre- and postradiation damages are listed in Table 18.2.

## 18.4 Conclusions

In this paper a review of some of the most relevant radiation damages aspects involving CMOS 28 nm bulk technology node have been presented. Specific electrical characterizations of irradiated devices have been presented for stand-alone single-finger MOST and analog amplifier operating in pixel read-out channels. Table 18.3 shows an intuitive resume of some of the most important aspect involving

**Table 18.3** Device-under-test rad damages resume

| Electrical parameter | nMOST | pMOST | Legend | |
|---|---|---|---|---|
| Leakage current—$I_{OFF}$ | ☹ | ☹ | ☹ | Reduced rad damages |
| Max driving current—$I_{ON}$ | ☹ | 😐 | ☺ | Significant rad damages |
| Threshold voltage—$V_{TH}$ | ☹ | ☹ | ☺ | Medium rad damages |
| Sub-threshold-swing—SubTh-SW | ☹ | ☺ | | |
| Majority carriers mobility—m | ☹ | 😐 | | |

n/pMOST after 1 Grad dose over exposition. nMOST suffer in terms of leakage currents increasing, threshold voltage negative shift, and efficiency loss. pMOST are more robust in front of leakage currents whereas they experience a dramatic loss in ON current (i.e., mainly due to majority carriers mobility degradation) and positive shift of the threshold voltage.

**Acknowledgments** This work has been carried out within the ScalTech28 experiment funded by National Institute for Nuclear Physics.

## References

1. P. La Rocca, F. Riggi, The upgrade programme of the major experiments at the large hadron collider. J. Phys. Conf. Ser. **515**(1), 012012 (2014)
2. J.R. Schwank, M.R. Shaneyfelt, D.M. Fleetwood, J.A. Felix, P.E. Dodd, P. Paillet, V. Ferlet-Cavrois, Radiation effects in MOS oxides. IEEE Trans. Nucl. Sci. **55**(4), 1833–1853 (2008)
3. T.R. Oldham, F.B. McLean, Total ionizing dose effects in MOS oxides and devices. IEEE Trans. Nucl. Sci. **50**(3), 483–499 (2003)
4. F. Faccio, G. Cervelli, Radiation-induced edge effects in deep submicron CMOS transistors. IEEE Trans. Nucl. Sci. **52**(6), 2413–2420 (2005)
5. F. Faccio et al., Radiation-induced short channel (RISCE) and narrow channel (RINCE) effects in 65 and 130 nm MOSFETs. IEEE Trans. Nucl. Sci. **62**(6), 2933–2940 (2015)
6. F. Resta, A. Pipino, M. De Matteis, A. Baschirotto, 28 nm integrated circuit for PIXel detector. Nucl. Instrum. Methods Phys. Res., Sect. A **904**, 140–148 (2018)
7. F. Resta, S. Gerardin, S. Mattiazzo, A. Paccagnella, M. De Matteis, C. Enz, A. Baschirotto, 1GigaRad TID impact on 28 nm HEP analog circuits. Integration **63**, 306–314 (2018)
8. D. Bisello, A. Candelori, A. Kaminski, A. Litovchenko, E. Noah, L. Stefanutti, X-ray radiation source for total dose radiation studies. Radiat. Phys. Chem. **71**, 713–715 (2004)
9. K. Terada, K. Nishiyama, K.I. Hatanaka, Comparison of MOSFET-threshold-voltage extraction methods. Solid State Electron. **45**(1), 35–40 (2001)
10. R.C. Lacoe, Improving integrated circuit performance through the application of hardness-by-design methodology. IEEE Trans. Nucl. Sci. **55**(4), 1903–1925 (2008)
11. W.M. Sansen, *Analog Design Essentials. Chapter 1*, vol 859 (Springer, New York, 2007), pp. 16–18
12. D.A. Johns, K. Martin, *Analog Integrated Circuit Design. Chpater 1* (Wiley, Hoboken, 2008), p. 28

13. A. Baschirotto et al., A fast and low noise charge sensitive preamplifier in 90 nm CMOS technology. J. Instrum. **7**(01), C01003 (2012)
14. F. Krummenacher, Pixel detectors with local intelligence: an IC designer point of view. Nucl. Instrum. Methods Phys. Res., Sect. A **305**(3), 527–532 (1991)

# Index


**A**
Active-switched-capacitor (SC) implementation, 26–28
Advanced multiphasing (AM), *see* Switched-capacitor (SC) converter
Ampere's law, 143, 150, 156
Analog beamforming, 218, 223
ASM330LHH
  performance, 247, 248
  phase delays, 247
  zero rate level, 247

**B**
Battery monitoring systems (BMS)
  charging profile, 144–145
  electro-mobility, 144–145
  Li-Ion battery charging phases, 145–146
  operations, 144–145
Beamforming architectures
  analog beamforming, 218
  digital beamforming, 219
  hybrid beamforming, 219
BMS, *see* Battery monitoring systems
Bottom-plate (BP), 113

**C**
Capacitive coupled instrumentation amplifier (CCIA), 155
Capacitor DAC (CDAC)
  CIFF structure, 24, 27, 29
  pipelined SAR, 46–47
Cascaded-integrator-feed-forward (CIFF) structure, 81
  active-SC integrators, 26–28
  CDAC, 24, 27, 29
  classic amplifiers, 30
  comparator input ratio, 30
  discrete-time, 24
  $\Delta\Sigma$ ADCs, 23–24
  dynamic amplifier, 30–31
  FIR and IIR, 26–28
  loop filter, 24–25
  noise issue, 30
  NTF, 25–27
  passive implementation, 28–30
  power consumption, 27
CDAC, *see* Capacitor DAC
Cellular base station (BTS), 6–8
Charge redistribution bus (CRB)
  multiphase soft-charging, 115–116
  SPCR, 114
Charge trap phenomena, 276
CIFF structure, *see* Cascaded-integrator-feed-forward structure
Common-mode rejection ratio (CMRR), 154, 155
Common-mode voltage range (CMVR), 155
Compressive sampling (CS)
  advantages, 17
  analog continuous-time domain, 15
  applications, 14
  approximation error, 17
  dynamic range, 16
  NUS implementation, 16–17
  PBRS functions, 15, 16
  requirements, 15
  RMPI architecture, 15–16


© Springer Nature Switzerland AG 2020
A. Baschirotto et al. (eds.), *Next-Generation ADCs, High-Performance Power Management, and Technology Considerations for Advanced Integrated Circuits*, https://doi.org/10.1007/978-3-030-25267-0

Compressive sampling (CS) (*cont.*)
Shannon's uniform sampling theory, 15
signal sparsity, 14
SNs senses, 14–15
THz range, 17
Continuous-time delta-sigma (CTΔΣ modulator
clock frequency, 61
clock-jitter problem, 77–78
feedback DAC
clock jitter, 72
current steering, 71
ISI, 71–72
FIR feedback (*see* Finite-impulse-response feedback)
loop filter, 72–73
maximum sampling rate and bandwidth, 59–61
multi-stage converters, 67–69
OS data converter, 61
quantizer
flash ADC, 69–70
high speed, 70
Schreier figure of merit, 59–60
single-loop modulator
architectural techniques, 62–63
building blocks, 61–62
ELD, 62, 64–65
energy efficiency, 66–67
maximum clocking frequency, 61–62
metastability errors, 62
stability theory, 62, 64
STF, 65–67
Continuous-time (CT) hybrid, 34–36
Correlated double sampling (CDS), 52
CS, *see* Compressive sampling
CTΔΣ modulator, *see* Continuous-time delta-sigma modulator
Current feedback instrumentation amplifier (CFIA), 155
Current sensing
automotive converters, 146–147
BMS
charging profile, 144–145
electro-mobility, 144–145
Li-Ion battery charging phases, 145–146
operations, 144–145
contacted current sensing
drawbacks, 147
error sources, 149
high-side sensors, 148
low-side sensors, 147–148
shunt resistor, 148–149
contactless current sensing, 149
closed-loop sensing configuration, 152
Hall and fluxgate magnetometers, 150–151
magnetic sensors (*see* Magnetic sensors)
open-loop sensing configuration, 151–152
PCB trace, 150
Rogowski coils, 150
motor control applications, 146
principles, 143–144
sense amplifier, 153–156
Current steering (CS), 71

**D**
Data-weighted averaging (DWA), 52, 53, 55
Dickson converter
MSC, 115–117
stage outphasing, 117–118
Digital beamforming, 219
Digital signal processing (DSP), 15
Digital-to-analog converter (DAC), 21–22, 36, 79
Digital-to-time (DTC) converter
building blocks, 9–11
current-mode circuits, 9
scalability, 13–14
time intervals, 9
Direct power injection (DPI), 259
Discrete-time (DT), 24, 34–36
Dosimetry circuits
battery level-shifter, 293, 294
C-sensor scheme, 289–291
current-to-voltage interface, 292
feedback resistance, 292
FG MOSFETs, 289
5-bit flash ADC, 293
$I$-to-$V$ interface, 290, 292
measurement results, 294–296
mirrored cascode opamp, 293
monolithic dosimeter, 290
DTC converter, *see* Digital-to-time converter
D-type edge-triggered flip-flop (DFF), 10–11
Dynamic base leakage compensation (DBLC), 287
Dynamic element matching (DEM), 36
Dynamic range (DR), 16, 61

**E**
Error-feedback (EF) structure
background calibration, 34
block diagram, 31–32

ΔΣ ADCs, 32
feedback summation, 33–34
FIR, 33–34
fully passive implementation, 32–33
NTF optimization, 31
PVT sensitivity, 34
Excess loop delay (ELD), 62, 64–65

**F**

Figures of merit (FOMs)
architecture front, 6
challenges, 4
conversion efficiency, 6
ISSCC and VLSI conferences, 4–5
linearity and noise performance, 6
non-CMOS technologies, 5
OSR, 4
performance objectives, 5
Schreier's FOM, 4–5
SNDR, 4
technology front, 4–6
FinFET process
measured performance, 102, 104
performance *vs.* sampling frequency, 102–103
single-tone output spectrum, 102–103
spur magnitudes *vs.* input amplitude, 102, 104
supply decoupling, 102
Finite-impulse-response (FIR) feedback
chopping
brute-force approach, 85
flicker-noise, 85–86
frequency, 87–88
low-bandwidth converters, 85
OTA, 86–87
semi-digital techniques, 87–88
shaped quantization noise, 87–88
compensation
challenges, 80
CIFF, 81
continuous derivative, 82
excess delay, 80
loop-filter pulse responses, 82–83
prototype's pulse response, 81–82
STF peaking, 83
without compensation, 83
conventional wisdom, 80
DAC, 79
design centering
closed-loop fitting, 85
loop filter, 83–84
open-loop fitting, 84
plain 1-bit quantizer, 79–80
single-bit modulator, 78–79
Finite-impulse-response (FIR) filter
CIFF structure, 26–28
EF structure, 33–34
FIR feedback, *see* Finite-impulse-response feedback
5G new radio (NR) standards
complex modulation, 220
massive MIMO and carrier aggregation, 220
mmWave 5G radio, 222, 223
phased arrays, 220
semiconductor technology, 220–222
Floating gate (FG) MOSFETs, 289

**H**

Hall elements (HEs), 253
Hall sensors, 157–158
Highly accelerated stress test (HAST), 172
High-resolution Nyquist-rate
I–ΔΣ
conversion step, 52–53
discrete-time, first-order modulator, 48
incremental modulator, 49
linear-exponential incremental ADC, 54–55
linear time-invariant system, 49–50
noise shaping, 49
oversampling, 49
slicing technique, 53–54
working principle of, 50–51
zoom ADC, 51–52
overview of, 41–42
SAR (*see* Successive-approximation-register (SAR))
Hybrid beamforming, 219

**I**

IC-PIX28, *see* Pixel detector in 28 nm technology
Incremental ΔΣ (I–ΔΣ)
conversion step, 52–53
discrete-time, first-order modulator, 48
incremental modulator, 49
linear-exponential incremental ADC, 54–55
linear time-invariant system, 49–50
noise shaping, 49
oversampling, 49
slicing technique, 53–54
working principle of, 50–51
zoom ADC, 51–52

INL/DNL effects, 94
Instrumentation amplifiers (INA's), 155–156
Insulated gate bipolar transistors (IGBTs), 168
Integrated fluxgate sensors (IFG), 158–160
International Traffic Arms Regulation (ITAR) restrictions, 280
Internet of Things (IoT), 14
  applications, 125
  challenge of, 126
  inductor-based buck converter, 134–135
  low-IQ PMIC
    analog core implementation, 128–129
    capacitor-less sub-regulator, 132
    LDOs, 129
    linear regulators, 132–134
    oscillator, 129
    supply monitor, pre-regulator, and bandgap, 130
    switching blocks, 131–132
    VDD, 128–129
  MLC, 134–136
  overview of, 126
  PMICs, 126–128
  SC converters
    efficiency measurement, 137–138
    interleaved circuit, 137
    theoretical power efficiency, 136
    two operation phases, 136–137
    unregulated intermediate supply, 135
  1S Li-ion batteries, 135–136
  2S Li-ion batteries, 134
  3-level buck topology, 139–141
Inter-symbol interference (ISI), 71–72
IoT, *see* Internet of Things
IoT applications, 215
ISO26262 single-point failure metric (SPFM), 258

**K**
Krummenacher feedback, 309

**L**
LFSR, 101
Long-range radar (LRR) systems, 231
Low-dropout regulators (LDO), 129, 132–134
Low-IQ PMIC
  analog core implementation, 128–129
  capacitor-less sub-regulator, 132
  LDOs, 129
  linear regulators, 132–134
  oscillator, 129
  supply monitor, pre-regulator, and bandgap, 130
  switching blocks, 131–132
  VDD, 128–129
LSM6DSO, 244, 245, 248

**M**
Magnetic sensors
  angle accuracy, 268
  applications, 253
  classification, 253, 254
  current sensing modules, 156–157
  DPI setups, 259
  ESD structure design, 260–262
  fluxgate sensors, 158–160
  Hall sensors, 157–158
  IEC 62132-4, 259
  ISO 7637-1, 259
  ISO-16750-2, 259
  magnetic amplification chain, 258
  magnetic signal path, 254–256
  measurement sequencing and digital corrections, 255–256
  offsets, 266, 267
  PHY technology
    analog ratio-metric operation, 263
    circuit diagram, 264
    cyclic redundancy check/checksum errors, 263
    emission performance, 266, 267
    Nibble bit detection error, 263
    output amplifier precision, 263–266
    pulse width modulation, 262
    SENT PHY architecture, 263, 264
    SENT protocol, 262
  principles, 155–156
  safety concept, 256–258
  sensor configurations, 254, 255
  stray-field robustness, 269–270
  supply system design, 259–260
  XFAB 0.18 um HV bulk technology, 270, 271
MEMS
  accelerometer, market segments, 243, 244
  applications, 243
  ASM330LHH
    performance, 247, 248
    phase delays, 247
    zero rate level, 247
  LSM6DSO, 244, 245, 248
  power consumption, 243
  for smart industry and smart driving, 245–246

STMicroelectronics manufactures, 243
testing and calibration, 245
ThELMA, 243
ultra-low power accelerometers, 245
waterproof pressure sensors, 245
Mismatch error shaping (MES) technique, 36–38
MM-wave applications, 216–217
automotive collision avoidance radar applications, 231
bit error ratio, 236
CMOS technology, 231
D-band integrated transmitter chip, 237
driving aid radar systems, 231
DSP techniques, 237
5G mm-wave, 229–230
frequency spectrum, 228
inter satellite links, 232
LRR systems, 231
market growth, 227
micro-satellites, 232
microwave circuit design, 236
MMIC, 231
mm-wave backhaul radios, 236
mm-wave imaging, 231
mm-waves technology market, 225, 226
mm-waves transmission, 227
point-to-point microwave links, 228, 229
power amplifiers, 238–239
propagation characteristics, 228
semiconductor technologies
economic considerations, 233–234
solid-state processes, 232
technology status and trends, 234–236
in telecommunication infrastructures, 225, 226
x-haul applications, 228
Multi-level switched-mode converter (MLC), 134–136
Multiphase soft-charging (MSC), 115–117
Multi-stage noise shaping (MASH) architecture, 67–69

**N**

Noise-shaping SAR (NS-SAR)
block diagram, 21–22
CIFF structure
active-SC integrators, 26–28
classic amplifiers, 30
comparator input ratio, 30
discrete-time, 24
$\Delta\Sigma$ ADCs, 23–24
dynamic amplifier, 30–31
FIR and IIR, 26–28
loop filter, 24–25
noise issue, 30
NTF, 25–27
passive implementation, 28–30
power consumption, 27
continuous-time hybrid, 34–36
discrete-time hybrid, 34–36
EF structure
background calibration, 34
block diagram, 31–32
$\Delta\Sigma$ ADCs, 32
feedback summation, 33–34
FIR, 33–34
fully passive implementation, 32–33
NTF optimization, 31
PVT sensitivity, 34
energy-efficient operation, 23
high-resolution, 22
implementation of, 23–24
MES, 36–38
OSR, 22–23
Noise transfer function (NTF)
CIFF structure, 25–27
EF structure, 31
I-$\Delta\Sigma$, 48–51
Non-uniform time sequence (NUS), 16–17
NS-SAR, *see* Noise-shaping SAR

**O**

Operational transconductance amplifiers (OTAs), 16, 23, 25, 86–87
Out-of-band gain (OBG), 72
Oversampled (OS) data converter, 61
Oversampling ratio (OSR), 23

**P**

Phased arrays, 216–217
Pixel detector in 28 nm technology (IC-PIX28)
CSPreamp, 309, 313
efficiency reduction, 313
$g_m$ reduction, 313
Krummenacher feedback, 309, 312
main performance, 313
radiation damages, 311–312
recovery time, 312
schematic diagram, 310
time-domain measurements, 311
X-ray facility and irradiation setup, 311
Point-of-load (POL) converter, 109–110
Power delivery network (PDN), 109–110
Power management IC (PMIC), 126–128

Power supply rejection ratio (PSRR), 73, 154
Process-voltage-temperature (PVT) variations, 30, 34
Pseudo-random binary sequence (PRBS), 15, 16
Pulse-Width Modulated (PWM) type signals, 146

**Q**
Quantization noise (QN), 49, 51

**R**
Rad-hard mixed-signal IC design
  CMOS technologies
    charge trap phenomena, 276
    cumulative effects, 276–278
    electron-hole pairs, 277
    intra-device leakage, 276, 277
    mobility, 277
    single event effects, 278–279
    STI, 276
    transconductance, 277
  dosimetry circuits (*see* Dosimetry circuits)
  extreme environments
    harsh environments, 275–276
    ionizing radiation, 274
    natural space radiation environment, 274–275
  microphotograph, 286
  RHBD techniques
    ADC converter, 283, 285
    ASET, 283
    auto-zeroed comparator, 285
    auto-zeroing technique, 285
    binary encoder, 284
    brute force solution, 288
    DBLC bandgap, 287
    design enhancement, 280, 281
    differential amplifier with split MOS and SC comparator, 288, 289
    digital domain, 281–283
    dual path design technique, 288
    ELT shape, 283
    flash ADC, 284
    ITAR restrictions, 280
    pipelined architectures, 284
    process enhancements (RHBP), 280
    radiation robustness, 286
    Rad-Pak®, 280
    RHBD at architectural level (RHBD-AL), 280
    RHBD at circuit level (RHBD-CL), 280
    RHBD at layout level (RHBD-LL), 280
    SAR architecture, 284
    shielded packages (RHBS), 280
    switched-capacitor comparator, 288
    switched-capacitor differential network, 289
    TID effect, 283
Radiation induced narrow channel effect (RINCE), 306
Random-modulation pre-integrator (RMPI) architecture, 15–16

**S**
SAR, *see* Successive-approximation-register
Scalable parasitic charge redistribution (SPCR)
  BP discharging/charging state, 113
  CRB, 114
  deep-trench capacitors, 114–115
  parasitic coupling, 113
  system overview, 113–114
  working principle of, 112–113
SC converter, *see* Switched-capacitor converter
Sensor nodes (SNs) senses, 14
Shallow trench isolation (STI), 276
Shunt current sensors
  drawbacks, 147
  error sources, 149
  high-side sensors, 148
  low-side sensors, 147–148
  shunt resistor, 148–149
Signal-to-jitter noise ratio (SJNR), 72
Signal-to-quantization noise ratio (SQNR), 53–55
Signal transfer function (STF), 61
  $CT\Delta\Sigma$ modulator, 65–67
  $I\text{-}\Delta\Sigma$, 48–51
Silicon-Limit, 193
Single cell (1S) Li-ion batteries, 135–136
Single event transients (SETs), 279
Size, weight, and power (SWAP), 15
SPCR, *see* Scalable parasitic charge redistribution
Stage outphasing (SO), 117–118
Standard-process (SP) nMOST, 1 grad total-ionizing-dose effects
  device-under-test geometry, 300, 301
  $I_{DS}$–$V_{GS}$ characteristics, 301, 302
  Keithley 707 switching matrix, 300
  nMOST electrical characterization
    $g_m/I_{DS}$ *vs.* TID dose, 304, 305
    $g_m/I_{DS}$ *vs.* $V_{GS}$, 304, 305

mobility, 309, 310
OFF current ($I_{OFF}$) behavior, 304, 306
SubThS reduction, 303
threshold voltage, 303
transconductance *(gm)* evolution, 304, 307, 309
pMOST electrical characterization
$\Delta V_{TH}$ current, 307, 308
$I_{DS}$–$V_{GS}$ characteristics, 305, 306
OFF current ($I_{OFF}$) behavior, 307, 308
RINCE, 306
saturation current with dose variation, 305, 308
SubThS *vs.* dose, 305, 307
TID tests, 300
Starved current buffer structure (SCB), 133
Successive-approximation-register (SAR)
building blocks, 42–43
coarse phase, 45–46
fine phase, 45–46
high-resolution, 43–44
mismatch error shaping, 43–44
pipelined SAR, digital calibration, 46–47
power consumption, 42
VCO-based comparator, 44
*See also* Noise-shaping SAR (NS-SAR)
Superjunction power devices
architecture design
dV/dt self-adjusting architecture, 199, 200
dV/dt self-adjusting regulating circuit, 204–205
dV/dt self-adjusting sensing circuit, 202–204
high voltage capacitance, 201, 202
$C_{GD}$ Miller multiplication factor, 195
charge balance rule, 193
dV/dt classification
active dV/dt, 199
dead time, 196
gate driver, 196
inverter-leg configuration, 196–198
motor drive stage and driver, 199, 200
negative active dV/dt, 198
passive dV/dt, 198
power-switching devices, 196
electromagnetic compatibility, 196
EMI, 196
gate-collector ($C_{GC}$), 194
gate-drain ($C_{GD}$), 194
hard-switching, 195
Io+ driving capability, 196
MOSFETs, dV/dt characteristics, 195
MOSFET structure, 194
non-linear depletion, 194, 195
power switches, 194
Silicon-Limit, 193
simulation results
low side active turn-on switching event, 206
LVsense signal, 206, 208
negative active dV/dt level, 206, 208, 209
positive partially passive dV/dt, 209, 210
positive totally passive dV/dt, 209, 210
600 V CoolMOS™ CFD7 SJ MOSFET, 205
zoomed portion of slow tail, 206, 207
SJ FET structure, 194
superjunction structure, 193
Switched-capacitor (SC) converter
continuously scalable conversion ratio
capacitors' voltage swing, 118–119
cornerstone phases, 118, 120
measured efficiency *vs.* output voltage, 120–122
micrograph, 120–121
theoretical efficiency, 120–121
voltage *vs.* phase diagram, 118, 120
design space, 110–111
efficiency measurement, 137–138
full integration, 110
interleaved circuit, 137
MSC, 115–117
POL, 109–110
SPCR
BP discharging/charging state, 113
CRBs, 114
deep-trench capacitors, 114–115
parasitic coupling, 113
system overview, 113–114
working principle of, 112–113
stage outphasing, 117–118
theoretical power efficiency, 136
two operation phases, 136–137
unregulated intermediate supply, 135
VCR, 111

**T**

TDCs converter, *see* Time-to-digital converter
Technological innovation, 8–9
Thick epitaxial layer for micro-gyroscopes and accelerometers (ThELMA), 243
Third-order full-integrator (IIR) based design, 26–28

Time-skew calibration block (TSCB), 101
Time-to-digital (TDCs) converter
  aperture plot, 11–12
  building blocks, 9–11
  current-mode circuits, 9
  digital temperature sensors, 13
  energy plot, 11–12
  in-depth examination, 13
  performance of, 12
  scalability, 13–14
  time intervals, 9
Two series (2S) Li-ion batteries, 134

U
Ultra-reliable mission critical applications, 215

V
Voltage-controlled delay unit (VCDU), 10–11
Voltage conversion ratio (VCR), 111

W
Wearable systems, *see* Internet of Things (IoT)
Wide bandgap integrated circuits
Wide bandgap semiconductors (WBGS)
  advantage, 168
  analog feedback, 172, 173
  applications, 174
  avalanche breakdown $E_c$, 167
  circuit topologies, 172
  digital noise margins, 173, 174
  doping control, 170
  driver circuit and packaged switch, 174, 175
  HAST, 172
  intrinsic concentration *vs.* temperature, 169
  ionization energy, 169
  linear voltage regulator, 174, 175
  minority carrier charge injection, 168
  MOSFETs, 170
  negative feedback techniques, 173
  noise margins, 173
  ohmic contacts, 171
  on-chip resistor divider networks, 173
  properties of, 167, 168
  radiation hardness, 169
  sheet resistance temperature dependence, 172, 173
  SiC BJT current gain, 172, 173
  SiC, ion implantation, 170, 171
  SiC MOSFET switches, 171
  silicon technology, 170
  space exploration, 169
  TTL ALU, 175, 176
  TTL inverter, 174, 175
  unipolar specific on-resistance *vs.* breakdown voltage, 168
  VLSI circuits, 175
Wide-bandgap transistors
  benchmarking
    based on $P_{loss}$, 184–186
    based on $Q$g, 183–185
    based on specific area, 185–186
    gate charge *vs.* $R_{\mathrm{dson}}$, 183, 184
    $R_{\mathrm{dson}}$ *vs.* breakdown voltage, 183
  gallium nitride, 179
  GaN gate drivers solutions, 189–191
  gate driver design, 179
  gate driving principle
    gate-source parasitic inductance, 189
    Kelvin gate connections, 189
    load current, 187
    Miller plateau, 187, 188
    off-to-on/on-to-off transition, 186, 188
    principle gate driver voltage waveforms, 186, 188
  gate loss ($P_g$), 180
  multiphase turn-on and turn-off schemes, 189
  output capacitance loss ($P_{Coss}$), 180–181
  output switching losses ($\mathrm{P_{sw}}$), 181
  $R_{\mathrm{dson}}$ loss, 181
  reverse recovery losses, 181–182
  semiconductors silicon carbide, 179
  SiC gate drivers solutions, 190–191
  slew-rate control, 186, 189
  switching slopes, 179
Wireless base stations
  implementation, 96–97
    architecture, 96–97
    channel, 98–99
    digital calibration, 99–101
    input buffer, 97
    sampling network, 97–98
  measurements
    measured performance, 102, 104
    performance *vs.* sampling frequency, 102–103
    single-tone output spectrum, 102–103
    spur magnitudes *vs.* input amplitude, 102, 104
    supply decoupling, 102
  requirements
    bandwidth, 94–96
    imperfections, 93–94
    receiver specifications, 92–93
    RF-sampling, 94–96
Wireless sensor networks (WSNs), 14
Wireless transceiver (TRX), 14

Printed by Printforce, the Netherlands